The Lithic Production System of the Princess Point Complex during the Transition to Agriculture in Southwestern Ontario, Canada

Chen Shen

BAR International Series 991
2001

Published in 2019 by
BAR Publishing, Oxford

BAR International Series 991

The Lithic Production System of the Princess Point Complex during the Transition to Agriculture in Southwestern Ontario, Canada

ISBN 9781841711928 paperback
ISBN 9781407353449 e-book

DOI https://doi.org/10.30861/9781841711928

A catalogue record for this book is available from the British Library

This book is available at www.barpublishing.com

BAR Publishing is the trading name of British Archaeological Reports (Oxford) Ltd. British Archaeological Reports was first incorporated in 1974 to publish the BAR Series, International and British. In 1992 Hadrian Books Ltd became part of the BAR group. This volume was originally published by John and Erica Hedges in conjunction with British Archaeological Reports (Oxford) Ltd / Hadrian Books Ltd, the Series principal publisher, in 2001. This present volume is published by BAR Publishing, 2019.

BAR

PUBLISHING

BAR titles are available from:

BAR Publishing
122 Banbury Rd, Oxford, OX2 7BP, UK
EMAIL info@barpublishing.com
PHONE +44 (0)1865 310431
FAX +44 (0)1865 316916
www.barpublishing.com

Acknowledgement

My interest in lithic analysis developed during the my graduate training at Tulsa University, Oklahoma, where I was engaged in a study of lithic artifacts from an Epi-Palaeolithic site in Japan. Don Henry and George Odell have inspired me a great deal with their contributions and commitments to the field of lithic analysis. Their common but somewhat different approaches (for example, Near Eastern versus North American lithic typology) have had great impacts on the research done in this book. I owe a deep debt of gratitude to George, who taught me use-wear analysis, for his insightful comments and humorous discussions on this work in its early stages.

The shift of my research focus from Near Eastern to North American lithic technology was made possible when I became a team member of the Princess Point research project at the University of Toronto. I would like to thank Gary Crawford and David Smith for offering me the opportunity to participate in the project and for encouraging me to take an interest in Ontario archaeology. Their generous help and enthusiastic guidance, as well as their friendship, stimulated the progress of this work over the years.

This book benefited from constructive and invaluable comments on early drafts from faculty members and colleagues in the Anthropology Department, University of Toronto. I thank Maxine Kleisdienst, Peter Storck, Ted Banning, and Tony Davis. I would also like to thank Bill Engelbrecht, W. Green, Charles Cobb, Rob MacDonald, and Susan Jamieson for their insightful comments on parts of the work.

Funding that enabled completion of this research was provided by a Social Science and Humanity Research Council (SSHRC) of Canada awarded to Drs Crawford and Smith. In addition, this work was also supported through the University of Toronto Open Fellowship, a SSHRC doctoral fellowship, an Ontario Graduate Scholarship, and an Archaeological Research Grant from Ontario Heritage Foundation. I thank these institutions/agencies for their financial support.

Gary Warrick and Jeff Bursy, then archaeologists at the Ministry of Transportation Ontario, generously provided archaeological materials and data from the HH sites for this study. I thank the Dinning and Riley Families for their permission to work at the Grand Banks site, the van Osch family for their permission to work at the Lone Pine site, and Murray Young for his permission to work at the Young 1 site. I also owe a debt of gratitude to Mr. Fred Moerschfelder and Mr. Jeff Bursy who taught me Ontario chert and flintknapping skills.

There were a number of people involved in my research, who provided invaluable insights and assistance. I am fortunate to have had such friends and colleagues with whom to discuss ideas that formed parts of this work. My first special thanks must go to my dear friend Alicia Hawhins, who went through all stages of this research with me (fieldwork, lab, and writing) sharing my excitement as well as frustration and offering her constructive comments on the early drafts. My thanks also extend to the Late Arne Carlson, David Lasby, and Trevor Ormerod for their participation in my use-wear experiments. Thanks are also extended to Mark Blackham, Vandy Bowyer, Chalton Carscallen, Gyoung-ah Lee, Ge Li, Andrew Martindale, James Quin, Julian Siggers, and Caroline Walker. I would also like to thank the following people who were involved in my research in the field, or in the lab, or both: Joanna Bandziorowski, the Late Andre Bekermen, Leigh Freeman, Demetra Georgakoupolis, David Goode, Heather McCallum, Zissis Parras, Lori Richards, Kathy Schoenberger, and Chris Watts. This book would not have apperred in its final form without the generous and responsible editing of Alicia Hawkins and Coraline Walker.

Last but not the least, my family deserves my deep appreciation for the emotional support and encouragement that I received during the process of this work. I thank my parents for their understanding and support while I have been away from our loving family. My infinite thanks to my wife, Bing Wang, who gave a great deal of help in various ways that made it possible for me to complete this work. My guilt and appreciation for her patience and hardship with our baby boy, Simon, while I was away in archaeological fields for several months a year, cannot be adequately expressed here. Without her understanding and companionship through some tough, stressful, and bitter times, this work would never have been completed.

Abstract

This study focuses on an examination of lithic production systems of Princess Point Complex of southwestern Ontario, Canada. The Princess Point Complex represents a transitional culture from the Middle to Late Woodland in the northeast North America (c.a. A.D. 500 - 1,000). The AMS dates on corn remains (*Zea mays*) recovered from this culture indicate the emergence of horticulture based on corn-production in the Lower Grand River Valley as early as A.D. 550. The analyses of lithic artifacts aim to reconstruct the pattern of Princess Point lithic production, and to explore the transformation of lithic production in relation to the emergence of food production in the study region.

The lithic data comes from a three-year field investigation at the Grand Banks, Lone Pine, and Young 1 sites (1993 - 1995). Over 2000 lithic samples were selected for both the typo-technological analysis and the use-wear analysis. The results of lithic analyses demonstrate that the Grand Banks lithic industry represents a generalized stone tool production. It suggests that the shift from specialized to generalized stone tool production, as a long-term technological change, is likely associated with the introduction of horticulture. While the subsistence shift may have brought about a series of changes in socio-economic structures, the transformation of lithic production might have been caused by, not one single factor, but an integrated relationship of external factors (increasing sedentism, use of localized resources, population increase) and internal factors (time-stress, productive organization). The study aims to stimulate research into lithic production in the context of agriculture in the study region, and to suggest new lines for further study.

Table of Contents

List of Figures

List of Tables

CHAPTER 1

INTRODUCTION

Over a century, the study of the transition to agriculture in prehistory has been one of the most hotly debated and intriguing subjects in archaeology and anthropology. To date, issues on the emergence of agriculture in different regions, especially places that are "non-centers of the origins," have been addressed in comparative and multi-leveled way from global perspectives (e.g., Bar-Yosef 1998; Harris 1996; Price and Gebauer 1995). Questions to be answered are no longer restricted to "where" and "when" wild plants/animals became domesticated or human beings started to practice domestication. Whereas searching for causes of the emergence of agriculture will be focus of research for a long time, archaeologists and anthropologists have started to conduct full-scale investigations into human behavior in association with social changes in such critical period. This study starts from this point, focusing on one aspect of social changes: *lithic technology*. It asks: what happened to the lithic production of a particular small hunting-gathering groups when they adopted sedentism and practiced domestication? Did the lithic production remain the same or was it transformed into a new form during the transition from hunting-gathering to food production, and how?

A few of recent studies on this topic have demonstrated that lithic technology had undergone dramatic changes in accommodation with a series of social-cultural changes due to adoption of food production by hunter-gathers in both Old and New Worlds (Henry 1989a; Odell 1996a, 1998; Parry 1986; Stafford 1999; Yerkes 1987). It is clear that these changes took place in very different forms at different cultural regions, thus investigations of such changes should have methodological and theoretic significance. In addition, these studies also indicate that lithic production in early farming societies is somewhat closely related to social organizations institutionalized for emergence of food production. That is to say, while changes in lithic production during the transition to agriculture is still likely influenced by techniques, mobility, and raw material availability, etc., other factors such as organization of production, control of raw materials, and network trade, have also taken into accounts for explanations. This book will present a similar study I conducted over the last decade, investigating changes in lithic production during the transition to agriculture in Southwestern Ontario, Canada.

In 1993, a multidisciplinary research project at the University of Toronto began investigating the transition from the Middle to Late Woodland in southwestern Ontario and the origins of horticulture in the Northeast Woodlands of North America. The Princess Point Complex (A.D. 500-1,000) of southwestern Ontario provides the primary focus of this research. To date, research has provided new insights into this regional culture complex ever since its first identification twenty years ago (Crawford and Smith 1996; Crawford et al. 1997, 1998; Shen 1995, 2000, in press; Smith 1997; Smith and Crawford 1995, 1997).

David Stothers (1977), who first systematically studied the culture, outlined a general scenario adducing that these people initiated food production, including the domestication of corn and other cultigens in the region. Although the traits of this cultural complex have been briefly described by Wright (1972:57-58), Mason (1981:319-324), and Fagan (1991:416-417), many issues concerning the nature of this culture still remain unexplored. These include chronology, settlement patterns, subsistence, and cultural interaction.

To date, the earliest AMS radiocarbon date on

maize (*Zea mays*) that comes from a Princess Point Complex site indicates that a shift to a mixed economy with horticulture, gathering, hunting, and fishing in the Lower Great Lakes region probably occurred earlier than previously thought (Crawford and Smith 1996; Crawford et al. 1997). Current research on the settlement pattern of this transitional Middle to Late Woodland culture also suggests that there was a significant change from short-term to year-around settlement organization in the Lower Grand Banks River, Ontario (Crawford et al. 1998; Smith and Crawford 1997). As a part of this research program, my study has been concentrated on lithic technology of the cultural complex in question, providing new insights into our understanding of lithic production system during the transition to agriculture in Ontario.

The Research Objectives

Lithic assemblages from Princess Point Complex sites are the subject of the study. Although it also deals with various aspects of human behavior in so far as they are associated with flaked stone tools, it is centered on an examination of the Princess Point lithic production system. My primary interests are the patterns of stone tool production and the transformation of lithic production concurrent with the introduction of horticulture in the region.

This study has descriptive and interpretative goals. At the descriptive level, it characterizes lithic production at the Princess Point sites currently under investigation. To systematically explore the lithic production pattern, I have designed this study to examine flaked stone tool manufacture and utilization through lithic typo-technological and use-wear analyses. At the interpretative level, the study will provide an explanation of changes in lithic technology during the transition to agriculture in the study region. At this level, attempts are made to examine a general trend of technological changes in lithic production from mobile hunter-gatherers to sedentary farmers. The theoretical framework used in this study is outlined in the next section. The specific objectives of this study are, in brief:

(1) to characterize the pattern of Princess Point lithic production through a detailed lithic analysis;

(2) to examine and explain the transformation of lithic production in relation to the beginning of food production in the study region.

Clearly, the study of the Princess Point lithic artifacts can provide perspectives on both material cultures in the context of the origins of agriculture and a regional manifestation of changes in lithic technology during this transitional period.

In order to properly interpret cultural affiliations from material data, I have adopted the lithic typological-technological classification system put forward by Henry (1989a:79-83), which has been developed and modified from Clarke's (1968) model of cultural classification. Three scales of diversity and complexity in lithic typology and technology correspond to the different levels of social groups in terms of time and space. They are "complex," "industry," and "assemblage" (see also Clark and Kleindienst 1974:73-74).

The term *complex* as defined in this study denotes the largest scale of lithic aggregations sharing general typo-technological characteristics. These are representative of a social-cultural system in a broad temporal and spatial range. The term is comparable to the techno-complex as defined by David Clarke (1968:357), and thus represents a general development or adaptive stage achieved by human beings (Clark and Kleindienst 1974:74). The temporal and/or spatial extent of a lithic complex, e.g., the Princess Point Complex, may encompass more than one lithic industry.

An *industry* is defined as an intermediate level of lithic typological and technological affinity, normally referring to entities of cultural groups in a relatively smaller range of time-space, for example, the Grand Banks Industry. Lithic industries within an industrial complex should share common lithic production strategies, although variability is expected to exist.

At the smallest scale of lithic artifact aggregation, an *assemblage* refers to the products of a stone

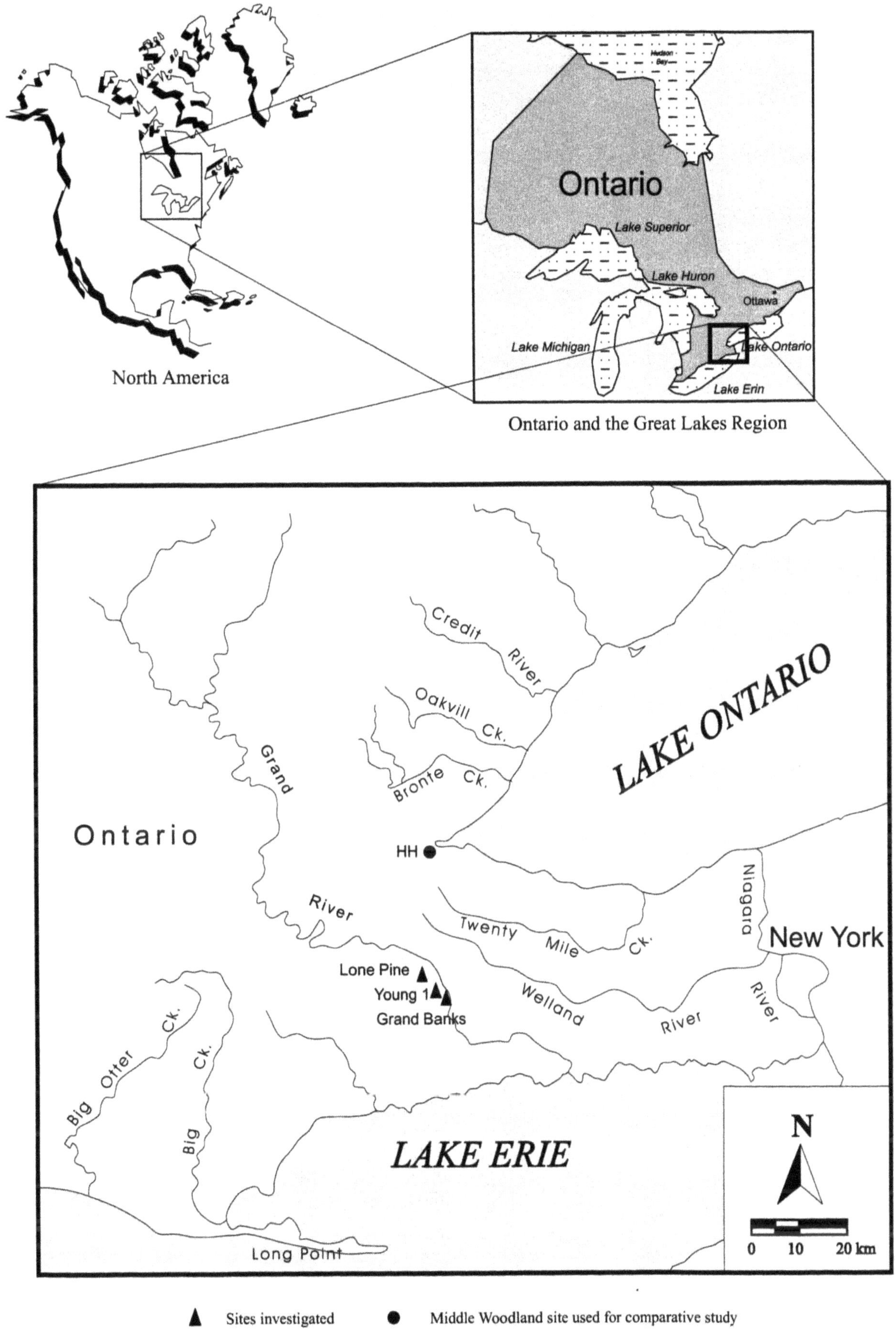

Fig. 1.1: Map of the Study Region.

tool production system that remains within a single archaeological context. This is a cluster of artifacts that share distinctive local technological and typological attributes, and represents a particular cultural occupation at a specific time and locality: e.g., the Grand Banks assemblage, the Young 1 assemblage, or the Lone Pine assemblage.

The temporal and spatial limits of this study should be noted here. The *study region*, in its narrowest sense, is the *Lower Grand River Valley* of Southwestern Ontario. However, the cultural inferences I make may apply to a broader region. This includes the northeastern shore of Lake Erie from Long Point to the Niagara Peninsula, and to the Credit River Valley at the western end of Lake Ontario (Fig. 1.1). Temporally, the study investigates the period from A.D. 500 to A.D. 1,000. This is the transitional period between the Middle and Late Woodland of North America, during which we see the emergence of food production in southwestern Ontario.

Study of the Princess Point Complex

The study of the Princess Point Complex of Ontario was initiated by David Stothers (1970, 1971, 1973, 1974a, 1974b, 1976, 1977), based on his extensive survey and test excavations in the Lower Grand River Valley during the later 1960s and early 1970s. One of his contributions was the discovery of maize (*Zea mays*) in early Late Woodland deposits, pointing to the early development of horticulture in the study region. Stothers utilized ceramic seriation to determine the temporal and spatial extent of this cultural unit. This culture complex, as he defined it, is characterized by its distinctive cord-marked pottery with internal bosses and circular exterior punctates and "Levanna" type triangular projectile points.

The recognition of this culture marked a developmental phase in the prehistory of southwestern Ontario: the initial period of the Late Woodland or the "transitional period" between Middle and Late Woodland traditions (e.g., Trigger 1985; Ellis and Ferris 1990). Recently Fox (1990) revised Stothers' synthesis of Princess Point to integrate new archaeological discoveries in Ontario (e.g., MacDonald 1986a, 1986b; Timmins 1985), and refined the scope of Princess Point both spatially and temporally. Two of the three regional foci of the Princess Point Complex, Point Pelee and Ausable, as defined by Stothers, are reassigned to the newly named the Western Basin Tradition (Murphy and Ferris 1990). According to Fox (1990) the Princess Point complex is confined to the lower valley of the Grand River (Stothers' Grand River focus). Fox and other researchers also argue that what Stothers refers to as the late phase of the culture should be excluded given the evidence from the Porteous village site near Brantford which is now taken as a manifestation of the early Glen Meyer stage (Noble and Kenyon 1972). They suggest instead that Princess Point ended at ca. 1,050 BP (A.D. 900), rather than at 1100 BP.

During the 1980s and early 1990s, salvage excavations in the region yielded nearly 80 Princess Point sites in the Lower Grand River Valley and its immediate surroundings. However, until a systematic investigation began in 1993, there was still little concentrated research on this cultural unit (Smith and Crawford 1993, 1994, 1995, 1997; Crawford and Smith 1996; Crawford et al. 1997, 1998).

It is apparent that a socioeconomic transformation took place during the Princess Point period. In southern Ontario, the early Late Woodland cultural complexes such as Pickering and Glen Meyer, Young Phase of the Early Western Basin Tradition, and Sandusky, were clearly demonstrated to be distinguishable in both physical and social contexts from Middle Woodland cultures such as Point Peninsula, Saugeen, Couture, and the Western Basin Middle Woodland (Ellis and Ferris 1990; Smith and Crawford 1993). These changes are indicated by: (a) an introduction of horticulture; (b) the appearance of longhouse village settlements; (c) an increase in population density; and (d) greater complexity in social organization. Despite these clear and important changes, little intensive research has been carried out on the cultures of the "transitional stage" in Ontario and New York (e.g., Riviere au Vase,

Sandbanks, Hunter's Home, and Princess Point), which remain poorly understood.

Secondary Origins of Agriculture

For present purposes, the term "agriculture" used in this study refers generally to a mixed economy with the production of domesticated crops such as maize, bean (*Phaseolus vulgaris*), cucurbit (*Cucurbita pepo*), sunflower (*Helianthus annuus* var. *macrocarpa*) and tobacco (*Nicotiana rustica*), combined with hunting, gathering, and fishing. The shift to this mixed economy is traditionally regarded as being completed in the study region about A.D. 1,100. Stothers (1977) suggests that corn was introduced to the Lower Great Lakes region before A.D. 800 (see also Stothers and Yarnell 1977), but this hypothesis was not confirmed until the new dates on corn from the Grand Banks site came to be available (Crawford et al. 1997). The new AMS radiocarbon dates on corn suggests that introduction of food production could have occurred in southwestern Ontario as early as A.D. 550.

In recent years, there has been a growing number of investigations of the origins of agriculture in eastern North America (Fritz 1990; Watson 1985, 1989; Scarry 1993; Smith 1987, 1989, 1992, 1993, 1995). These studies point to areas such as the Midwest and Southeast U.S. as being centers of indigenous or primary domestication. However, domestication in the Northeastern Woodlands (including southern Ontario, central and southern New York, Pennsylvania, and New England) is still considered to have been the result of "secondary origins" (Cowan and Watson 1993; Crawford 1999; Crawford and Smith 1997; Smith and Crawford 1997). Thus, transition to agriculture in Ontario occurred through diffusion or migration, or both.

Incorporated into his study of the Princess Point Complex in the 1970s, Stothers proposed a northward migration model to account for the spread of agriculture in Ontario (Stothers 1977; Stothers and Yarnell 1977). This model suggested that Princess Point people came from somewhere south or southwest of Ontario (Ohio, Illinois, and Indiana), and that they were culturally linked to the Hopewellian culture. Climatic changes in the mid-second millennium B.P. resulted in warmer and moister living conditions. This allowed people to successfully grow the corn which had been brought from further south.

This climatic amelioration model for the origins of agriculture in Ontario was supported, in the 1970s, by two lines of evidence. Firstly, the Hopewell culture relied upon corn as its subsistence base. The use of this productive resource was thought to have resulted in the growth of localized populations and eventually in the demise of the Hopewellian Interaction Sphere (Stothers 1977). Secondly, the lack of evidence for cultural continuity between the southern Ontario Middle Woodland and Princess Point suggested that southern Ontario was colonized by corn-based horticulturists from elsewhere (Stothers 1977; Stothers and Yarnell 1977).

This model, however, is in large part refuted by recent research on plant domestication in the Eastern Woodlands (Fritz 1990; Crawford 1999; Crawford and Smith 1997; Smith 1989). It has been recognized that Middle Woodland plant husbandry in the Midwest was not based on corn. Wymer (1987) provided a case study from southern Ohio, where early Late Woodland cultigen assemblages are qualitatively indistinguishable from the Middle Woodland assemblages (including the absence of maize). Muller (1987), in attempting to explain cultural changes in the Lower Ohio Valley, suggested that corn was not a prime mover, and proposed that the new technologies used to adapt to new resource procurement strategies lessened the interdependence of Middle Woodland peoples. This, in turn, led to changed settlements which facilitated population growth. Crawford and Smith (Crawford and Smith 1997; Smith and Crawford 1997), in favor of a multifactorial approach, suggest incorporating research into demography, sedentism, and competition/cooperation in the study of agricultural origins in the Lower Grand River Valley. Following Muller's model, they consider that the increased localization of populations and the correspondingly increased pressure on localized wild resources have a great potential to lead to the development and intensification of horticulture. Crawford and Smith further hypothesize that the Princess Point Complex may represent the localization phase, setting the stage for the adoption of corn horticulture (Crawford et al. 1997; Smith and Crawford 1997; Crawford et al. 1998; Bowyer 1995).

Lithic Study

Previous studies of Princess Point lithic assemblages have been restricted to categorized descriptions. These are mostly found in Stothers' dissertation (1977), but are also scattered among other reports of archaeological surveys and excavations (e.g., Timmins 1992a; Parker 1994). From 10 archaeological sites originally assigned to the Princess Point Complex, Stothers (1977) provided 9 types of tools from the flaked stone artifacts, in addition to cores and "chipping detritus." He mentioned that the predominant tool types are triangular projectile points and scrapers. Other types such as drills and gravers appeared infrequently. More recently researchers have noted that "informal" tools are a significant component of Princess Point lithic assemblages (Timmins 1992a; Parker 1994; Murphy n.d.; Lennox and Morrison n.d.). Fox (1990:175) summarized the Princess Point lithic assemblages as constituting a "limited range of bifacial chert tools and a variety of cutting and scraping tools produced on flakes." However, no detailed analysis of lithic artifacts from any reported Princess Point site has been conducted. The present understanding of Princess Point lithic production by local archaeologists has been confined to statements about: (1) the use of Onondaga chert found along the north shore of Lake Erie; and (2) the presence of thin and finely flaked triangular projectile points known as Levanna points (Bursey 1995:47). The fact that small-sized triangular points replaced the notched points which predominated during this period has been considered as a result of technological change from the spear-and-dart to bow-and-arrow in northeastern North America (Blitz 1988; Christenson 1986; Hall 1980; Kelly et al. 1984; Morse and Morse 1990; Shott 1993).

Shortly after A.D. 500 a dramatic change in lithic technology of the Eastern Woodlands from standardized to unformalized core reduction occurred (Johnson and Morrow 1987; Jeske 1992). Unfortunately, a similar change has not yet been fully documented in southwestern Ontario. Whatever the case, we are also uncertain as to factors affecting these changes and further study needs to be conducted.

As far as methods of lithic analysis are concerned, there are two research problems. First, the lithic classification currently used for the study of Middle-Late Woodland lithic assemblages from Southern Ontario is confusing. Given the fact that previous studies of lithic artifacts have emphasized the morphology of "diagnostic" artifacts, lithic classification has been performed to aid in the descriptions of shaped tools. Some researchers have successfully established "types" for Paleo-Indian and Archaic lithic industries (e.g., Lennox 1986; Ellis and Deller 1988; Deller and Ellis 1992). However, during the Middle and Late Woodland, there was a recognized decrease in the number of bifacial tools and other formal tools and an increase in flake tools. Thus, the term "utilized flakes" is poorly defined, and yet still commonly used in classification systems. This often refers to "informal tool" categories. It is obvious that the use of this term in archaeological reports is to aid in functional interpretations. I argue that defining and using the term "utilized flakes" as a designated tool type often results in misuse of lithic classification. Young and Bamforth (1990) have presented experimental data to underscore this point, and these works is also supported by my own experimental and use-wear analytical data. A detailed discussion of this criticism will be presented in Chapter 3.

Second, there is a lack of understanding of stone tool use in this crucial period. To date, there exist no publications of lithic analysis of Woodland materials from southern Ontario in which functional interpretations are made on the basis of detailed use-wear analysis*. Some lithic analysts continue to doubt the reliability of use-wear analysis, and advocate the conventional use of morphological forms to inform the evaluation of tool use implications (e.g., Ellis and Deller 1988:123). However, in the last decade, practitioners of use-wear analysis have demonstrated confidence in this analytic method, and have successfully applied this method to archaeological

* Some use-wear research on Ontario Paleo-Indian lithic industries has been conducted by Dr. J. Tomenchuk (pers. comm.).

research in all parts of the world (Tomenchuk 1985; Shea 1987, 1991; Grace 1989, 1990; Odell 1996a; Rousseau 1992; Lewenstein 1987; Jensen 1994; Sievert 1992; Unger-Hamilton 1988, 1989; Levi-Sala 1996).

Given these specific problems in previous studies of lithic artifacts from the region, this research will integrate both typo-technological and use-wear analysis. Within a specified theoretical framework, this study will focus on the lithic production system -- a chain-of-relationships that is constituted by human involvement in stone tool manufacturing and utilization.

The Lithic Production System: A Theoretical Consideration

Lithic production, according to Ericson (1982, 1984), is defined as a process of lithic material modification with the intent to *form* and *use* a particular object. It is assumed that the manufacture and utilization of stone tools in a social system is closely linked to the investment of human energy involved in production and decision-making (Ericson 1984:3). The basis of the study of lithic production is to emphasize the *patterns* and/or *processes* of lithic material modification within a given society. The lithic production may be patterned by the way humans place production within constraints of both the physical environment and social relations. Thus, a lithic production system reflects a chain-of-relations of production which involves: raw material procurement in relation to access to and/or exchange of resources and products; manufacture and use of lithic products in relation to lifeway strategies; subsistence and social activities associated with tool utilization and maintenance such as specialization and labor allocation; and social organizations reflected by spatial patterning of lithic production, exchange or trade of lithic products, etc.. An understanding of lithic production organization, through studies of stone tool manufacture and utilization, is key to the reconstruction of social organization in a dynamic society.

It should be noted that the study of the process of lithic material modification has long been of interest to archaeologists. Recently, the concept of lithic technological organization has been elaborated by researchers using extensive archaeological records (Johnson and Morrow 1987; Nelson 1991; Carr 1994). This approach provides a framework for inquiring into the variability of prehistoric stone tools and strategies of manufacturing, using, transporting, and discarding tools and the materials needed for their manufacture and maintenance (Nelson 1991:57). The organization of lithic technology investigates the correlation between human uses of stone tools and their physical environments (e.g., resources). Although this approach is also designed for exploring economic and social variables influencing stone tool-use, most case studies in both Johnson and Morrow's (1987) and Carr's (1994) volumes have given a great deal of weight to variables such as mobility and availability of raw materials. For example, according to Kelly (1988:717, also see 1992), it seems that "mobility plays a large part in determining the organization of hunter-gatherer lithic technology." While some argue that different settlement strategies (mobile vs. sedentary) have the greatest effect on technological organization (e.g., Parry and Kelly 1987), others emphasize the availability of raw materials in determining the strategies of stone tool-use (Andrefsky 1994; Bamforth 1986). However, as Torrence pointed out (1994), a reduction in settlement mobility is not necessarily a reason for a high level of competition over local resources, and a more likely reason for such competition is a shift in the nature of food-getting and/or in territoriality and the ownership of resources. This clearly implies that the study of lithic production should consider constraints of social relations.

The concept of the organization of lithic technology provides a great deal of potential for probing past behavior in relation to stone tools. However, its applications at present are far from satisfactory. One of its weaknesses, in my opinion, is the over-emphasis on external factors as reasons for changes in lithic technology. As yet, little attention has been paid to internal factors such as human decisions, labor intensification, productive organization, or ownership. Human adaptive strategies in prehistory emphasized survival (e.g.,

Jochim 1981); changes in subsistence may have been conditioned by external factors, but determined by internal factors. Cultural change must be determined by the relationships of external and internal factors, and such changes represent the balancing of physical and social constraints. In the example of the transition to food production, an alternative view is that change in the social relations of production affected how food was produced, and food production in turn provided an economic base for building a new social structure (e.g., Bender 1975, 1978, 1981).

Recent studies of the political-economic system in small scale societies, using ethnological and archaeological evidence, have yielded new insights concerning the emergence of social inequality in the context of agricultural intensification (papers in Upham 1990 and Price and Feinman 1995; Upham 1983; Green and Sassaman 1983). Notably, lithic studies within such theoretical frameworks also attempt to explicate social complexity (Johnson 1996; Nassaney 1992, 1996; Rosen 1996; Yerkes 1989; Arnold 1987, 1993). The fundamental differentiation between hunter-gatherer societies and agricultural communities lies in centralized control over resources in the latter (Bender 1975; Wolf 1982). One of the consequences of the emergence of food production in human social life was a modification in the structure of society (e.g., Bender 1975, Price and Gebauer 1995). As Bender (1975) has pointed out:

> Hunter-gatherer societies tend to be structured on a flexible kinship base and to exhibit strong fissionary tendencies. ... Agricultural societies, on the other hand, tend to be organised on a more permanent corporate basis. This is necessary because there is a great need for long-term stability to permit co-operative land-clearance, defence, etc. Even in the early stages of food-production, mixed farming (with both herding and crop-growing) may require community organisation. ...
>
> Within the corporate structure of the agricultural society the family operates as a semi-independent economic unit and this may lead to slight differences in 'wealth' and some social stratification. Such differences may stimulate trade: more important members of the community will want to underline their position by obtaining 'luxury' goods, such as fine stones, pigments or shells. ...
>
> Inter-community exchange is probably more significant in this respect. It is less important in very uniform ecological areas where everyone will tend to produce the same range of food, and more important where the terrain is diversified and there is a variety of farming systems associated with various crops or animals. Specialisation will encourage exchange. ... Where such exchanges are controlled by particular members of the community, their prestige and power will be considerably enhanced (Bender 1975:8-9).

Here, Bender considered several variables of social relations in early farming societies: the organization of production, surplus accumulation, specialization, and inter-community exchanges. To this list, I would add ritual activity in a given society as an important element of social relations. In his recent papers, Hayden (1990, 1992) has argued specifically that it is the competitive and feasting aspects of Big Man rivalry that are the driving forces behind food production. Ritual performance would also likely increase the prestige and power of certain community members. Social relations between people and within societies are based on, among other things, resource control, ownership, and the value ascribed to materials.

Therefore, the reconstruction of the organization of lithic production in food producing societies should provide some insight concerning social structures. This is not to say that the study of the lithic production system can solve all the problems in the quest to understand social relations. As a matter of fact, discussions of social variables are always limited by the archaeological record. However, to propose such an alternative approach (the study of the lithic production system) to lithic analysis is to provide a new avenue which may lead to a better understanding of the relations between the external (physical) and internal (political, economic, ideological) factors which influenced cultural change in general, and changes in lithic technology in particular.

Table 1.1: Some Common Tools Used in Proto-historic Iroquoian Societies.

Tools	Worked Objects	Working Purpose	Working System
wooden spade [1,2,3]	soil	planting seed	horticulture
wood shovel [3]	soil	hoeing & digging	horticulture
shoulder-blade of a deer [3]	soil	hoeing	horticulture
stone axe [2,5]	tree/wood	clearing land	horticulture
stone chisel [5]	tree/wood	clearing land	horticulture
wooden spear [2]	fish	killing	fishing
bone "harpoon heads" [2]	fish	killing	fishing
arrow-and-bow [2,4,5]	animal	killing	hunting
stone & bone scraper [1]	animal skin	fleshing skin	clothing
wooden mortar and pole [1,4,5]	corn	grinding corn	food preparation
flint knife and blade [3,5]	meat	cutting and carving meat	food preparation
stone hatchet [3,4]	wood	making wooden utensils	woodworking
beavers' incisor teeth [3]	wood	making wooden utensils	woodworking
stone gouge [3,5]	wood	making wooden utensils	woodworking
turtleshell [3]	wood	making wooden utensils	woodworking

1: Trigger 1987:32-45
2: Trigger 1969:26-41
3: Waugh 1916
4: Tooker 1967:72
5: Morgan 1901

Ethnological Evidence of Stone Tool Use

To illustrate the above point concerning explorations of social relations through analysis of lithic artifacts, let us take a quick look at the ethnographic evidence drawn from proto-historic Iroquoian societies, concentrating on the use of stone tools. The purpose of this is to examine the patterns of stone tool use and the social dynamics involved in stone tool production in existing ethnological data. Princess Point is hypothesized to be in some way ancestral to the Ontario Iroquoians (Trigger 1981, 1985; Smith and Crawford 1995, Crawford and Smith 1996). Recent investigations of the Princess Point Complex have established cultural links between Princess Point and Glen Meyer, an early stage of the Ontario Iroquoian Tradition (Wright 1973). As Smith and Crawford (1995:68) argued, "the roots of many aspects of later Iroquoian culture can be traced directly to Princess Point."

If this is the case, a study of proto-historic Iroquoian tool-use may shed light on the organization of production with respect to flaked stone tool manufacture and utilization. It should be noted, however, that proto-historic Iroquoian societies exhibit major differences in socio-political organization from preceding societies. Ethnographic analogy should be approached with caution, but given that the ethnographic evidence comes from the same study region as the archaeological materials, this approach may be useful.

Proto-historic Iroquoian societies have been extensively studied through numerous ethnohistoric records (Tooker 1964, 1985; Trigger 1969, 1976, 1978, 1985; Heidenreich 1971, 1990). As Trigger notes (1981), however, most ethnohistoric information takes the form of broad generalizations, with little in the way of supporting case studies. Ethnographic or ethnohistoric observations concerning Iroquoian lithic technology are rare, for stone technology was not a priority of European concerns during the sixteenth and seventeenth centuries (Fox pers. comm.).

In his very early study of the Iroquoian tribes, Lewis H. Morgan made the following observations:

> Metallic implements were unknown among them, as they had not the use of metals. Rude knives of chert were used for skinning deer, and similar purposes. For cutting trees and excavating canoes, and corn mortars, in a word, for those necessary purposes for which the axe would seem to be indispensable, the Iroquois used the stone chisel, Uh'-ga-o-gwät'-hä. In cutting trees, fire was applied at the foot, and the chisel used to clear away the coal. By a repetition of the process, tress were felled and cut to pieces. Wooden vessels were hollowed out by the same means. Fire and the chisel were the substitutes for the axe. The chisel was usually about six inches long, three wide, and two thick; the lower end being fashioned like the edge of an axe. Stone gouges in the form of a convex chisel. were also used when a more regular concavity of the vessel was desired. Stone mortars for pounding corn, grinding mineral paint, and for pulverizing roots and barks for medicines, were also among their utensils (Morgan 1901:9-10).

This clearly indicates that stone tools were used in a relatively wide range of activities in historic Iroquoian societies, including food production and food preparation. As agriculturists, Iroquoians used chisels or axes (celts) to cut trees for clearing land. Tools made of chert were used for meat cutting, as well as for corn grinding and pounding. Chert tools were also used in making wooden utensils, such as dishes, spoons, etc. (Waugh 1916:14-15, 65, 71).

However, these historic documents also suggest that stone tools were not used exclusively as implements in daily activities. Table 1.1 lists some examples of tool types that were used in different subsistence related activities. Wooden tools, bone tools, and shell tools were part of the Iroquoian tool kit. Wooden spades or shovels were used for hoeing the land and planting seeds, while wooden spears and nets were employed for fishing. Bone tools such as scrapers and awls were used for the manufacture of clothing. Hunting was carried out using the bow-and-arrow, and traps and animal drives were also commonly used (Trigger 1969; Tooker 1964). But "an important cutting material throughout a very wide area was flint or chert" (Waugh 1916:71).

In addition to being used for clearing land, hunting, and preparing food, stone tools made of chert or flint played an important role in manufacturing wooden tools and utensils. "Spoons, bowls, arrow shafts, snowshoes, sleds, clubs, and suits of armour were fashioned out of wood, using stone tools" (Trigger 1987:44). Hennepin also remarked that "when the Savages are about to make Wooden Dishes, Porringers or Spoons, they form the Wood to their purpose with their Stone Hatchets, make it hollow with their Coles [sic] out of the Fire and scrape them afterward with Beaver's Teeth to polish them" (Hennepin 1903:103, cited in Waugh 1916:65).

How was lithic production organized in these societies? Although ethnographic accounts do not summarize this particular topic, the implications of earlier observations clearly indicate a division of labor in Huron society (Tooker 1967; Trigger 1969, 1987; Heidenreich 1971, 1978). The women were responsible for horticulture and collecting, and were engaged in food preparation and other house-work. The men made most of the wooden tools and utensils with stone tools, and also manufactured stone axes, adzes, arrow points, scrapers, and drills. "Large tools such as adzes and axes were pecked and ground out of a fine-grained granite and hornblende schist, while the smaller tools such as scrapers, points, and drills were usually chipped out of chert" (Heidenreich 1978:383). Many tools were manufactured from small nodules of chert, and the bipolar technique was observed (Trigger 1976:44).

Heidenreich (1978:382) pointed out that: "although there was apparently no full-time specialization in manufacturing or economic pursuits, it is obvious that some individuals were more adept at making pottery, pipes, or flint items than others. Each extended family, comprising a household, looked after its own requirement." In horticultural activities each family cleared their own field (Trigger 1969, 1976; Tooker 1964). This implies that the household was an important productive unit and may have co-ordinated the acquisition of raw materials and the production of tools for food production.

The ethnographic evidence also suggests some kind of spatial patterning in lithic production. Morgan recorded that : "arrow-heads of chert, or flint, were so common that it is scarcely necessary to refer to them. Occasionally they are found with a twist to make the arrow revolve in its flight. ... It is not uncommon to find the places where these arrow-heads were manufactured, which is indicated by the fragments of chert which had been made by cleavage" (Morgan 1901:10). He implied that arrow point manufacturing was a kind of specialized productive activity, and that this production was carried out in specific locations.

The ethnographic accounts also suggest that the men produced a wide range of ornamental objects of stone or other materials for trade. Biggar (1930:453) mentioned that the Natives of the Gulf of St. Lawrence area had given a Spanish fisherman "deer and wolf skins" for "axe and knives and other trifles" (see also Heidenreich 1990:478-479). It should be noted that the items exchanged might be metal tools of European manufacture. However, it also implies that acquiring stone tools through trade, or producing stone tools for trade, might have occurred in earlier periods.

Stone tools, especially small flint knives, were commonly used as butchering tools in ritual ceremonies. Waugh described the following event:

> Dog's flesh was formerly consumed on special occasions and as a ceremonial observance. Dog feasts, in fact, are said to have been offered to 'Aireskoui,' the Sun, who was also the god or 'demon' of war, this observance securing success in war or hunting as well as the satisfactory interpretation of dreams and the recovery of the sick. The burning of the white dog at the Mid-winter Festival may be a survival of this. Stags and deer were sometimes offered in the same way (Waugh 1916:133).

Lithic production therefore appears to have played an important role in the economic life of historic Iroquoian people, as well as in their social lives. Reports of investigations of Iroquoian archaeological sites regularly include numerous tool types (Wright 1973; Latta 1976; Williamson 1990; Ramsden 1990; Dodd et al. 1990). These tools, which include axes, adzes, triangular or notched projectile points, scrapers, drills, etc. may function as their names imply and as described in ethnographic documents, but this has, by no means, been empirically documented. Furthermore, the lithic production systems of northern Iroquoians have yet to be reconstructed.

Research Design and Research Questions

Having established the theoretical framework of this study with supports from the ethnographic record, I will now outline my research procedures for archaeological interpretations of the lithic production system (Fig. 1.2). According to this research design, an Princess Point lithic production system is reconstructed using three levels of interpretations. The goal of level 1 interpretations is to characterize the pattern of lithic production. The characterization of lithic production includes both manufacture and utilization. Discrete and metric attributes and edge damage patterning of flaked stone artifacts are first studied through typo-technological analysis and use-wear analysis. These archaeological observations provide the grounds for descriptions of both core reduction strategy and use-tasks through data manipulations. Interpretations of the manufacture and utilization of flaked stone tools is carried out on the basis of an understanding of the core reduction strategy and of use-tasks.

Core reduction strategy represents human choices and a set of actions in tool manufacture. These technological dynamics include raw material procurement and selection, techniques for reducing core nodules (core reduction) and for producing appropriately sized and shaped flakes (blank production), and selection of retouch methods for tool modifications. Since the impact of such actions can directly affect the morphology of stone tools, the reconstruction of a core reduction strategy can be accomplished through observation of discrete and metric attributes of lithic

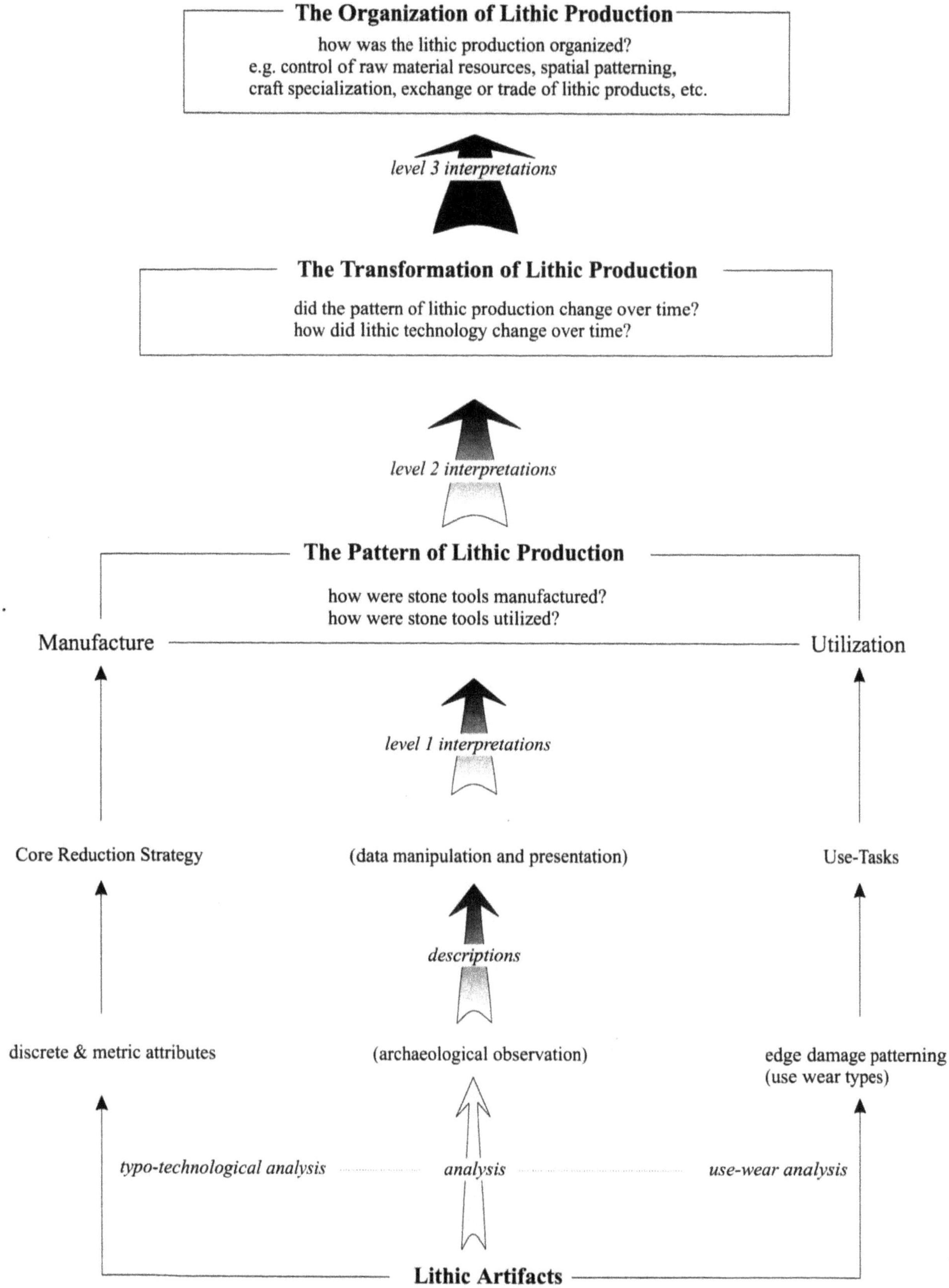

Fig. 1.2: A Diagram for the Archaeological Study Leading to Interpretations of the Lithic Production System.

objects.

On the other hand, the objective of the study of use patterning of lithic objects is to reconstruct the activities involved with the use of stone tools. Activities are, arguably, components of the *behavioral chain* within various working systems (Schiffer 1976). By a working system, I mean a complex of related activities sharing a general goal; for example food production, food preparation, hunting, or fishing, etc.. From the ethnographic data, we know that a single stone tool can function in the same manner for similar tasks in different working systems. In other words, the reconstruction of activities in various working systems requires an understanding of "the dynamic relationships among the participating elements" (Schiffer 1976:49). Therefore, tool-use patterning is not sufficient to interpret actual activities in working systems (Kleindienst pers. comm.). In an effort to reconstruct tool utilization, use-wear analysis is appropriate and has been applied in this study. The purpose of use-wear analysis is to identify edge damage caused by use, and inferences concerning tool utilization are made through the interpretation of tool motions and contact materials. The term, "use-task" (UT), is used in this study, instead of activity. A UT is represented by a single combination of tool motion and contact material (e.g., scraping seasoned wood).

At the second level, transformation of lithic production is interpreted on the basis of an understanding of lithic production. The focus at this stage of research is to examine the possible changes in patterns of lithic production over time and the factors that influenced these changes. While it should be noted that the causes of changes in lithic production are more complicated than expected, explanations regarding the transformation of lithic production during the transition to food production need to be considered within the dynamics of social structure. For example, on the one hand, increased population would cause productive labor allocation; on the another hand, established land-ownership and household economic activities would be one of factors leading to a generalized tool production (for detailed arguments see Chapter 9). Time constraints arising from intensified field horticulture would likely also re-form the pattern of lithic production in early farming societies. To an extent, this study is suggesting that the transformation of lithic production may possibly occur as a response to the emergence of food production in the study region.

The highest level of interpretation (level 3 in the Fig. 1.2) in the reconstruction of the organization of lithic production is, of course, the most difficult to approach through archaeological research. The interpretations at this level are concerned with social organization, cultural interactions on a regional scale, and culture change. The goals at this level of interpretation are to explore issues such as: how and why a particular pattern of lithic production is present in a given cultural system; how lithic production could be affected by physical conditions (environment, settlement, subsistence, etc.) and by social factors (economic organization, resource accumulation and control, etc.) or vice versa; what role lithic production plays during cultural changes, and so forth. Therefore, the reconstruction of a lithic production system should be based on all existing data and the comprehension of such data on a regional basis, which at present is clearly beyond the scope of this study.

Recently, the political-economic approach to lithic analysis has been used to explore the above issues. For instance, Nassaney (1992, 1996) claimed that the incipient elite of the prehistoric southeastern United States may have tried to increase their power by controlling access to lithic resources and/or to the organization of lithic production. Using lithic data from the Plum Bayou culture in central Arkansas (c.a. A.D. 700 -- 950), he concentrated on raw material distributions, spatial patterning of lithic production, and intentional thermal alteration, to elucidate three aspects of social relations: control of access to lithic resources, labor allocation, and labor intensification. The results are intriguing, if not totally satisfying. In his study of Early Formative sites in Mesoamerica, Parry (1987) compared lithic data from different house structures and found that lithic production was carried out by every household. Biface production was specialized and carried out only in a few households, while every household produced its own flake tools.

While hard labor and a higher energy input into production may be demanded in order to create a surplus, craft specialization provides opportunities for product accumulation and exchange. Craft specialization is always linked to greater cultural complexity; it may also be related to controlling access to resources and exchange (Arnold 1987, 1992; Johnson 1996). Lithic production is by no means limited to manufacturing tools necessary for food production. Other lithic products were also fabricated for exchange, as ornamental objects, and for entertainment. The skill of fine flintknappers would have been passed on to younger generations.

Ritual practice played an important role in the enhancement of individual authority. Until recently, little attention has been paid to the use patterning of stone tools in religious ceremonies (Sievert 1992; Odell 1994a). Functional studies of lithic artifacts from mortuary sites indicates that specially-made points, blades, or other stone tools may have served specific functional purposes in ceremonies, such as bloodletting. Other aspects of the role of stone artifacts in ritual practice, such as spatial layout of stone products and offerings, should also be considered as part of the organization of lithic production.

However, as preliminary research concerning the Princess Point Project, this study emphasizes the description of the Princess Point lithic industry and the first two levels of interpretation. Given the data currently available, the reconstruction of the organization of Princess Point lithic production is not possible. The study of an organized lithic production system, which contains a series of social variables, requires large scale and comprehensive evidence. Although this study is limited in scope, and, therefore, unable to offer such interpretations, an understanding of the social variables that may influence the form or patterns of, and the transformation of, lithic production, is a step towards a comprehensive understanding of the lithic production system.

With this research design, three basic research questions arise:

(1) What kinds of core reduction strategy are represented in the lithic production system of the Princess Point Complex, and how were stone tools manufactured?

(2) What kinds of tool use-tasks of flaked stone tools are represented in the lithic production system of the Princess Point Complex, and how were stone tools used?

(3) Were there any changes in core reduction strategies and tool use-tasks during the transition from the Middle to Late Woodland cultures in southwestern Ontario, and how did lithic production change over time?

Summary

In this chapter, I have outlined the goals and the theoretical framework of this research, and reviewed ethnographic data concerning the possible organization of lithic production. On the basis of these arguments, a research design for studying lithic production systems has been outlined. A series of research questions have been articulated.

In the next chapter, I provide background on the relevant fieldwork of the Princess Point Project, and on the physical setting of the study region. Chapter 3 presents the methods of lithic analysis. I explain how attribute analysis can be used to arrive at meaningful interpretations, and how these attribute analyses will be manipulated. A use-wear experiment will be documented. A sampling strategy for lithic analysis is offered at the end of the chapter.

Chapters 4 through 7 comprise presentations of analytical results for the four main artifact classes. Although many tabulations are used to assist in the presentation of data, a summary at the end of each chapter contains an extended discussion concerning the issues derived from data analysis. Chapter 8 offers a comparative study. It focuses on intersite and intrasite comparisons of tool-use pattern in the study region. These results furnish the first detailed study of the Princess Point lithic industry. The interpretations presented in the concluding chapter (Chapter 9) emphasize the transformation of lithic production. An explanation of the shift to a generalized lithic production during the Princess Point period in relation to food production is offered.

CHAPTER 2

ARCHAEOLOGICAL INVESTIGATIONS: FIELDWORK AND ECOLOGICAL SETTING

Fieldwork

Initiated in 1993, the Princess Point Project investigates this particularly culture, with four main objectives (Smith and Crawford 1995:56):

1. to clarify the temporal and spatial range of the Princess Point Complex (*distribution* and *chronology*);
2. to reconstruct the economy in terms of the origin of food production (*subsistence*);
3. to investigate both regional and intra-site *settlement patterns*; and
4. to examine the *material culture* of the Princess Point Complex.

After five years of fieldwork and nearly a decade of research, the project goals have been well achieved. This cultural manifestation is now better understood than before. My following introduction of the project will be brief due to a number of publications are already in place. Readers who are interested in the provisional findings of this multidisciplinary research are encouraged to refer to Crawford 1999; Crawford and Smith 1996, Crawford et al. 1997, 1998, Shen 1995, 1999, 2000; Smith 1997, Smith and Crawford 1997, Walker et al. 1997.

The first three seasons of fieldwork were conducted in 1993-1995. This included field surveys of the Credit River Valley and the Lower Grand River Valley, test excavation of Lone Pine and Young 1, and full-scale excavation of Grand Banks. The objective of the first field season was simply to determine scales of site sizes and artifact density at three archaeological sites thought to be Princess Point. Based on the findings of the first season, our team shifted to concerntrate on the Grand Banks site. The methods of investigation employed at each site are slightly different due to the different site situations. Deposits were screened through 1/8" mesh and both wet and dry screening were employed. Diagnostic artifacts such as rim sherds, projectile points, and pipes, were plotted *in situ*. A large quantity of soil samples was taken for flotation. For each 5 cm arbitrary level approximately 20 liters of soil were collected for each one-meter square. Features such as pits and hearth floors were completely collected for flotation (for detailed discussions on method of fieldwork, see Smith and Crawford 1997).

Lone Pine (AfGx-113)

Lone Pine site was first discovered during the early 1990s. Presence of diagnostic pottery sherds led to recognize the site as a Princess Point site. The site is located at a fork in Roger's Creek, a tributary of the Grand River, and approximately 2 km from the Grand Banks site, at an elevation of 190 meters above sea level. This upland site is in a woodlot that has been logged but not ploughed, and is therefore relatively undisturbed. Both controlled surface collection and excavation of the site were carried out. Nearly fifty 1 m^2 test units were excavated and four areas were expanded (Fig. 2.1). These expanded areas were chosen for investigation due to: (a) the presence of a high density artifacts on the surface (Area A and D) and (b) the presence of hearths and pits (Area B and C). A systematic surface collection was carried out. A total of 933 surface artifacts were individually plotted. Areas of four N-S trenches and two E-W trenches were laid out for surface collection.

Cultural deposits at Lone Pine are relatively shallow, averaging only 15 cm in thickness over a base of dark brown silty Haldimand clay. About 73,000 artifacts were recovered during the two seasons' test excavations and surface collection

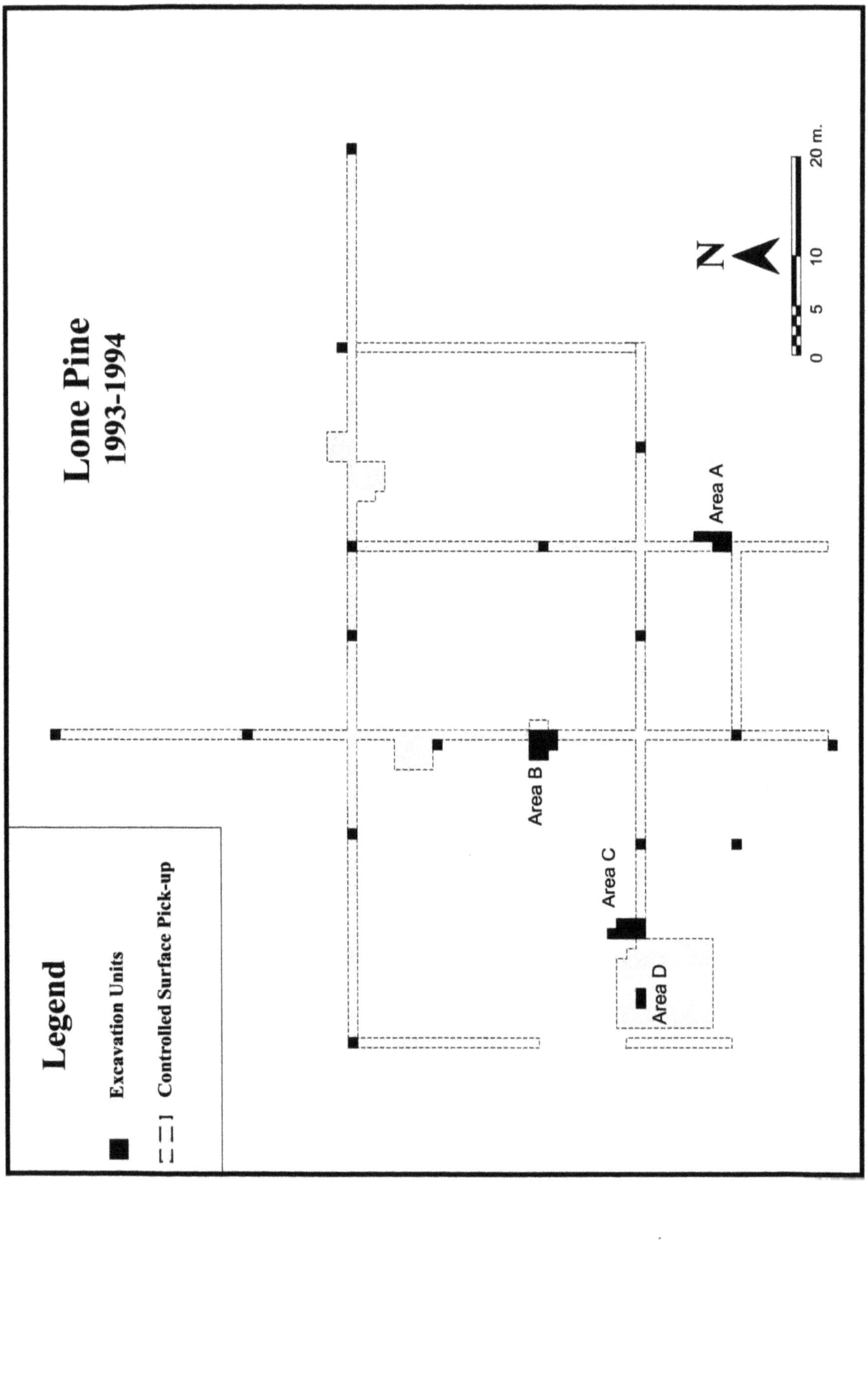

Fig. 2.1: Map of the Lone Pine Site (AfGx-113).

(Photo 1 & 2). The greatest concentrations of artifacts were in Areas B and C where the hearths were located. The two hearths are large and oval in shape. Hearth 1 in Area B is about one meter in diameter, and fired clay is 12 cm thick at the deepest point, indicating its intensive use during the occupation. Two small pits, and an abundance of artifacts, including most of the ceramic smoking pipes, were found in association with Hearth 1.

Maize was recovered from the flotation samples. Two radiocarbon dates on kernels have yielded calibrated dates of AD 885-1060 and AD 1165-1295. Both a preliminary study of ceramic seriation by Bekerman (1995) and a comparative study of ceramic pipes from the study region by Smith (1996) have placed the Lone Pine site in the early Late Woodland period in Ontario. Thus it could be an early Glen Meyer site. However, it is possible that the late Princess Point and early Glen Meyer overlap in time (Smith 1997).

Given the data available to date, it is difficult to determine whether Lone Pine was a seasonal or year-round, long or short-term occupation. "On the one hand, Lone Pine may be a village; i.e., a large, year-round, multi-year but short-term occupation. On the other hand, it may represent small, seasonal camps used over a long period of time, and thereby covering the entire plateau area" (Smith and Crawford 1995:62). Despite the fact that only limited test excavations were carried out at Lone Pine and no post-moulds were found, Crawford and Smith favor the interpretation that it is a village settlement similar to the Porteous site (Noble and Kenyon 1972), indicated by the size of the hearths and the density of artifacts (Smith and Crawford 1995, 1997). For the purposes of this dissertation, I define Lone Pine as a permanent village site immediately succeeding the Princess Point and dating to ca. A.D. 1,000.

Young 1 (AfGx-6)

Young 1 was first identified as a Princess Point site by Stothers during his Grand River survey (Stothers 1974). It is situated on a knoll on the first terrace of the Grand Valley, about half a kilometer from the Grand Banks site. The site is primarily of interest because of its physical setting, situated between the floodplain of the Grand River where Grand Banks is located, and the upland area where Lone Pine is located. It was hoped that test excavation of this site would yield information that would help to confirm or refute Stothers' model of Princess Point settlement patterns as being one of seasonal movement (Stothers 1977).

Unfortunately the site is on cultivated land and has been badly disturbed by ploughing. Two one-meter squares and three half-meter test units were excavated during the 1993 field season. A cultural layer could not be clearly distinguished from the plough zone which was 25 to 45 cm thick. Because of the disturbed nature of the site and the very dry and clayey nature of the soil, artifacts were mainly recovered by shoveling and screening through 1/4" mesh. The clayey soil also made the process of flotation almost impossible, but soil samples of about fifty liters were taken and processed.

Nearly 2,400 artifacts were recovered from 3.5 square meters. Although this is a very limited artifact sample compared to Grand Banks and Lone Pine, it includes cord-wrapped stick marked pottery. So far, no diagnostic artifacts indicate cultural affiliations other than Princess Point. However, no radiocarbon dates have been obtained. Stothers (1974a) interpreted Young 1 as a Princess Point short-term seasonal camp, based on his assumption that such first terrace settlements on the top of knolls by river banks were used as temporary places when basecamps were flooded. At present we have no further faunal or botanical evidence to point to precise seasonality and/or scheduling of the site. We have no direct data to disagree with Stothers' assumption that Young 1 was a short-term settlement. However, density of artifacts from our test-excavation yield 687 per one square-meter, suggesting that there is potential use of site as relatively long-term settlement. Extensive excavation is necessary to confirm this view.

Grand Banks (AfGx-3)

The Grand Banks site was first identified and tested by Stothers in the 1970s (Stothers 1977). This well-preserved site appears to represent an extensive floodplain occupation. According to

Stothers, Grand Banks might have been a seasonal camp site occupied from the late spring to early fall. Stothers believes that the occupants of the Grand Banks site would have dispersed into upland areas in the winter due to annual flooding and ice rafting (Stothers 1977:123). This model is being challenged by data recovered through excavations at Grand Banks and by a geomorphological survey of the Grand Banks bar (Crawford and Smith 1996; Crawford et al. 1998; Walker et al. 1997).

In the course of three field seasons, a total of 81 square meters was excavated at the Grand Banks site (Fig. 2.2). We intended to place our test units in the general area of the site where Stothers carried out his test excavations. Unlike Stothers, who excavated directly on the river bank, we worked some 10 meters away from the bank, initially placing 2 one-meter square test units in the southern part of this area (Area A), and 4 test units 20 meters to the north (Area C). We soon

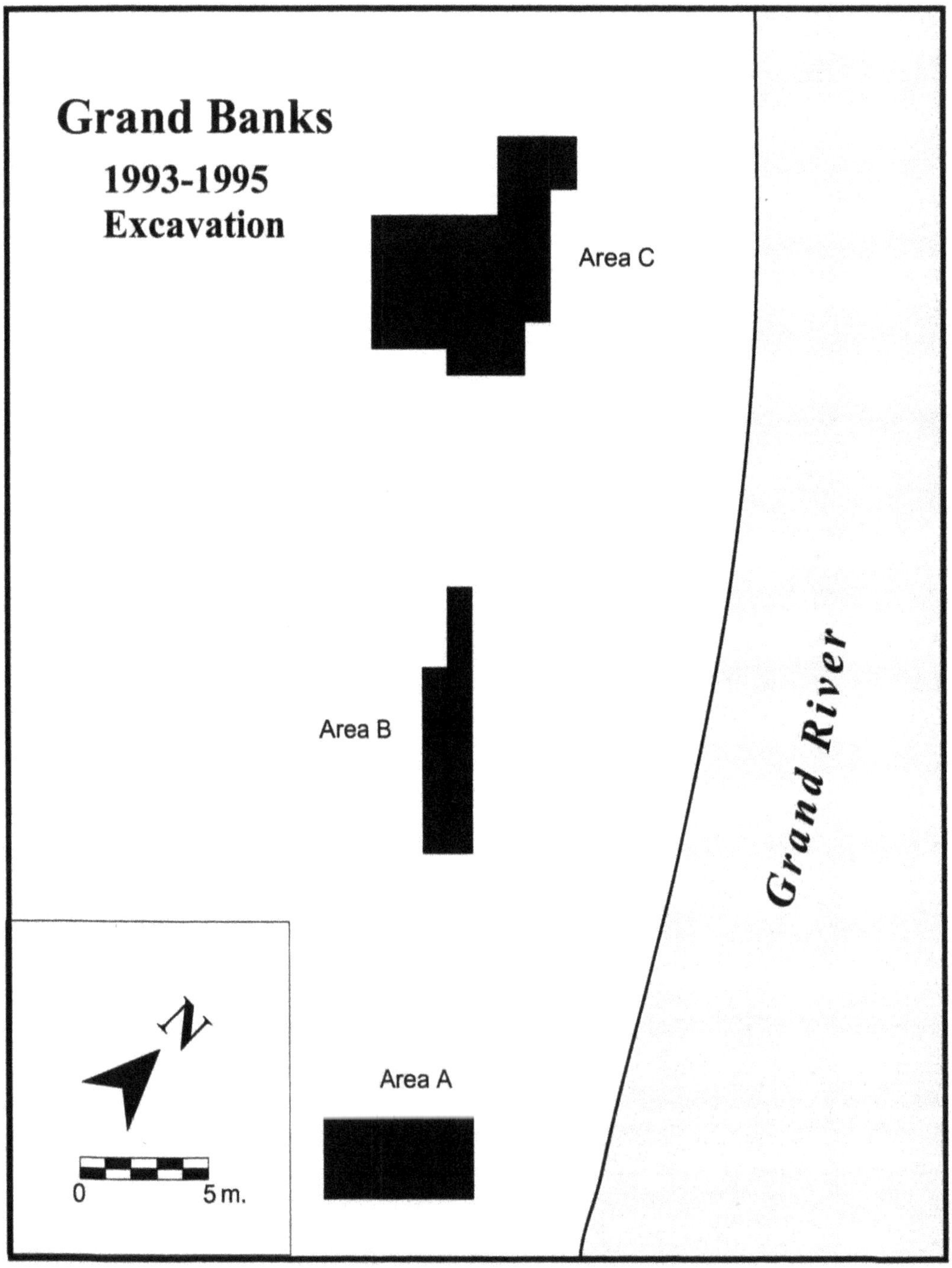

Fig. 2.2: Map of the Grand Banks Site (AfGx-003).

discovered that the stratigraphic profiles from the two areas were completely different. In Area A two cultural layers are separated by accumulated alluvium and reach a depth of at least 2 meters below the surface. Area C shows no indication of similar cultural deposits and non-cultural "subsoil" is reached at a depth of only 75 cm. In both areas, however, there is evidence that the top 50 cm has been ploughed.

In order to clarify our understanding of the stratigraphy, we placed a six meter trench (Area B) between the two areas. We also expanded Area C in order to clarify the patterning of post moulds. Fortunately, the stratigraphy of Area B clearly links Areas A and C. Details will be given in the following section as they are critical to our understanding of site formation processes at Grand Banks. Radiocarbon dates were obtained on corn remains found during the first two seasons and provided dates for this multi-component site. The lower cultural layer represents a Late Archaic or Early Woodland occupation and the upper one represents a Princess Point occupation. Given this, the goals of the third field season were to simply enhance our knowledge of settlement patterns and to increase the size of the artifact sample. During the 1995 season, we expanded all three excavation areas. Due to our primary interest in the Princess Point component, we only dug through the Princess Point deposits, and left the lower cultural layer untouched.

Nearly 56,000 artifacts (not including bone and plant remains) were recovered from Grand Banks (Photo 3 & 4) and a number of post moulds, hearths, and pits were exposed and excavated. About half the artifacts recovered from the site are ceramics, most of which are typically Princess Point products. Ceramic pipes are rare at Grand Banks. A large amount of soil was taken for flotation. During the 1993 season, nearly 2000 liters of soil were processed, and abundant plant remains were recovered (Bowyer 1995).

The results of our fieldwork indicate that the Grand Banks site formation processes may differ from those originally suggested by Stothers. That the Grand Banks site represents a spring to fall seasonal campsite is now open to question.

Grand Banks Site Formation

A team from the Department of Geography at the University of Toronto carried out a geomorphological survey of the lateral bar on which the Grand Banks site lies. Approximately 150 core samples were taken from the entire bar (1000 x 200 m) including a portion of the archaeological site. Walker and Desloges (Walker 1995) conclude that:

> The Grand Banks lateral bar site represents a long-term, slowly accreting landform that has evolved under semi-confined, low-energy channel conditions. Alluvial soil development supports this belief. There is no evidence, however, to support catastrophic stripping and subsequent re-deposition of alluvium, nor is there any indication that channel migration and lateral accretion have contributed significantly to bar development (Walker 1995:56-57).

If this statement is correct, we can assume that during the Princess Point occupation the Grand Banks site was not subjected to regular catastrophic floods and therefore the occupants of the site may not have been affected by the flood plain conditions. Grand Banks need not have been seasonally abandoned. As Crawford and Smith (1998) point out, the geomorphologic evidence supports the notion that the flood plains were stable to the extent that year-round occupations were feasible.

A diagram of the stratigraphic profile of Area B is presented in Figure 2.3. It is of note that two dark brown (10YR3/2) cultural layers (Paleosols) are present in southern parts of the excavation units. A radiocarbon date from Stothers' 1970s excavation dates to 3010+/-65 b.p. cal. 1379 - 1126 B.C. This was rejected by Stothers (1977) as too old. The dating of the Paleosol I (the lower cultural layer) is supported by our own conventional radiocarbon assay on charcoal dated to 3120+/-80 b.p. cal 1520 - 1150 B.C. (Beta-75096). The closeness of this date to Stothers' early date is not coincidental: Stothers' early date must have come from Paleosol 1 which we believe he either ignored or misidentified as Princess Point.

The Paleosol II (the upper cultural layer) has been dated by three AMS radiocarbon dates, all on corn remains recovered from Area A: 1250+/-80 b.p. cal. A.D. 780 (TO-4585); 1500+/-150 b.p. cal A.D. 570 (TO-5308); and 1570+/-150 b.p. cal A.D. 535 (TO-5307) (see also Crawford and Smith 1997; Smith 1997). These dates confirm that the later occupation at Grand Banks falls into the early to middle phase of the Middle to Late Woodland transition period (A.D. 500 - 900); support for this also comes from ceramic seriation (Bekerman 1995). Most importantly, these dates are significant for two reasons: (1) this represents the earliest date of maize growing in Ontario; and (2) it may push back the initial date for this culture from A.D. 650 to approximately A.D. 500.

Radiocarbon dates enable us to sort out the chronology of the two paleosols; however, the relationship between Areas A and C still need explaining. The excavation of Area B clarified this concern (Fig. 2.3). In the south portion of the trench, the two paleosols are separated by the sterile alluvium similar to the one shown in Area A. But the Paleosols do not exist at the extreme north portion of the trench, as well as in Area C. In the center of the Area B, the Paleosol 1 rises rapidly to join the Paleosol 2, and further north they appear to be mixed and disturbed by ploughing. According to Ian Walker (1994 personal communication), who carried out the geomorphological survey at the site, Area C may have been a small knoll. If this assumption is correct, the paleosols may have been truncated, either through extensive land-use or erosion. There is supportive evidence from fieldwork observations for this assumption. The confusing pattern of post moulds and features of different ages in Area C may be interpreted as multiple components. In addition to Late Archaic/Early Woodland and Princess Point materials, a couple of large rock clusters attest to modern usage of this area. A radiocarbon date of charcoal from a post in this area indicates that it was recent: 140+/-70 b.p. (Beta-75094).

Although our excavations at Area C were recovered a large quantity of artifacts (about 24% of the total), interpretation of this material is difficult in light of the apparent disturbances. With this in mind, I selected flaked stone artifacts from the Paleosol II only in the Areas A and B for the purpose of this study.

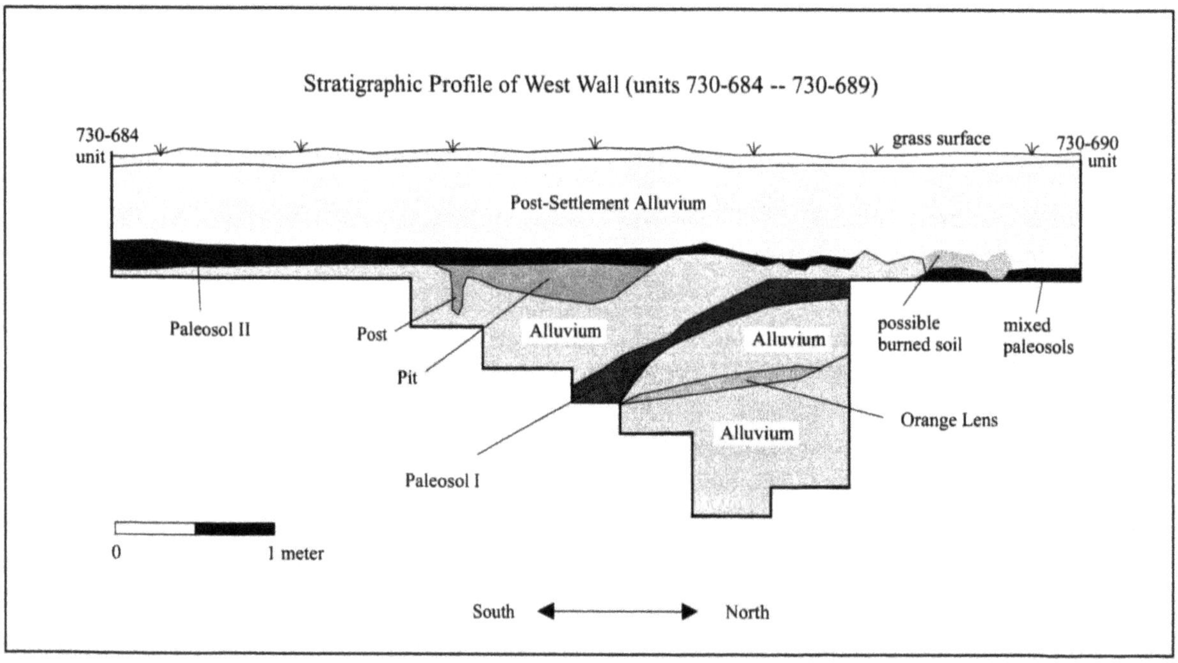

Fig. 2.3 The Grand Banks Stratigraphy (Area B).

Ecological Setting of the Site

The time period under discussion spans from roughly A.D. 500 to 1,000. During this time, the physical and biological environment of the study region did not differ greatly from today (Karrow and Warner 1990). During the Middle to Late Holocene, ecological conditions of southwestern Ontario were relatively stable. The following discussion will focus on the physical setting of the study region, the palaeoclimate, the natural resources available, and the regional bedrock formations.

The Physical Setting and Palaeoenvironment

The three sites under investigation are all located in the riverine environment of the Grand River Valley. The Grand River is the largest river in the Lake Erie drainage system in Ontario (Chapman and Putnam 1969:139-147). The river itself is over 400 km long and, with its many tributaries, drains approximately 6000 km^2 of southwestern Ontario. The river flows onto a low lying plain along the north shore of Lake Erie. According to Chapman and Putnam (1969), the Grand River divides into two parts at the city of Brantford. In its upper part, the river and its tributaries mostly flow in spillways formed in the till plain, whereas the lower Grand River flows in a channel across the lake plain. The river has a history of flooding, and many lateral bars have formed in the lower Grand River (Chapman and Putnam 1969).

These floodplains have provided suitable conditions for human settlement throughout the post-glacial era. According to information on file at the Ontario Ministry of Culture, Tourism and Recreation, more than 700 archaeological sites are located in the Lower Grand River Valley, of which as many as 50 sites yield evidence of Princess Point occupations (Fig. 2.4).

Floodplain dynamics and their relationship to human settlement have been studied by Crawford and his colleagues in some detail (Crawford et al. 1998). They identified two stable periods at Grand Banks which likely represent regional equilibria (see above). During these periods, two paleosols developed at the Grand Banks site. The more recent paleosol, PII, dates from at least 1,500 years ago to the nineteenth century, while the earlier palaeosol, PI, appears to date from ca. 3,200 to 2,800 years ago. It was likely that during the Princess Point occupation there was no major threat of flooding.

The Lower Grand River Valley is part of a humid and warm region of southern Ontario. The average mean temperature is about -10°C in January and 25°C in July, with average annual precipitation ranging between 65-900 mm. The valley is part of a region of Ontario which has an annual growing season of about 150 frost free days (Brown et al. 1974; Presant and Acton 1974).

The Holocene pollen record from Southwestern Ontario indicates the vegetation history through changes in tree populations. The generalized pollen diagram (see Karrow and Warner 1990 Table 2.1) suggests that elm, maple, and beech populations had increased steadily during the period between 8,000 to 1,000 BP, but decreased sharply around 1,000 BP. Spruce, red/jack pine, and white pine populations, which were prevalent in the Early Holocene, significantly decreased during the Middle to Late Holocene. There occurred a climatic shift to cooler and wetter conditions during A.D. 350-400 (Bryson and Wendland 1967; Baerreis and Bryson 1965), which are referred to as the Scandic Episode. Stothers (1977; see also Stothers and Yarnell 1977) claimed that this climatic amelioration is responsible for the introduction of corn in the Lower Great Lakes Region. The palaeoclimatic trends during this period, as suggested by Campbell and Campbell (1989), indicate a short warm period from A.D. 900 to 1,250, followed by a gradual cooling of about 2°C which lasted until about A.D. 1,800. The climate then rapidly warmed and remained at modern range, which are similar to those of the period from A.D. 900 to 1,250. These small climatic fluctuations over the last 1,100 years are clearly shown on pollen diagrams from Crawford Lake and Gignac Lake, Ontario (McAndrews 1988; Burden et al. 1986) and Marion Lake, Michigan (Bernabo 1981; also see Campbell and Campbell 1989:15).

Soil types in this area are light-textured

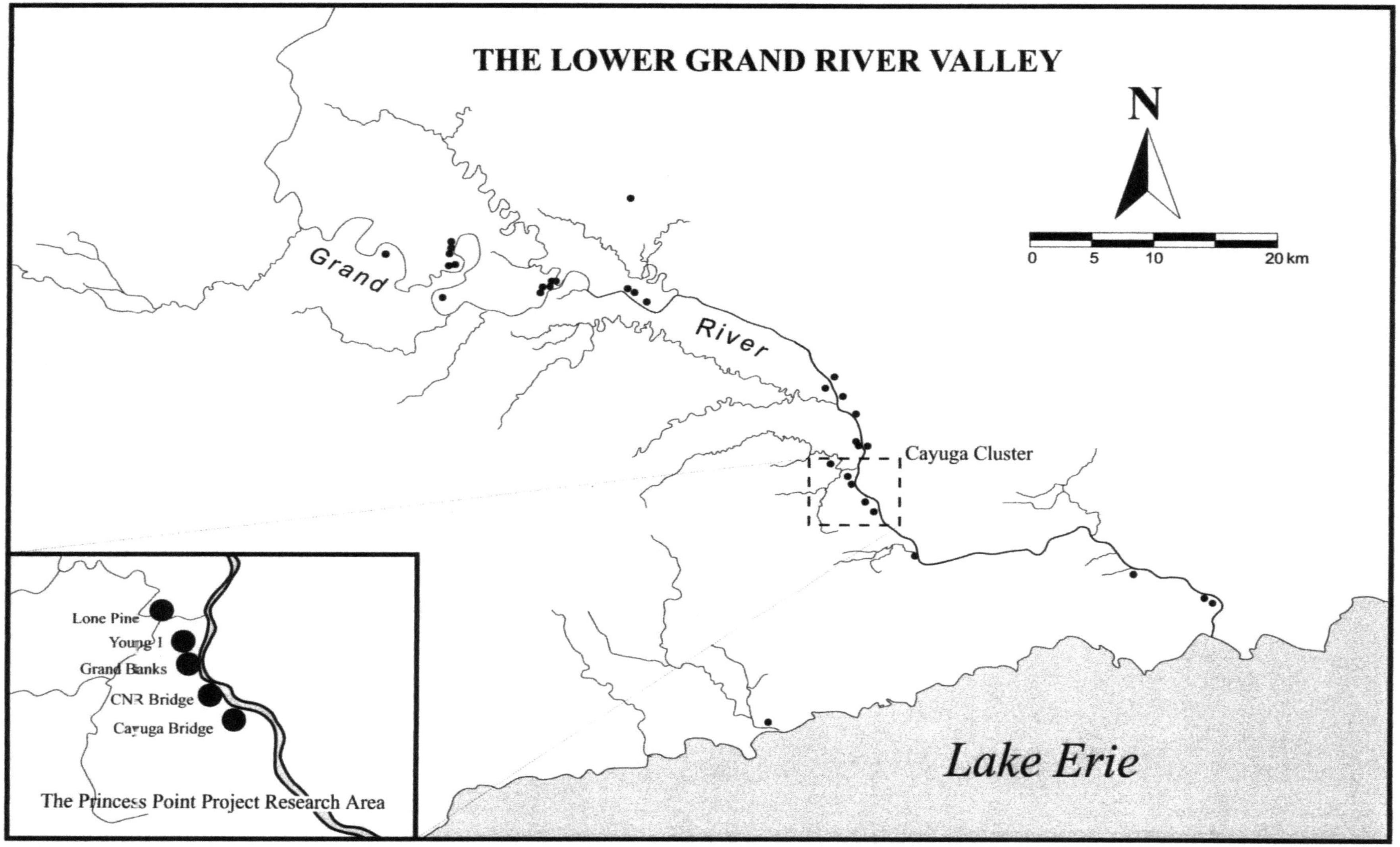

Fig. 2.4: Map of the Lower Grand River Valley showing the Princess Point Site-Clusters.

Haldimand clay and Oneida clay loams. These soils have developed under a deciduous or mixed forest vegetation; in the study region these are represented by the Carolinian Biotic Zone and the Deciduous Forest Region. The Haldimand clays are considered favorable for fruit crops. With this ecological setting, the floodplain of the lower Grand Valley provided favorable physical conditions for agricultural activities during the first millennium AD.

Natural Resources

The Lower Grand River Valley falls into the Carolinian Biotic Region (Dice 1943), which is characterized by climax hardwood forest dominated by nut-bearing trees. Oak, elm, maple, ash, hickories, and beech are present. According to McAndrews and Boyko-Diakonow (1989), from the Middle Holocene to A.D.1,000 the pollen record from Crawford Lake, southern Ontario, is dominated by beech, maple, elm, oak, and birch. This mixed hardwood forest shifted slightly to a pine-oak-birch dominated forest at ca. A.D. 1,400, and this is probably in part attributable to clearing of the forest and to old field succession (McAndrews and Boyko-Diaonow 1989). However, it is more probable that the cooling of the climate c.a. A.D.1,100 would have caused the death of beech trees while allowing oak and pine trees to grow (Campbell and Campbell 1992). After A.D. 1,850, pollen diagrams clearly indicate a dramatic increase in grass and ragweed, reflecting the clearing of forest for agricultural purposes. This serial ecological succession might generate favorable habitats for deer and other animals, as well as useful plant resources.

The preliminary study of archaeological plant remains, recovered by flotation from the Grand Banks site, provides a basis for reconstructing the past environment (Bowyer 1995; Crawford 1997). Three important cultigens are identified: maize, chenopods, and sunflowers. A number of nut species were found, including acorn (*Quercus sp.*), hickory (*Carya sp.*), and butternut (*Juglans cinerea*). In addition, abundant remains of several fleshy fruits were also recovered. They include American nightshade (*Solanum americanum*), bramble (*Rubus sp.*), ground cherry (*Physalis sp.*), and strawberry (*Fragaria virginiana*).

Although, to date, there is no detailed faunal analysis available from the study period, preliminary studies of faunal remains from Lone Pine indicate that the Lower Grand River Valley provided a habitat for white-tailed deer, gray squirrels, red squirrels, and raccoons (Cabeceiras 1994; Knight 1993). According to Campbell and Campbell (1989), based upon their analysis of Neutral faunal assemblages, the animal species habituated to forests, such as moose, snowshoe hare, wolf, black bear, and chipmunk increased in importance after A.D. 1,450, while the field species (eastern cottontail, woodchuck, grey squirrel, mouse, and short-tailed weasel) decreased. Campbell and Campbell attribute this change to climatic deterioration, which is reflected by a shift in subsistence strategy at Neutral sites. Fish from the Grand River and its tributaries were also very important in local diets. Identifiable fish remains from Lone Pine comprise a high percentage of the total faunal sample analyzed. Migratory birds were likely hunted, although we have no evidence for this as yet.

Bedrock Formation and Raw Materials

The geological formations of the Lower Grand River Valley consist mainly of Paleozoic carbonate and fine sedimentary rocks with occasional outcrops of Silurian dolostone and Devonian cherty limestone and sandstone (Chapman and Putnam 1969). The Lower Grand River Valley area is dominated by two chert formations of the Devonian Group: the Bois Blanc and Onondaga formations (Eley and von Bitter 1989; Leudtke 1992; Quin 1996). The Onondaga formation, consisting of a variety of Onondaga chert, has a fine grained structure and a conchoidal fracture pattern. It is an ideal raw material for making tools. Another chert type, Haldimand chert from the Bois Blanc formation, is regarded as a poorer quality chert with only a medium to fine crystalline structure and is thus considered a less easily workable material (Eley and von Bitter 1989:19). The two kinds of chert are distinguishable on the basis of color. While all Onondaga cherts are medium to dark grey and black, Haldimand chert is light grey to white in colour (Quin 1996; Eley and von Bitter 1989; Parker 1986). These two formations run along the west side of the Grand River to the Lake Erie shore at which point they run due east, essentially paral-

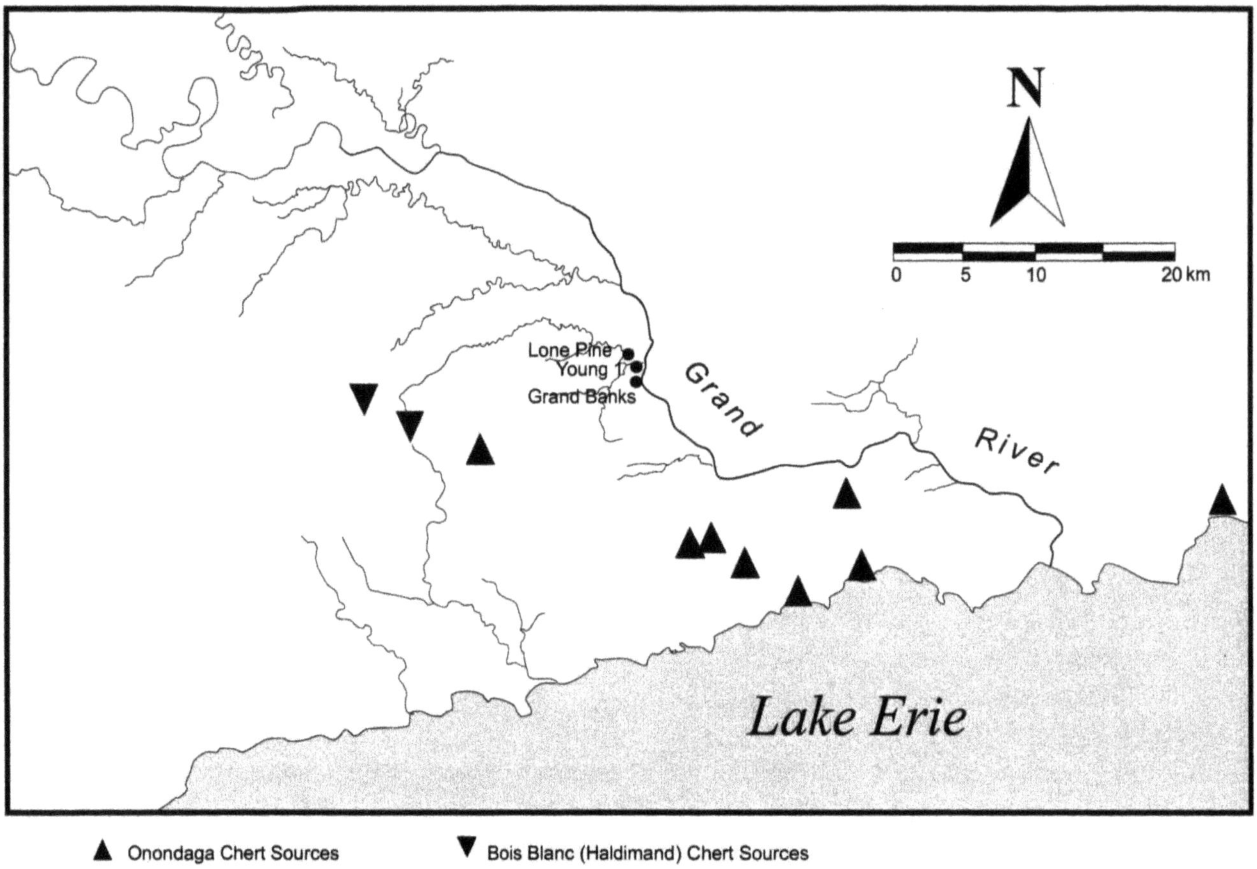

Fig. 2.5: Map of Chert Sources.

leling the shoreline to the Niagara River. Ormerod (1994), in his preliminary study of the Lone Pine assemblage, observed that only these two types of raw material were used at this occupation site. However, other types of chert, such as those from the Ancaster and Dundee formations, are found in the study region and were probably also used by the inhabitants in the Lower Grand River Valley.

The inhabitants of the Lower Grand River Valley likely used mainly Onondaga cherts, as they were both plentiful and easily accessible. In a field survey of chert outcrops and quarries located between the north shore of Lake Erie and the Grand Banks site, we identified 8 sites as Onondaga chert sources and 2 as Haldimand chert sources (Fig. 2.5). They are, on average, 15 km from the Grand Banks site. Two outcrops of Bois Blanc chert were also visited. These two outcrops are closer to Grand Banks, within 10 km of the site. Quin (1996) has conducted a study of Princess Point lithic raw material sourcing, using Instrumental Neutron Activation Analysis. Due to a limited archaeological sample, at present, Quin is not able to identify the exact chert sources which the Princess Point people utilized. However, his analysis demonstrates the potential for sourcing identification, and, from the data available, confirmed that the cherts used at the Grand Banks site are from the Onondaga sources along the north shore of Lake Erie (Quin 1996).

Palaeoenvironmental reconstruction suggests that the Lower Grand River Valley was a resource-rich area during the Princess Point period. The physical setting of the archaeological sites provided the prehistoric inhabitants in this region with possibilities for a mixed economic strategy of farming, hunting, fishing, and gathering (Crawford et al. 1998). It is not possible at present to determine the specific site type of Grand Banks or Lone Pine; to do so, detailed paleoethobotanical and faunal analyses are required. However, given abundant resources and favorable climatic conditions, prehistoric use of the sites could have included a variety of economic activities. Based on this assumption, stone tool production at these sites would be expected to be a generalized pattern in tool manufacturing and utilization (e.g., Bleed 1986). The rationale for such expectations will be summarized in the following chapters.

The evidence cited above indicates that during the period from A.D. 500 - 1,000 there were changes in the food procurement system and settlement patterns of the people occupying southwestern Ontario. There were likely corresponding changes in social organization. Researchers have found evidence for lithic technological changes during this period in the Northeast as well as during the shift to agriculture (Fitting 1975; Mason 1981; Grinffin 1983; Parry and Kelly 1987; Yerkes 1987; Jeske 1990; Ellis and Ferris 1990). If lithic technology did change accordingly, then how so? In order to address this question, it is first necessary to characterize the Princess Point lithic assemblages. Subsequently inferences regarding the Princess Point lithic production system can be addressed.

CHAPTER 3

METHODS OF LITHIC ANALYSIS

Bearing in mind that stone tools comprise only a part of the prehistoric tool-kit, flaked stone tools have been exclusively chosen for the purpose of this study. The lithic analysis includes artifact classification, attribute observations, use-wear examination, and data manipulation. I began by devising a classification scheme that has been used to catalogue the collection. This proposed typology is somewhat different from the lithic classification scheme currently used for Woodland lithics from Southern Ontario. The attribute observations made on the assemblages fall into two categories: (1) typo-technological characteristics and (2) edge damage formations as determined by use-wear analysis. The selection of attributes, their interpretive significance, as well as the methods used to make behavioral inferences, are presented in the following sections. Sampling strategies and data manipulation are also outlined.

Typology: Artifact Forms

Studies of archaeological typology have demonstrated that artifact classification is essential to archaeologists. Difficulties arise, however, as a result of diverse approaches to typology (Chang 1967; Whallon and Brown 1982; Klejn 1982; Dunnell 1986). My purpose here is not to provide a lengthy review of the Spaulding-Ford debate of the early 1950s (Spaulding 1953, 1960, 1982; Ford 1954a, 1954b). Adams and Adams (1991) insist that there can never be an "all-purpose" typology. Drawn from their explicit conceptual reviews, Adams and Adams (1991) propose that to be practical and cost-effective, a classificatory system should be made based on specific research needs and logical relations amongst objects (types). In accordance with these requirements, to make and use a typology is simply to describe the form of artifacts on the basis of, for example, their morphology or modification alone.

This idea is not new. In practice, archaeologists have employed this methodology for at least fifty years. Krieger (1944), in his classic article, pointed out that classification must "provide an organizational tool" in order to demonstrate "historical meaning in terms of behavior patterns" (Krieger 1944:272). Integrated with technological studies of lithics, typological studies have been of considerable use in the reconstruction of cultural history in the Old World (cf. Henry's (1973, 1989, 1995) works on Jordanian prehistory). What is new about Adams and Adams' classification theory is that a typology should be regarded as a conceptual system that comprises a comprehensive set of *mutually exclusive types*, and based upon a set of common criteria dictated by the purpose of the typologist. They further state that "each type is a category created by the typologist, into which he or she can place discrete entities having specific identifying characteristics, to distinguish them from entities having other characteristics, in a way that is meaningful to the purpose of the typology" (Adams and Adams 1991:91).

This definition, in my opinion, places typology into a practical classification framework, meaning that any archaeological research can start with a working typology as long as it is developed with reference to a specific purpose (e.g., Read and Russell 1996). In this case, the purpose of my working typology is to establish a relatively standardized framework whereby types of stone artifacts can be categorized for future analysis and comparison. Following Adams and Adams (1991), a *typology* is a particular kind of classification made specifically for the sorting of analytical objects into mutually exclusive categories (sometime called types). The types assigned to the same lithic class must be logically related. In this regard, the system of lithic classification generally used in southern Ontario archaeology is problematic because, for example, "utilized flakes" are included as a *type*.

A Critique of the "Utilized Flakes" Category

Two general categories: tools and debitage (non-tool), are in current lithic classification systems used in Ontario archaeology. Tools are normally further subdivided into "formal tools" including bifacially and unifacially retouched types, and "informal tools" mainly referring to utilized flakes. In most survey and excavation reports, the separation of utilized flakes from other reduction byproducts is purposeful as it helps to assist in tool function interpretations. While this may appear to simplify the process of functional analyses of lithic artifacts, it in fact causes typological confusion. I would suggest an alternative typology that excludes "utilized flakes" (Shen 1999). I dismiss the use of this term for following three reasons:

First, current definitions of utilized flakes are ambiguous and inconsistent. Although it is commonly recognized that utilized flakes are those flakes bearing use-damage on flaking edges, the criteria used to define utilized edge(s) remains unclear. Storck has used the designation "utilized flake" or "worked flake" for those that "exhibit a small amount of unifacial marginal retouch along either a single edge or a short segment of one edge" (Storck 1974:6, see also 1978:37). Lennox characterized "utilized flakes" as those specimens showing "use retouch resulting in the removal of a series of small flakes (less than 2 mm) from one or more edges" (Lennox 1993:6, see also 1986:228). Apparently minimal secondary retouch of margins has been regarded as an indication of flake utilization. While Storck considered the shapes of used margins (straight, convex, or concave) as evidence of use, Lennox used size of flaking scars as a criterion to distinguish use retouch from deliberate retouch. Recently, Storck modified his definition of utilized flakes to emphasize the size of the edge damage/retouch, stating that "utilized flakes also exhibit very small, irregularly spaced flake removals along one or more edges, but the flake scars are so small as to suggest they were formed by use wear rather than by deliberate pressure retouch" (Storck 1997:84-85).

Inconsistencies in defining utilized flakes also exist among other researchers. Some search for evidence of marginal retouch (Deller 1976:14), others for use marks in the form of nibbling along the edge (Finlayson 1977:179), and yet others, for continuous flaking scars (Timmins 1992b:241; Bursey 1994:51). An assumption underlying these definitions is that edge modification on utilized flakes was caused by tool utilization. However, microscopic examinations of flake edges suggest damage may caused by forces other than use. In process of lithic analysis, I believe that it is useful to distinguish edge-damaged flakes from unmodified flakes since the former may have a greater possibility of being utilized than the latter. Thus, in practice, such edge-damaged flakes seem more important than unmodified flakes. However, in a typology that embraces a strong morphological aspect, a group of flakes with use traces could be drawn from the unmodified, retouched, or shaped flakes. Therefore, the edge-damaged flakes should not be considered "informal tools" in the first place.

Secondly, utilized flakes are not morphologically distinguishable types, but result from demonstrable use-wear which may occur on a variety of flake types. If we follow the above definitions, it is clear that "utilized flakes" may include any flake types bearing edge damage. Lennox (1986) and others (e.g., Timmins 1992b) further divide utilized flakes into various sub-types according to the morphology of unretouched flakes, for instance, utilized primary flakes or utilized secondary flakes, in order to determine the possible use preferences of unmodified flakes at any given occupational sites. This practice suggests that utilized flake is not a *mutually exclusive category*, but a group of typed flakes bearing worn edges. Thus, there may be "utilized primary flakes," or "utilized biface thinning flakes," or "utilized blades," and so forth. Typologies in Ontario which include utilized flakes appear to be based on a biased assumption that all lithic artifacts fall into one of two groups: used (tools) and unused (debitage flakes) artifacts. However, use-wear analysis demonstrates that not all artifacts classified as tools on the basis of morphology were used, and those which were used may not have been used in the manner implied by their names (e.g., a scraper may not have been used for scraping) (Odell 1981). Therefore, in this study I treat all shaped, edge-damaged, and unretouched flakes as use-potential tools, and not as utilized pieces unless their use-wear is identified.

Thirdly, the identification of flakes as utilized in the absence of proper equipment and training produces highly inaccurate results. Although Fox once correctly pointed out that separating these utilized pieces depends on actual use retouch and use polish, no further explanations of how such "non-magnification" approaches should be carried out have been offered (Fox 1979:68). Tomenchuk, who conducted a well-controlled experiment on edge-damaged types, noted: "it is my experience that at present, no criterion or technique exists that is infallible in distinguishing between manufacture and utilization damage" (Tomenchuk 1985:508). Methodologically, to determine whether or not a flake was actually used, one must rely on careful observation of use traces, through microscopic examination of a combination of microchipping fracture and abrasions. Yerkes and Kardulias correctly point out: "microwear analysis has helped change our view of lithic typology, especially as it relates to functional categories of artifacts such as 'utilized flakes'" (Yerkes and Kardulias 1993:103).

To illustrate this point, we carried out an experiment similar to the one performed by Young and Bamforth (1990) at the University of Toronto in early 1995. Three graduate students, each with some lithic analysis experience, volunteered to participate in a blind test. They were given a set of 45 pieces that consisted of both used and unused flakes. Each was instructed that only a portion of these flakes was utilized by me for various tasks, while the others were intentionally retouched, or not modified at all. Participants were asked as if they were cataloging the assemblage so as to determine which flakes were qualified as utilized flakes. One of them was also permitted to use a 10x hand-lens. The goal of this exercise was to determine how well a practitioner could judge whether a flake had been utilized, and what factors would obscure that judgment.

The results, presented in Table 3.1, indicate that about half the specimens were misinterpreted. Coincidentally, the scores of these three participants are the same: 25 out of 45 were identified correctly. Use of the hand lens did not make a difference in the results. During the test, I observed that the participants regarded some edge-retouched or abrasion flakes as utilized flakes. Most unretouched and unused flakes (#12-17) were easily picked out. However, it should also be noted that in this experimental set there were no naturally or culturally damaged items as a result of washing, transporting, trampling, or trowel retouching. These types of damage are common in archaeological samples. In the utilized experimental samples, almost all those used on soft materials such as meat, hide, and the like (#25-28, 31, 34, 38, 43), were incorrectly identified. Polish, rounding, and microchipping fractures resulting from these activities can only be observed under a microscope. By contrast, flakes used on hard materials such as bone, antler, and dry wood left clear chipping scars on the utilized edge which were easily detected by the blind testers. However, three experimental flakes used to bore either hard materials or soft materials (#23, 25, 34) were totally misinterpreted by the three practitioners because the utilized portions of these flakes were limited to their tips, leaving the lateral edges undamaged.

These results match those of Young and Bamforth's (Young and Bamforth 1990:Tables 2 and 3) experiment. Their test involved nine archaeologists and eleven used and unused flakes and minimally retouched flakes. The results are also "not very encouraging: only 36 out of 144 possible inferences, or 25 percent, were correct." (Young and Bamforth 1991:404). Their experiment, supplemented by ours, suggests that simply sorting utilized flakes by distinguishing edge damage is highly inaccurate. Macroscopic identification of used edges on flaked stone artifacts should be carried out with extreme caution if it must be carried out at all (Young and Bamforth 1991:408).

It should be noted that, while I strongly suggest the use of "utilized flakes" as a categorized type in lithic classification be eliminated, the term is still useful describing tool use-patterning in general. In this study, I use "utilized flakes," or "utilized pieces," to refer only to objects which are determined to have been used through microscopic examination. In another words, "utilized flakes" represents pieces of all types of lithic artifacts, not a single type.

A Working Typology

In my study of lithic artifacts, I exclude the term "tool" and "utilized flake" from lithic classification (Shen 2000, in press). This system suggests that all lithic artifacts from an archaeological assemblage can be divided into four classes: *cores, formed types, unretouched flakes,* and *flak-*

Table 3.1: Blind Tests on Utilized Flakes.

Specimen #	Raw Material	Edge Modification	Utilization	A	B	C
1	Onondaga	retouched	none	T	T	F
2	Onondaga	retouched	none	T	F	F
3	Onondaga	retouched	none	T	T	T
4	Onondaga	retouched	none	T	T	F
5	Onondaga	retouched	none	F	F	F
6	Onondaga	retouched	none	F	F	T
7	Haldimand	retouched	none	F	T	T
8	Haldimand	retouched	none	T	T	T
9	Haldimand	retouched	none	T	F	F
10	Haldimand	retouched	none	F	F	F
11	Haldimand	retouched	none	F	F	F
12	Haldimand	not retouched	none	T	T	T
13	Onondaga	not retouched	none	T	T	F
14	Haldimand	not retouched	none	T	T	T
15	Haldimand	not retouched	none	T	T	T
16	Onondaga	not retouched	none	T	T	T
17	Onondaga	not retouched	none	F	T	T
18	Obsidian	not retouched	none	F	F	T
19	Obsidian	retouched	none	F	T	F
20	Obsidian	retouched	none	F	T	F
21	Obsidian	not retouched	none	T	F	F
22	Obsidian	not retouched	scraping hard wood	T	T	T
23	Obsidian	not retouched	sawing hard wood	F	F	F
24	Obsidian	not retouched	scraping hard wood	T	T	T
25	Onondaga	not retouched	drilling fish skin	F	F	F
26	Onondaga	not retouched	sawing hard wood	T	F	F
27	Onondaga	not retouched	drilling pig hide	T	F	F
28	Onondaga	not retouched	scraping fish	F	F	T
29	Onondaga	not retouched	cutting pork meat	F	F	F
30	Onondaga	not retouched	cutting fresh wood	T	T	T
31	Onondaga	not retouched	scraping carrots	F	F	F
32	Onondaga	not retouched	sawing antler	T	T	T
33	Onondaga	not retouched	cutting fresh wood	T	T	T
34	Onondaga	not retouched	boring carrots	F	F	F
35	Onondaga	not retouched	scraping dried hide	T	T	T
36	Onondaga	not retouched	slicing dried hide	F	F	T
37	Onondaga	not retouched	sawing dried bone	T	T	T
38	Haldimand	not retouched	cutting carrots	F	F	T
39	Onondaga	not retouched	scraping soaked bone	T	T	T
40	Haldimand	not retouched	sawing dried bone	T	T	F
41	Onondaga	not retouched	scraping hard wood	T	T	T
42	Onondaga	not retouched	scraping fresh wood	F	T	T
43	Onondaga	not retouched	scraping fresh bone	F	F	F
44	Onondaga	not retouched	scraping dried hide	T	T	T
45	Onondaga	not retouched	scraping sweet potato	F	F	F
Correction Rate				55.60%	55.60%	55.60%

*: RM=Raw Material; A, B, C= scores for three experimenters; T=True; F=False

ing debris (Fig. 3.1).

Formed Types

Use of the term "formed type" is proposed here to replace the term "tool" category. An object may be a "tool" because it is manufactured according to a particular design specification for a particular purpose, or because it is utilized as such. Some designs reflect their functions, such as projectile points, scrapers and drills. These objects exhibit regular, patterned alterations of edges and surfaces as a result of deliberate retouch. Their intentional modification for specific use – their morphology – makes them "tools," as this word is commonly understood and used by Ontario archaeologists engaged in lithic classification.

A problem occurs, however, when the word "tool" is used to refer to objects classified according to functional criteria as well. In a functional typology, this same word "tool" may refer to artifacts that have been used, regardless of their shape. Although shaped artifacts have a particular use-related function or functions, the study of artifact function is methodologically distinct, depending as it does on use-wear analysis. A morphological "scraper" might not always have been used for scraping whereas, conversely, a flake of no distinctive shape might have been used as a scraper. To avoid confusion in terminology, a morphological scraper and other type-specific objects should not, perhaps, be considered as belonging to the more general category of "tool," but rather to some other-named conceptual category, such as the "type collection," a term proposed by Odell (1985, 1996, 1998). Adoption of one of these alternative terms for "tool" would make it possible for the term "tool" to refer solely to objects that are interpreted as having been utilized.

The formed type, therefore, is defined here as a cluster of pieces that display obvious and intentional modification through retouch into certain forms. Intentional modification refers to deliberate retouch, and shows regular patterned alterations on the edges or surfaces of lithic objects. Formed types such as projectile point, biface, uniface, scraper, drill, etc., are referred to only on the basis of physical appearance, rather than according to the functional identity of the pieces. It is worth noting that pieces with macro-fractures, which may have been caused by natural forces, human trampling, or direct utilization, should not be considered as variations of formed types. The same holds true for those specimens displaying a physical similarity to the type forms, but which lack evidence of intentional retouch. As I mentioned earlier, manufacture retouch is often difficult to distinguish from use or non-use edge damage. To set a commonly acceptable rule, I deliberately defined retouch based upon: (1) the presence of at least three or more contiguous flake scars in a similar pattern oriented roughly perpendicular to the edge; and (2) the maximum length of the individual flake scar, which should be at least 2 mm.

Investment of human energy in the modification process makes it likely that most formed type artifacts had particular use-related function(s). However, the study of artifact function is methodologically different, and should be addressed independently of typology. Lithic artifacts might not be modified for functional reasons only: rather, some modification reflect manifestations of style. Reconstruction of cultural preferences in retouch techniques among certain prehistoric groups has been documented through the study of retouch styles among certain Near East and African lithic industries (e.g., Marks 1976, 1977; Henry 1989a, 1985; Close 1978, 1989). Thus, analysis of retouch attributes must provide a focus for the class of formed types.

Artifacts assigned to the formed type class are characterized by either surfacial retouch or edge retouch. These pieces are usually named according to morphological typologies set forth by Bordes (1961) and others (e.g., Clark and Kleindienst 1974; Tixier 1974; Bar-Yosef 1970; Henry 1973). The following descriptions of eleven types are applied only to the lithic assemblages studied in this research.

Projectile Point: An object that has bilateral symmetrical edges converging to a point with bifacially or, rarely, unifacially flaked is termed a projectile point (Justice 1987). Basal modification is the most distinctive characteristic used to discriminate this type from others.

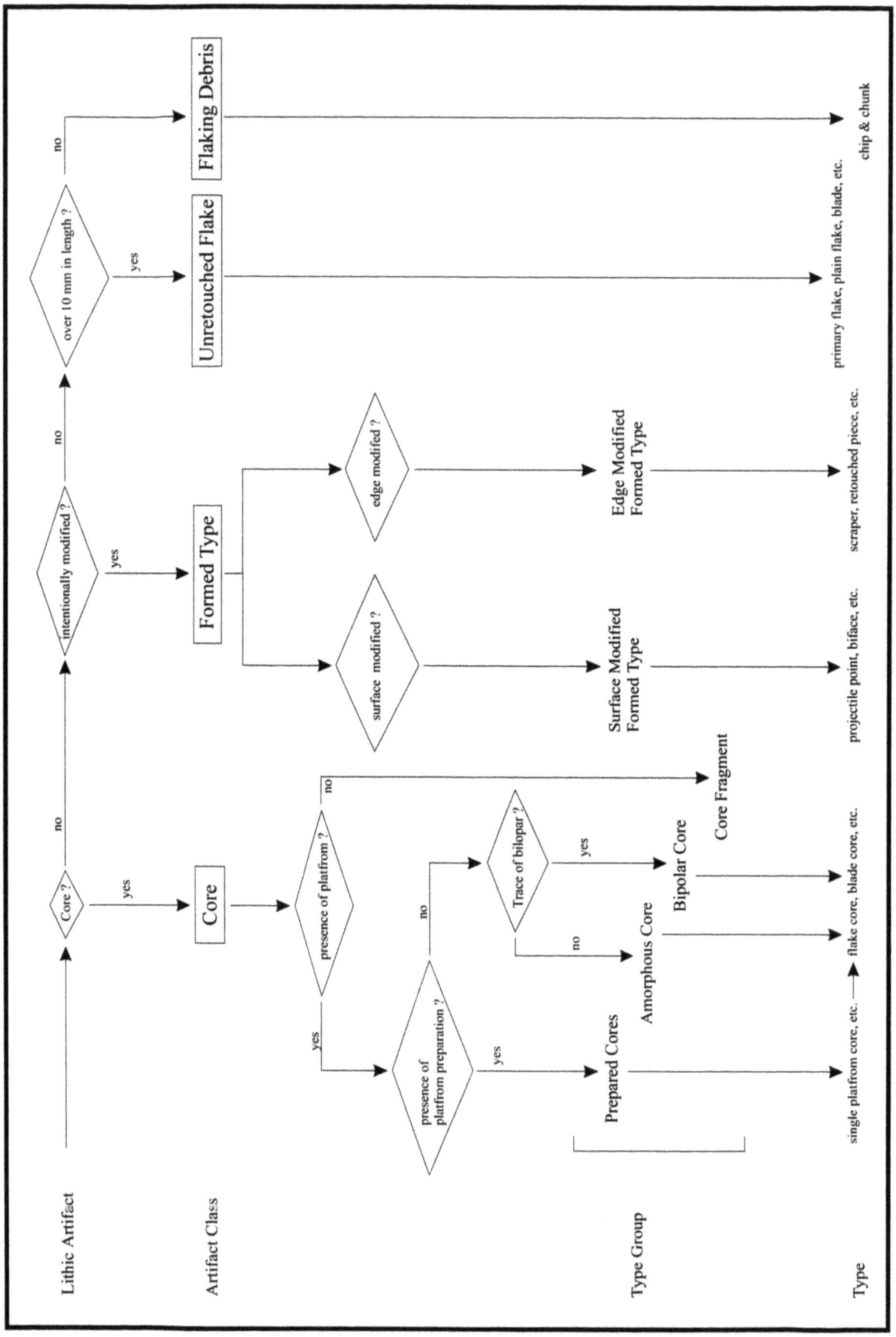

Fig. 3.1: Flow-Chart Indicating Lithic Classification Applied in This Study.

Biface: An object that is bifacially flaked, and cannot be classified otherwise is termed a biface. The item is surface retouched for the most part, rather than restricted to the periphery of the pieces. Biface subtypes are defined according to the reduction sequence outlined by Callahan (1979), except that his stage 1 (procurement) is excluded. Stages 1-3 here are his stages 2-4. The stage 1 biface, or Callahan's "initial edging biface" (Callahan 1979:10), is characterized by roughly symmetrical and widely spaced scars along the margin. The piece often displays a denticulate or curvy outline. The width-thickness ratio of a complete stage 1 biface is 2:1 or greater (Callahan 1979:10). The stage 2 biface, or "primary thinning biface," shows minimal edge projection and irregularities, and the central ridge is less sinuous than that of the previous stage. A complete stage 2 piece has a width-thickness ratio of 3:1 or 4:1, and an edge angle restricted to between 40° and 60°. A stage 3 biface has a flattened cross-section with a w/t ratio in excess of 4:1 and edge angles consistently in the 25° - 45° range. The edge of the stage 3 biface is quite straight and scars are closely spaced. One of the principal features distinguishing this stage from the previous one is that flakes frequently extend past the center of the piece and undercut previously produced flake scars from the opposite margin.

Uniface: A uniface is surfacially retouched, either on the dorsal or ventral side, and usually on the more convex surface. The flaking scars are usually invasive, extending past the central line, though scars sometimes remain along the margin.

Scraper: A scraper can be either unifacial or bifacial and displays a steeply retouched edge. One surface of the scraper is usually flat, while the other is convex. Scrapers are distinguished from other edge retouched types in that they usually possess steeper retouch and continuous modification along one or more edges. Two major subtypes of scrapers are defined according to the location of the steep edge: endscrapers and sidescrapers.

Retouched Piece: A retouched piece possesses intentional retouch, but its edge is not significantly modified into a further classifiable form. Based on the location and orientation of the retouch, retouched pieces are divided into unilateral-obverse (dorsal side), bilateral-obverse, unilateral-inverse (ventral side), bilateral-inverse, and alternated (both sides) retouched pieces. Partially retouched pieces are those having a retouched edge length comprising less than one-fifth of the edge.

Drill: A drill has a pronounced, roughly parallel-sided projection or bit. This bit is bifacially retouched and is rhomboid-to-circular in cross-section. Usually a drill is considered to be a bifacial tool. However, I consider drills to be an edge modified formed type because the bifacially retouched bit is actually an edging modification. It may become a surface-modified formed type only if the non-bit part of the object exhibits evidence of surface retouch.

Notch: A notch, as named, possesses notched retouch on the edge. More commonly, they are called *spokeshaves* in the lithic assemblages in the Northeast. Subtypes of the notch are defined according to the number of notched retouches (single, double, or multiple) (also see Tixier 1974).

Burin: A burin is defined by the technological characteristic of a blow transverse to an edge, often transverse to the piece itself. The number of blows and blow locations form different subtypes of burins. Tixier's classification (1974) of burins, derived from that of Bordes (1961), is commonly used and adopted here.

Perforator: A perforator, sometimes called a "graver," is characterized by the presence of a short and broad projection, in contrast to a drill. To qualify as a perforator, the projection must exhibit intentional modification on both sides of the tip.

Truncation: A truncation exhibits marginal, regular, and continuous retouch, truncating either a proximal, a distal, or a lateral part of an object. The modification normally exhibits abrupt retouch on the edge.

Microlith: Retouched pieces, geometric (triangle, lunate, and trapeze/rectangle) and non-geometric forms, are made on bladelets with special techniques.

Cores

A core, as commonly recognized, is an object from which flakes have been removed. Johnson (1987) points out that, to determine lithic tool production systems, one should start with the distinction between amorphous core and prepared core technology. Defining morphological types of cores depends on the appearance of core striking platforms and core facets.

Unlike cores from Old World lithic assemblages, cores recovered from the Northeastern Woodlands tend to be amorphous, which renders them difficult to classify. "Amorphous core" is commonly used to describe the core morphology recovered from this part of the world (Johnson 1986, 1989; Custer 1987; Patterson 1987). In contrast, cores which present a regular shape such as cylindrical and pyramidal cores are classified as "prepared cores" (Johnson and Morrow 1987). For the purposes of comparison I adopt these terms in this study.

Prepared cores are ones whose platforms are formed and maintained in order to remove predetermined flakes (e.g., blades). A principle feature that distinguishes this type of core from others is the presence of a platform or platforms from which most flakes (including blades) were removed in a consistent orientation. This type of core corresponds to single platform and opposed or right angle platform (two platforms) cores in the Old World (e.g., Clark and Kleindienst 1974). The prepared core technology often requires specific procedures to make standardized products, such as blades or Levallois flakes. Although very few specialized cores, if any at all, are recognized in Northeastern Woodland lithic assemblages, this type of core may represent a reduction technology that, presumably, exemplifies skillful and controlled flake removal.

Amorphous cores, in contrast, show minimal platform preparation. Flakes are removed from different directions, and as a result, cores are irregular in shape. The counterpart of this type of core in the Old World may be multi-directional cores or formless cores (Clark and Kleindienst 1974). Amorphous core technology may often indicate unstandardized production which, as argued by some authors (Johnson 1986, 1989; Custer 1987; Parry and Kelly 1987), resulted in a predominance of "expedient" flake tools.

Bipolar cores are those that produced bipolar flakes with bipolar reduction. This technique has been discussed by Binford and Quimby (1963), Crabtree (1972:42), Forsman (1975), Hayden (1980), Barham (1987), Casey (1993), and Kuijt et al. (1995). Cores are struck from one end, while the other end is placed on an anvil. They are often "triangular, rectangular, or pie-shaped, with characteristic platform crushing and flake removal on one or both ends" (Kuijt et al. 1995:118). Flakes and blades may be removed from both platforms on one or both surfaces (Clark and Kleindienst 1974:91).

Core fragment is listed as a separate subtype here because core fragments normally represent a large portion of the core assemblage. A core fragment exhibits only a partial striking platform, if any at all. One surface of the core fragment does not have any negative scars, indicating that the pieces are "blow-off" byproducts of core reduction. These pieces are relatively large blocky chunks, but they are not big enough to continue flake removal.

Unretouched Flakes

Often this group of lithic artifacts is referred to as "debitage" or "debris" in Ontario archaeological classification. They include "all waste material generated by humans in lithic reduction" (Shott 1994:70). This study defines *unretouched flakes* as a group of flakes that have been intentionally removed from cores but display no evidence of edge-retouch. These flakes are distinguished from flaking debris in that they have the potential to be modified into any of the formed types, but can also be used directly as unmodified pieces. They could be often chosen as tool blanks, so that unretouched flakes are sometime called "basic tools" (Casey 1993). The complete unretouched flakes, which are relatively large in size (>10 mm), normally display three surfaces: a dorsal surface, a ventral surface, and a striking platform surface unless the platform is linear or pointed. Corresponding to the sequence of core reduction, unretouched flakes can be further divided into the following types:

Primary flakes are those displaying at least one-third cortex or natural surface coverage on their dorsal surfaces. This includes both conventionally recognized "primary flakes" and "secondary flakes," which are also called "decortication flakes."

Core trimming flakes exhibit evidence of core preparation, sometimes called "core rejuvenation flakes." At least one side of the dorsal surface functioned as a striking platform in earlier uses of the core.

Bipolar flakes are byproducts of bipolar core technology. Thus, they often display opposing bulbs of percussion and force waves (ripples) on the ventral surface that run in opposite directions.

Biface thinning flakes are byproducts of biface production. Thus, they exhibit traits indicative of a bifacial edge on the striking platform. Most scar facets on the dorsal surface usually run transverse to the long axis of the flake. Platform angles are often acute and lipping or, rather, a diffuse bulb of percussion is present.

Plain flakes are those which do not fall into other categories. Most of them may have been intentionally removed to be used as tool blanks. These pieces possess no cortex, and are often called "tertiary" or "interior" flakes.

Blade/bladelets are also known as "prismatic flakes" in northeastern North America. They have roughly parallel edges with at least one ridge running the length of the blade/bladelet. Specific measurements are applied to them in that both have to be at least twice as long as they are wide. According to Tixier (1974), bladelets are distinguished from blades by their small size, being less than 50 mm in length and less then 12 mm in width. Since neither blades nor bladelets are a major component of unretouched flakes assemblages from late Woodland cultures, I have chosen to lump them together for analytical purposes.

Others include those types that do not regularly appear in the sampled assemblages. These include byproducts of special production such as burin spalls and microburins.

Flaking Debris

Flaking debris refers to waste flakes that are usually by-products of "blow-offs" during percussion flake removal, or tiny flakes resulting from pressure flaking. Flaking debris contains only two major subdivisions: chips and chunks.

Chips are defined as small-sized pieces, with a maximum length of less than 10 mm. *Chunks* refer to pieces with an irregular and angular form, making it difficult to determine their dorsal and ventral sides. Chunks may exceed 10 mm in size whereas chips, by definition, may not. On the other hand, chips exhibit distinguishable ventral and dorsal surfaces, but chunks do not. The distinction between a chunk and a small-sized core fragment is sometimes unclear, but principally, the latter exhibits negative flaking scars while the former does not.

Unfortunately, due to the nature of flaking debris, waste pieces have not been given adequate attention until recently (Ahler 1989a, 1989b; Patterson and Sollberger 1979; Stahle and Dunn 1982, 1984). Baumler and Downum's (1989) experimental study of small-size flaking debris has demonstrated that by-products of unifacial retouching (e.g., scraper manufacture) are morphologically distinct from those produced by flake core reduction. The former activities, therefore, can be isolated and identified in the archaeological record.

Typo-Technological Analysis: Manufacture

Flaked stone tool manufacture generally consists of two main steps, excluding raw material procurement. The first step is to obtain proper flakes (blanks) to be used directly as tools or to be modified into tool types. This process is a form of *core reduction*. Methods of trimming nodules and reducing core size can vary; for example, bifacial reduction, flake core reduction, and bipolar core reduction have been well recognized and discussed (see Amick and Mauldin 1989). The second step, if necessary, is to fabricate the blanks into shaped objects by secondary retouch either on the surface or on the edge of the blanks. This process is referred to as *tool production* in this study. Thus, two main objectives of the typo-technological analysis are to determine: (1) what kinds of core reduction technology were employed at sites, and (2) how stone tools were produced.

An *attribute* is a measurable or observable quality of an artifact and may be either a continuous or a discrete variable. A flake can bear numerous attributes, thus selection of such attributes should be subject to the analytic purposes of the researcher and should be tested through lithic experimentation (Amick et al. 1989; Shott 1994, 1996a; Odell 1989a). Due to time and financial constraints, an experiment that I have designed to test attributes of lithic manufacture has yet to be carried out*. However, relevant experiments by others have been consulted for this study and are listed in Table 3.2.

It is of note that the attributes selected and tested by each experimenter are different, but overall their results are very informative. Shott (1994) suggests a set of effective attributes for debris analysis, including weight, cortex, dorsal scar count, platform angle, and platform facet. Odell (1989a, Fig. 10 &11) derived 13 discriminating attributes. He found, in an attempt to distinguish flake core reduction from bifacial core reduction, that attributes such as bulb scar shape, distal termination, and dorsal perimeter scarring are particularly useful (Odell 1989a). Similar conclusions are found in Tomka's experiment (1989). Bulb scar, platform facet, lateral profile and margin, have been found to be indicators of different loading forces for flake detachment (hard- and soft-percussion and pressure) (Henry et al. 1976; Hayden and Hutchings 1989), whereas cortex, dorsal scar count, and platform attributes are used in discriminating early from late stages of core reduction (Shott 1995, Mauldin and Amick 1989, Magne 1985). Kuijt et al. (1995), following Sullivan and Rozen's (1985) flake typology, demonstrated that besides bipolar core and flake morphology, the breakage of unretouched flakes can be the best indicator of bipolar core reduction. Casey's experiment demonstrates that predominant flakes, removed by bipolar reduction, exhibit crushed platforms, hinge and snap distal terminations, and flat or diffused bulb shapes (Casey 1993).

As far as metric data are concerned, analysis of size distribution has gained much attention recently (for lengthy review see Shott 1994). Several models have been proposed in order to recognize bifacial core reduction at archaeological sites, such as Patterson's log-linear model (Patterson 1981, 1982, 1990), Stahle and Dunn's weibull distribution model (Stahle and Dunn 1982, 1984), and Ahler's mass analysis (Ahler 1989a, 1989b). While these studies are at a preliminary stage and may yet provide more convincing results, formal analysis of metric variables may still be relevant. In the experiment cited above, Odell (1989a) found that maximum length and maximum width are the best attributes for discriminating flakes removed by flake core reduction from bifacial flakes. Width at midsection, on the other hand, is an effective indicator of bifacial reduction stages. Weight, thickness, and platform size relative to flake size, are capable of distinguishing pressure application from percussion (Henry et al. 1976, Hayden and Hutchings 1989).

Size comparisons between various unretouched flakes and tool blanks of formed types can also contribute to core reduction and tool production strategies (Henry 1989b; Shen 1992). In addition, the overall size of unretouched flakes and tool blanks, and other discrete attributes such as lateral profile, distal shape, alignment of lateral edges, and cross-section, have been useful in describing the morphological characteristics of lithic products. Their differentiation among archaeological assemblages can yield, to some degree, cultural inferences (Henry 1973, 1995; Shen 1992).

Deriving cultural inferences using attribute

* Application for funding for a controlled experiment of lithic core reduction technology at the University of Toronto in 1995 was submitted to the Ontario Heritage Foundation. Unfortunately, it has been put on hold due to cancellation of the grant program.

Table 3.2: Experimental Sources Consulted in this Study.

Sources	Reduction Technology	Attributes
Ahler (1989a 1989b)	bifacial reduction	size grades
Barham (1987)	bipolar reduction	raw material, flake type, spatial distribution
Baumler and Downum (1989)	core and unifacial reduction	completeness, size grades
Casey (1993)	bipolar reduction	bulb shape, platform type
Henry et al. (1976)	bifacial reduction	bulb scars, weight, maximum thickness
Kuijt et al. (1995)	bipolar reduction	completeness
Magne (1985, 1989)	bifacial reduction	platform, dorsal scar count, cortex,
Mauldin and Amick (1989)	bifacial reduction	cortex, flake size, platform size, scars count, completeness, weight,
Odell (1989a)	bifacial and flake-core reduction	distal termination, dorsal perimeter scarring, dorsal scar removal, bulbar scar, weight, striking platform thickness, width at midsection
Patterson (1981, 1982, 1990)	bifacial reduction	size distribution
Patterson (1995)	core reduction	heat treatment
Shott (1996a)	bifacial reduction	platform, cortex, size distribution
Stahle and Dunn (1982, 1984)	bifacial reduction	size distribution
Tomka (1989)	bifacial and flake-core reduction	cortex, Dorsal scar count, size, platform facet, platform grinding, platform cortex

analysis may not be as simple as many lithic experimenters expect, however. For instance, removal of flakes by soft hammer percussion may produce lipping flakes, but not all lipping scars are produced by soft hammer percussion (Clark and Kleindienst 1974). Flakes with high dorsal scar counts may be the result of later reduction stages. However, due to a decrease in flake size during core reduction, dorsal scar counts may not necessarily increase (Mauldin and Amick 1989).

To avoid such frustration, I adopted an alternative approach to the interpretation of attributes: *paired attribute analysis*, as applied by Tomka (1989) and Mauldin and Amick (1989). This method cross-tabulates attribute classes and plots pairs of attributes in order to verify the reliability of core reduction inferences. As Mauldin and Amick demonstrate, although cortex coverage might be expected to decrease during the core reduction process, a combination of flake size and cortex coverage may provide a more accurate representation of the reduction stage. Similarly dorsal perimeter scarring might be a good indicator of bifacial reduction, but this inference can be enhanced by also considering the multiple facet platform and the curved lateral profile. Flakes that have lipping bulbs, which also have relatively small-sized platforms and/or expanded lateral margins, are almost always produced by soft hammer percussion, whereas those with large platform flakes with crushed facets are probably removed by hard hammer percussion. Using this method of paired attribute analysis, or even multiple attribute examination, interpretations of core reduction and tool production should be reliable.

Secondary retouch, either on surfaces or on edges, is a necessary and important procedure in tool modification, and information on tool production can be further retrieved through examining the retouch techniques employed. Since there is more than one way to make a type of tool, the choice of retouch methods by the knapper reflects manifestations of style (Close 1978, 1989; Henry 1973). It is believed that the comparison and analysis of retouch attributes yields information that helps to identify a lithic industry, or a specific cultural group. Inizan et al. (1992:67-72) summarize seven common retouch variables, derived from earlier works of Bordes (1961, 1969) and Tixier (1974), that are worth considering. Among these, retouch type, position of retouch, and pattern of retouch are highly regarded as stylistic indicators (Henry 1973:56, Close 1989:12).

The orientation of the artifact chosen when observing attributes was dorsal side up with the proximal end of the artifact towards the observer

(Fig. 3.2). In cases where the dorsal and ventral sides have been modified, as with bifacial types, the side on which the catalogue number was written is treated as the "dorsal" side. While recording the exact locations of retouch units, I found Odell's 8-polar co-ordinate system (PC) useful (Odell 1979). This PC system places samples to be analyzed on a polar coordinate grid divided equally into eight portions (Fig. 3.3), and the PC section is used to refer to different sections of the lateral edges of the sample. Distortion occurs in the observation of elongated pieces, and, to avoid this, an alternate system is employed. In this system, the dividing lines of the PC section covers an equal amount of the sample perimeter.

Use-Wear Analysis: Utilization

Concepts, Methods, and Definitions

The study of stone tool use has long been of interest to archaeologists. Over a century ago, Sir

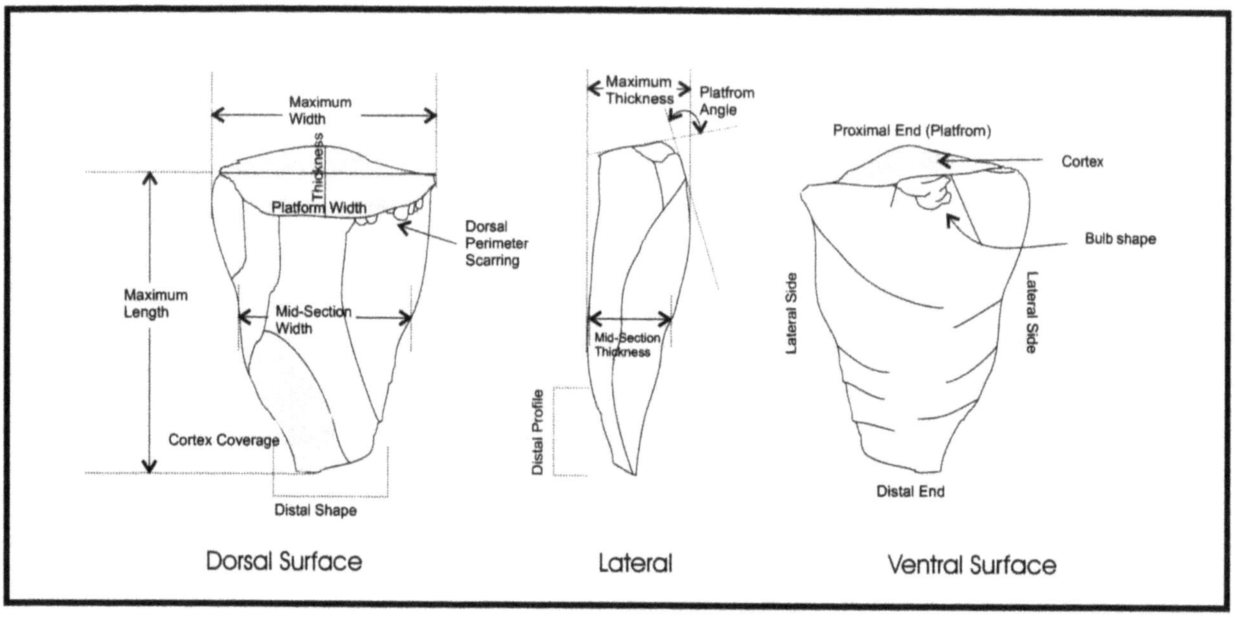

Fig. 3.2: The Orientation of Artifact Observation.

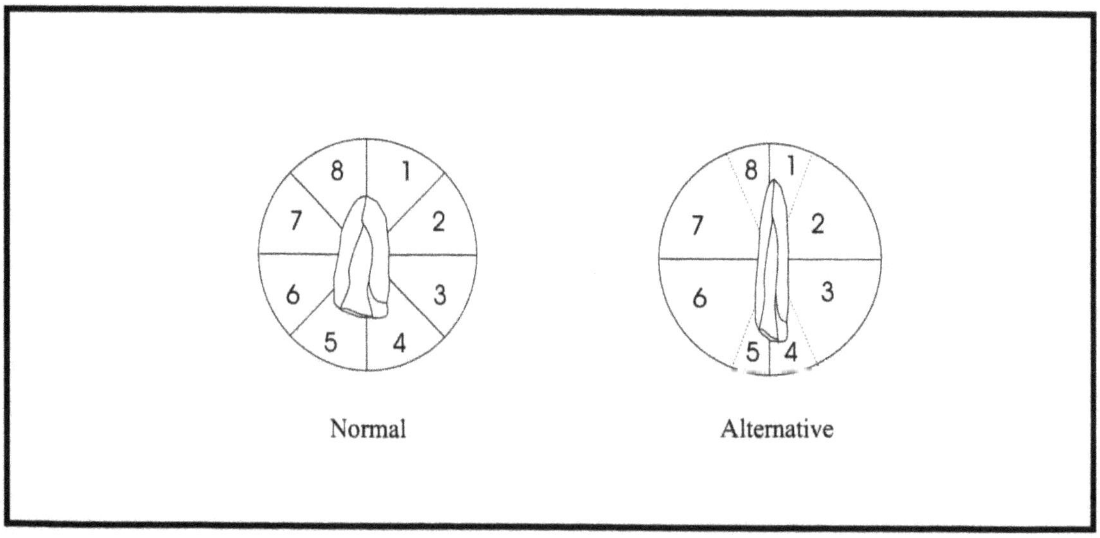

Fig. 3.3: The 8-polar Co-ordinate System (adapted from Odell 1979).

John Evans' published *Ancient Stone Implements, Weapons and Ornaments of Great Britain* (1872). Since the introduction of Semenov's translated monograph *Prehistoric Technology* (1964), use-wear analysis has become an essential component of lithic artifact analysis in North America. A historic review of this particular method is not appropriate here; anyone who is interested is referred to Tringham et al. (1974), Odell (1975, 1977), Hayden (1979), Odell and Vereecken-Odell (1980), Keeley (1974, 1980), and Kamminga (1982). Use-wear analysis, as a means to study stone tool edge damage caused by use, has three analytic forms according to Grace (1990): edge-damage analysis, edge-wear analysis, and micro-wear analysis. Edge damage analysis is based on the morphological attributes of used edges and macro-wear. Examples of this type of wear analysis are seen in Schiffer's work (1976) on the Joint lithic assemblage, Parry's work (1987) on Formative Oaxaca, Mexico, and Casey's work (1993) on the Kintampo Complex in Northern Ghana. These studies have concentrated on edge sizes, edge angles, shapes of edges, and degree of modification and finishing, to arrive at "use-potential categories." Edge-wear analysis, using reflective-light microscopes (low-power technique), considers the overall configuration of micro-fractures and abrasion to obtain information about tool motions and the range of contact materials in terms of their hardnesses (e.g., Odell 1977, 1996; Lewenstein 1987; Shea 1991). Micro-wear analysis concentrates on use-polish distributions and formations using incident-light microscopes or SEM (high-power technique) (e.g., Keeley 1980; Vaughan 1985; Unger-Hamilton 1993; Yerkes 1987). Using micro-wear analysis, more precise information about specific contact materials can be obtained. As Grace claims (1990), it is useful to integrate the three forms of use-wear analysis as a hierarchical analytical method.

Traditionally, the analytical context of stone tool use studies has been functional analysis. "Functional analysis," as it is currently practiced among most lithic analysts, is restricted to retrieving evidence of the way in which stone tools articulate with other materials in order to provide interpretations of prehistoric behavior (Odell pers. comm.); that is, the physical use of stone tools. However, clarification of the differences between *function* and *use* has been needed since the First Conference on Lithic Use-Wear at Burnaby, British Columbia, in 1977 (Hayden 1979:57-62). In general, "function" has a much greater range of social-cultural significance in human behavior than "use," a point which was made sixty years ago by Ralph Linton:

> The terms function and use have been employed interchangeably even by certain members of the functional school, but the author feels there is a very real distinction and that there will be constant confusion unless this is made clear. The difference between use and function is most obvious in the case of material expressions of culture such as tools and utensils. Thus the primary use of an axe is for chipping, that of a spade for digging, but any one will feel the inappropriateness of applying the term function to such utilizations. **The use of any culture element is an expression of its relation to things external to the social-cultural configuration; its function is an expression of its relation to things within that configuration**. Thus the axe has a use or uses with respect to the natural environment of the group, i.e., to chop wood. It has functions with respect to both the needs of the group and the operation of other elements within the cultural configuration. ... The function of a trait complex is the sum total of its contribution toward the perpetuation of the social-cultural configuration (Linton 1936:404) (emphasis mine).

Thus, to Linton, function refers to the *interrelationship* of individuals and nothing else, while use refers to the *utilization* of a trait within a cultural system. From an evolutionary perspective, Dunnell considers the *functional* characteristics of an object to be those which "directly affect the Darwinian fitness of the populations in which they occur" (Dunnell 1978:199). He believes that functional elements should be taken into account with respect to evolutionary process, and that "function" should be considered separately from "use" of tools in archaeological contexts. In the context of this study, therefore, I consider that a study of stone tool edge damage is a means of retrieving information on how tools were manipulated in direct contact with other materials; that is, tool use-patterning, not function (see Hayden 1979:57-62).

According to Schiffer, "use" is defined "as the minimal instance of behavior directed towards completion of a task" (Schiffer 1975:250, see also 1979:19). Although I am in general agreement with this definition, I have altered the terminology slightly. In Schiffer's definition, a task to be accomplished may comprise a series of "uses." I define a "use task" (UT) to express a particular tool motion with one kind of contact material, i.e., scraping seasonal wood, on a limited use-segment of a stone tool. By definition, any particular tool can exhibit several different use-tasks if different segments of the edge were employed. In this study, the term "use" is used to describe tool utilization in general; thus tool use-patterning is represented by a series of UTs determined from use-wear study.

In order to present possible use-tasks of flaked stone tools at the Princess Point sites, use-wear is essential. Ormerod (1994) carried out edge-damage analysis of flaked stone tools as a part of the Princess Point Project. He applied Schiffer's method of use-attribute type flow-charts (Schiffer 1976, Tables 9.5-9.8) to determine the use-potential categories of the Lone Pine lithic assemblage. Ormerod's edge damage analysis demonstrated that most Lone Pine tools were employed in light-duty activities, with a small portion of flake tools capable of working in medium-duty activities. No heavy-duty tools were noted. These results encourage me to ascend to the next level of use-wear analysis because samples from the Lone Pine lithic assemblages have been selected in this study.

Therefore, the study of stone tool use patterning in this research is approached through the low-power use-wear analysis. Selected artifacts have been examined through a stereoscopic microscope with reflective lighting. Two microscopes were used: a Nikon SMZ-10 series with a magnification range between 6.5x and 40x (at the University of Tulsa), and a Nikon SMZ-1 series with a magnification range between 8x and 35x (at the University of Toronto). These magnifications have further been enlarged with the assistance of a pair of 20x eyepieces and a 2x objective. Thus, edge-wear images can be enlarged up to 160x.

Apart from my personal training by Dr. Odell at the University of Tulsa, there are two reasons for choosing the low-power over the high-power technique. First, for the last fifteen years, low-power use-wear analysis has proven to be a reliable means of providing use-task assessments of stone tools. As a result, it has been employed in a great number of studies (Tomenchuk 1985; Lewenstein 1987; Shea 1987, 1991, 1995; Rousseau 1992; Odell 1996a; Phillips 1988; Sievert 1992). For years there have been criticisms and arguments over whether or not the low-power approach could provide accurate information on polish and striation wear derived from abrasive uses (Richards 1984; Odell 1985, 1990). This situation evolved, to my knowledge, due to an over-emphasis on microfractures as discriminating wear types in Odell's early work (1974, 1977, 1981b). Odell's motivation in using the low-power technique was to call attention to edge fractures at a time when only polish and striations seemed comparatively reliable indicators of use (Keeley 1980). As we have learned, the low-power approach is capable of detecting polish and striations. Abrasion is in fact an important criterion when assessing edge wear determinations (Odell pers. comm.). In order to demonstrate this, this study will present an additional set of experimental and archaeological data.

Second, when it comes to the study of *use-patterning*, it is necessary to examine a large number of artifacts. Stone tool use-patterning cannot be generated from examination of less than one or two dozen of lithic samples, no matter how accurate their individual assessments are. Given a limited time frame, the low-power use-wear analysis is advantageous in this respect. With my present skill, a worn piece takes on average about 15 minutes to analyze. This means that about 25-35 pieces can be examined within a single working day. During a two-year period, over 2,000 artifacts were processed. A further advantage of the low-power use-wear approach is that it has great potential in its application in various types of research. When large samples are examined, the results of edge wear determinations can be integrated with results of the typo-technological analysis. In this way, a comprehensive understanding of overall lithic production and use can

Table 3.3: List of Edge Wear Variables.
(adopted in part from Odell (1977, 1996))

Microfracture

Scar Termination; following terminology agreed upon at the First Lithic Use-Wear Conference (Hayden 1979:133-135)
1. feather
2. hinge
3. step
4. snap

Scar Pattern; based on the configuration of scarring
1. uni-directional
2. bi-directional
3. perpendicular
4. crushed

Scar Size; according to visibility of majority of scars
1. small: visible under greater than 10x magnification
2. medium: visible under 10x magnification or less
3. large: visible by the naked eye

Scar Distribution (Surface); based on appearance of majority of scars on surface
1. bifacial even
2. bifacial, concentrated on Ventral side
3. bifacial, concentrated on Dorsal side
4. unifacial on Ventral side
5. unifacial on Dorsal side

Scar Distribution (Marginal); based on configuration of scar distribution on lateral margin
1. scattered, uneven
2. close
3. run-together
4. denticulated
5. alternating
6. rolled-over

Abrasion

Rounding
1. light: if the basic outline of the margin can still be seen or projected
2. moderate
3. heavy: if the rounding has obliterated the general outline of the edge itself

Polish; depending on the degree of reflection from the surface
1. absent
2. matte
3. bright
4. incipient

Striation; according to the orientation to the edge
1. absent
2. parallel to edge
3. perpendicular to edge
4. diagonal to edge

be gained.

Edge wear variables have been repeatedly stressed by almost every use-wear analyst. Although the definitions often vary slightly from one analyst to another, I have no intention of significantly modifying these recognized variables. Hayden's volume (1979) has standardized most terms we are using at present (Table 3.3). Three types of edge wear variables are commonly discussed by use-wear analysts: determination variables, microfracture variables, and abrasive variables. While determination variables are edge wear assessments that use-wear analysts use to ascertain tool use, the other two kinds of variables are those on which use-wear analysts are dependent to make edge wear judgments.

Determination variables consist of four basic types: employed unit, employed location, tool motion, and contact material. The *employed unit (EdU)* is a discontinuous portion of the artifact where use wear is shown. The concept was first developed by Knudson (1973, in her own term: employable units) and later by Odell (1979, also see 1996:37). Obviously each worn artifact could have more than one unit. Given the fact that utilized pieces may have several EdUs which involved different use-tasks, the employed unit rather than the utilized pieces, is treated as a basic analytic unit. The *employed location* is the location of the worn portion on flake/tool edge. Employed location is expressed using the 8-polar co-ordinate system illustrated in Fig.3. *Tool motion* refers to the physical action in which stone artifacts are employed as devices to modify other objects. Some closely related actions, such as cutting and sawing, slicing and carving, scraping and shaving, etc., are not clearly distinguishable either in real life or in terms of the edge wear they produce (although sometimes they may be). I have therefore lumped these pairs together as single tool motions, respectively. *Contact material* refers to the substance that is contacted during the process of tool motion. These substances are divided into 8 relative hardness groups: soft animal (SA), soft vegetal (SV), medium-soft vegetal (1M), medium animal (MA), medium-hard vegetal (2M), hard animal (1H), very hard animal (2H), and very hard inorganic (3H). Exact materials can only be inferred in light of the nature of resistance grades of the materials.

Microfracture refers to fractures and edge damage caused during tool utilization, whereas *abrasion* refers to impressions and marks left by pressure from contact during use. Microfracture patterns are usually described in terms of four variables: *scar termination*, *scar pattern*, *scar size*, and *scar distribution*. Abrasion comprises three important formations: *edge rounding*, *polish*, and *striation* (Table 3.3). The presence of microfractures in different scarring combinations gives substantial clues to wear types (Tomenchuk 1985; Odell and Vereecken-Odell 1980; Odell 1981; Shea 1991). However, abrasion is critically important and an excellent indicator of the presence of use wear, and can be detected and discriminated under 200x magnification (Lewenstein 1987; Shea 1991; Rousseau 1992; Odell 1996). Recent research on polish formation suggests that polish appearance and distribution alone are insufficient to interpret tool use (Levi-Sala 1993, 1996). A complex combination of microfracture and abrasion presence is very helpful in determining tool motions and worked materials.

Since raw material and flake edge morphology have the most impact on the scarring patterns and polish, an understanding of fracture mechanics is also necessary for accurate determination of tool use. For instance, even if both were employed for cutting or sawing on the same worked material, a thin and straight edged flake with a flat surface would have relatively different microfracture patterns from a dull and curved edged flake with a convex dorsal side. The former would likely display an even distribution of bifacial scarring, whereas the latter has unifacial scars on the dorsal side with the appearance of having been used for scraping. In this case, abrasion, especially polish and/or striation, is a crucial indicator. On these two tools, polish is likely formed on both sides of the flake surface in a linear appearance near the edge (Photos 6 & 8). This kind of polish formation easily eliminates any possibility of transverse motions like scraping or shaving (Photo 10).

However, we still encounter difficulties in determination of tool use from edge wear attributes.

This is especially true of the low-power approach. The high-power analysts have been developing correlations between polish formation and tool use in order to provide a one-to-one discriminating attribute study (Grace et al. 1988; Grace 1989, 1993). Through controlled experiment and using the low-power technique, Tomenchuk (1985) has for the first time successfully shown correlation between edge wear type and use-fractures. This indicates the powerful potential to objectify the use-wear variables. However, practical problems remain; Tomenchuk's system is largely hampered by extensive utilization of mechanical engineering as an application in archaeological research. Shea (1991) used a 9-digit coding combination of edge wear attributes in a computer simulation. However, his study was still based on a foundation of experimental judgments. Thus, at present, use-wear analysis still relies on what Odell refers to as *interpretative states*, meaning tool use assessments dependent on correlations between observed use-wear and wear produced experimentally (Odell 1996a:38). Therefore, the reliability of interpretative assessments should be, and must be, ensured by an understanding of a combination of microfracture and abrasive variables and intensive experimentation.

Experimental Study

Use-wear experiments are essential to use-wear practitioners. Knowledge of tool use is founded on a corpus of experimental work involving a large variety of tool motions and worked materials (Brink 1978; Keeley and Newcomer 1977; Tringham et al.. 1974; Odell 1977; Keeley 1980; Odell and Odell-Vereechen 1980; Shea 1991; Vaughan 1985; Richards 1988; Lewenstein 1987). There is no doubt that use-wear experiments have provided lithic analysts with a great understanding of how tool edge can be damaged through use, and that this has enabled them to arrive at interpretations of use patterns of certain tool types. For instance, Brink (1978) carried out an experimental study to examine the function of endscrapers found at a prehistoric site in west-central Alberta. Ahler (1971) examined the function of projectile points recovered from Rodgers Shelter, Missouri, through a series of experiments on replicated points. More recently, Rousseau (1992) has demonstrated how a particular shaped tool type, so-called "key-shaped unifaces," might have been employed at prehistoric sites in the Interior Plateau of British Columbia. A wide range of use-wear experiments on retouched and unretouched flake tools, generally employed by most use-wear practitioners, allowed for the recognition that unretouched flakes, in addition to shaped artifacts, were used in a wide variety of activities (Keeley 1980; Yerkes 1989; Shea 1991; Odell 1996).

The common procedure for such experiments consists of the following steps: (1) manufacturing a set of flaked tools, commonly from raw materials similar to ones found at archaeological sites; (2) selecting a range of materials to be worked, generally in different categories of hardness such as hides, meat, wood, bone, and the like; (3) performing a series of different motions (scraping, cutting, drilling, etc.) on target materials with selected tools, and recording the process in detail; and (4) examining the used tools under designated microscope(s) in order to recognize wear variables caused by performed actions. Ideally, for purpose of comparison, specimens are examined prior to being used as well as after use. In order to ensure reliability and credibility, a blind test usually takes place after the experiments have been carried out.

However, problems in experimental use-wear exist. First of all, methods used in the experiments vary among use-wear analysts due to their different research objectives . While some researchers have concentrated on the formation of polish (Kelley 1980; Vaughan 1985; Richard 1988; Grace 1989), others have focused on microfracture variables (Brink 1978; Tringham et al. 1974; Odell and Odell-Vereechen 1980). This deviation is being overcome in some recent experiments, which have studied both abrasion and microfractures caused by use (Shea 1987, 1991; Lewenstein 1987; Levi-Sala 1996).

Second, most of the aforementioned experiments were not controlled. Although these experiments could be used to educate the eyes, they are not replicable. In most instances use-wear experiments have served as a basis from which to judge the quality of use-wear interpretation. This requires use-wear practitioners to conduct their own experiments almost every time, especially prior to the examination of a new archaeological collection. The advantage of this is to provide a use-wear analyst with a wide range of reference collection. The disadvantage is that such repeated efforts are costly and time-consuming. Because these experiments are uncontrolled it is not possible to compare experiments nor to judge their reliability.

Third, each experiment is also limited to a certain range of tool motions and contact materials. Again, because affected by the individual analyst's research objectives, these vary between experimenters. Credibility of use-wear interpretations would be questionable if an analyst performed a limited range of experimental actions. For instance, Rousseau's study on the function of the "key-shaped unifaces" is based on a series of ten experimental tools, which were mostly worked on wood materials. As a result, his interpretation of these artifacts as "woodworking tools" may be questioned. Conventionally, a minimum of 50 or so tasks in a set of experiments is required for a fair coverage (Odell pers. comm.). However, no matter how many tasks are performed, replication of all possible prehistoric uses could never be achieved. To minimize this limitation, I studied a large number of experimentally produced samples (see below), in addition to my own 49 experimental tasks, in order to cover a fairly wide range of wear types.

Although use-wear experimentation is problematic, the efforts made by most use-wear analysts should not be dismissed. We are, however, better able to generalize tool use patterning in a static condition. Therefore, the aims of my use-wear experiments are (1) to understand and recognize the formation and appearance of edge wear types caused by various actions; (2) to evaluate the tool use assessment derived from similar experiments conducted by other researchers; and (3) to ascertain reliability of tool use assessments based on the edge wear variables. Thus the experimental study entailed three steps prior to archaeological examination. They included a microscopic study of a large number of experimental collections at the University of Tulsa, use-wear experiment at the University of Toronto, and a blind test.

The experimental collection (TU) consisted of 178 pieces made by the students of Dr. G. Odell. One hundred and one different use-tasks were carried out using these samples, employing 26 tool motions on 65 different contact materials. This collection was examined for wear types and later used as a reference for comparison with my own experimental specimens (ES). The use of this comprehensive collection allowed me to carry out this use-wear study without carrying out extensive experiments. It was an effective and speedy way to get acquainted with a large number of wear types in order to be able to recognize and compare them.

My experimental procedures followed the general guidelines used in most conventional use-wear experiments conducted by others (mentioned above). However, my experiment was strictly conducted in the manner described by Odell and Odell-Verrecken (1980) because the results of my experiment were to be directly compared to that of the study collection. As usual, the work being done during the experiment was purposeful. The reader is advised that the force applied were not controlled. The use-tasks of stone tools in the experiments certainly do not replicate aboriginal conditions, but represent common practices in reality.

The material from which most of the experimental tools were fashioned consists of a dark grey, fine-grained Onondaga chert that was acquired from quarry sites along the north shore of Lake Erie. It was selected because the Onondaga chert in our archaeological samples. Locally available Haldimand chert was also used. Obsidian, quartz, and an unknown fine-grained flint, were also used for a small number of tools (8 in total). Although these types of raw material were not found in the Princess Point lithic assemblages, I utilized them for observations regarding the edge wear variability. All tools used in the experiment were produced by direct hard and soft hammer percussion. The selected flakes were either used directly or used after minor edge retouching. Forty-two flakes were utilized for 49 use-tasks (7 pieces have two employed units) (Table 3.4).

Every piece was examined microscopically before use, to observe the edge damage or current

Table 3.4: Use-Wear Experiment Task List.

Task	ES Numb	RM	Tool Motion	Resistance Grade	Exact Worked Material	Time	Stroke
1	ES01	Haldimand	sawing	very hard animal	dried cow limb bone	15 min	1450
2	ES02	Onondaga	sawing	very hard animal	dried cow limb bone	18	1920
3	ES03	Onondaga	cutting	soft animal	fresh pork meat	22	1980
4	ES04	Onondaga	scraping	medium animal	fresh pig limb bone	15	1650
5	ES05	Onondaga	cutting	medium soft vegetal	fresh pine-tree wood	30	1700
6	ES06	Onondaga	scraping	medium soft vegetal	fresh pine-tree wood	30	1925
7	ES07	Onondaga	cutting	medium soft vegetal	fresh pine-tree wood	15	1230
8	ES08	Onondaga	cutting	medium animal	dried pig hide	20	2100
9	ES09	Onondaga	slicing	medium animal	dried pig hide	10	400
10	ES10	Onondaga	scraping	medium animal	dried pig hide	25	1875
11	ES11	Onondaga	cutting	medium animal	dried pig hide	20	2600
12	ES12	Onondaga	drilling	medium animal	dried pig hide	15	1200
13	ES13	Onondaga	scraping	medium animal	dried pig hide	20	2500
14	ES14	Onondaga	scraping	medium hard vegetal	dried hackberry wood	20	2400
15	ES15	Onondaga	sawing	medium hard vegetal	dried hackberry wood	22	2800
16	ES16	Onondaga	scraping	soft vegetal	sweet potato	30	3450
17	ES17	Onondaga	cutting	soft vegetal	carrot	25	5750
18	ES18	Haldimand	cutting	very hard animal	soaked red-tailed deer antler	20	2400
19	ES19	Onondaga	scraping	hard animal	cooked pig limb bone	20	1700
20	ES20	Onondaga	scraping	soft vegetal	carrots	15	2000
21	ES21	Onondaga	projecting	soft vegetal	carrots	18	2000
22	ES22	Onondaga	boring	soft vegetal	carrots	8	800
23	ES23	Onondaga	scraping	soft animal	fresh hide	8	1000
24	ES24	Onondaga	slicing	soft animal	fresh hide	7	1000
25	ES25	Onondaga	boring	medium animal	fresh pig limb bone	8	800
26	ES26	Obsidian	scraping	medium hard vegetal	dried apple-tree wood	15	2100
27	ES27	Obsidian	sawing	medium hard vegetal	dried apple-tree wood	20	2600
28	ES28	Obsidian	scraping	medium hard vegetal	dried apple-tree wood	23	3100
29	ES29	Other Flint	cutting	medium animal	fish	12	660
30	ES30	Onondaga	slicing	medium animal	fish	15	900
31	ES31	Onondaga	cutting	medium animal	frozen pig feet	16	2100
32	ES32	Onondaga	scraping	soft animal	fresh pig hide	20	1900
33	ES33	Onondaga	penetrating	soft animal	fresh pig feet	15	520
34	ES33	Onondaga	shaving	soft animal	pig hide & fat	25	1200
35	ES34	Other Flint	scraping	hard animal	soaked cow limb bone	15	1670
36	ES34	Other Flint	drilling	hard animal	soaked cow limb bone	7	600
37	ES35	Onondaga	cutting	medium hard vegetal	dried hackberry wood	20	1800
38	ES35	Onondaga	boring	medium hard vegetal	dried hackberry wood	25	2100
39	ES36	Onondaga	sawing	medium hard vegetal	dried hackberry wood	15	800
40	ES36	Onondaga	boring	medium hard vegetal	dried hackberry wood	11	600
41	ES37	Quartz	sawing	medium hard vegetal	dried hackberry wood	21	1600
42	ES38	Other Flint	cutting	soft vegetal	grass	22	2500
43	ES39	Onondaga	shaving	medium soft vegetal	fresh pine-tree wood	25	3300
44	ES39	Onondaga	shaving	medium soft vegetal	fresh pine-tree wood	15	1600
45	ES40	Other Flint	cutting	medium hard vegetal	dried hackberry wood	5	320
46	ES40	Other Flint	graving	medium hard vegetal	dried hackberry wood	8	550
47	ES41	Onondaga	chopping	medium hard vegetal	large unknown-tree trunk	15	800
48	ES42	Onondaga	digging	hard inorganic	mud with small rocks	15	2300
49	ES42	Onondaga	graving	hard inorganic	sandstone	15 min	800

Table 3.5: Recording Sheet for Experimental Specimen Observation.

Part I: Experimental Process Record

1. Experimental Specimen (ES#) ________ 2. Raw Material ________
3. Worked Material ________ 4. Resistance Hardness ________
5. Retouched before experiment ? (yes/no)
   If yes, retouch units and comment ________
6. Microscope scanning before experiment ? (yes/no)
   If yes, comments ________
7. Action: ________ 8. Motion defined ________
9. Process measurement:
   stroke/minute ________ total time worked ________ total stroke ________
10. Employed unit ________ 11. Prehensile unit ________
12. Remarks ________ 13: Drawing working position of tool

Part II: Analytic Recording

1. Microscopic examination: scanning at ________
   edge damage record at ________ polish record at ________
   striation record at ________ photography taken at ________
2. Edge damage unit and comments ________
3. Microfracture:
   scar termination ________
   scar pattern ________
   scar size ________
   scar distribution ________
4. Abrasion
   rounding ________
   polish ________
   striation ________
5. Comparison with study collection (TU) ________
6. Remarks ________

status of edge formation. Both recording sheets for edge wear observations before and after flake utilization were used (Table 3.5). Twelve tool motions and 19 types of contact material were employed for fulfilling these 49 use-tasks (see below). Wear produced by my own experiments was immediately compared to wear observed on the TU collection specimens. The length of time and number of strokes was recorded.

The observation and comparison of edge wear on the experimental specimens confirm that wear types display homogeneous patterns under similar conditions. The following sections provide a brief description of the common features of edge wear resulting from the experiments.

Tool Motions

Twelve common tool motions were applied to utilized tools in my experiment. Edge wear was observed and compared with that of the studied TU collection. Common characteristics of microfracture and abrasion are summarized below.

- *Cutting* is a type of longitudinal motion that is unidirectional and parallel to the edge of tools;
- *Sawing* is a type of longitudinal motion that is bi-directional and parallel to the edge;
- *Slicing/carving* is variation of longitudinal movements when the tool is held at or near a right angle to the worked surface, moving either unidirectional or bi-directional along the edge.

All the above tool motions tend to penetrate the worked material by applying force directly on an edge opposite the working edge. Observation of edge wear on experimental samples confirms some conclusions made by other use-wear experimenters. Cutting usually produces scarring on both sides of the working edge. This scarring is characterized by a unidirectional orientation. Striations are parallel to the edge, if present. Polish is likely on both surfaces of a working edge. Sawing (the 2-way motion) provides a similar scarring pattern, but exhibits bi-directional scarring. In slicing or carving, scarring tends to be located more on one surface than on the other because of the angle of movements. Striations, if present, are often unifacial and usually slanted, or diagonal to the edge.

- *Scraping* is a transverse movement of a tool being pulled towards the operator across its working edge.
- *Shaving* is a movement of a tool also towards the operator, but with short and quick strokes transverse to the working edge.
- *Planing/whittling* is a transverse tool motion of a tool that is pushed away from the operator. The tool in contact with the contact material remains connected throughout the length of movement.

All the above tool motions shave off material with the working edge of the implement held at roughly a right angle to the direction of use. All such tool motions produce mostly unifacial scarring. Polish appears on the side which is in contact with the material worked. Usually, planing or whittling produce more abrasive wear on the surface than scraping and shaving do. In scraping and shaving, scarring is clumped on the working edge which sometimes shows evidence of crushing. Edge-perpendicular striation is diagnostic of the working edge.

- *Penetrating* is the tool motion involving the tip of a tool.

Wear caused by a penetrating motion is rather complex compared to that described above. Because of forces directly applied on the tip, scars are usually snapped off, or crushed. Unidirectional orientation of the scarring from the tip area inwards can be seen. A useful indication of such wear is more abrasion on both dosral and ventral surfaces rather than on the edge.

- *Drilling* is a circular motion producing both rotary pressure to the sides of the drill and direct pressure downward into the contact material.

The typical wear of this kind is bi-directional scarring, symmetrical along the axis of the EdU. Rounded and crushed edges are expected. Polish is limited to the edges because the edges are in contact with the material.

- *Graving* is a tool motion which involves the tool tip. Graving has elements of either, or both, longitudinal and transverse motions, depending on the particular case.

A useful indication of graving wear is that unifacial scarring appears opposite to the surface where abrasion such as polish and striation occur.

- *Chopping* is a downward movement into the main body of the material using relatively strong force.

The working edge of a chopper usually exhibits large bifacial scarring, asymmetric abrasion, and edge-oblique striations, if present. Scars are usually hinge- and step- terminated.

- *Digging* is a similar movement to chopping, but involves the tip of the tool.

Digging results in a heavily rounded tip, or the blunting or crushing edge. Polish is diffused and scars are large in size and feather- or step-terminated.

Contact Materials

Materials that were worked were selected according to different hardness categories. The method is, of course, subjective, but is still valid in its representation of relative resistance grades used by other use-wear experimenters. The following materials were selected for this study.

- *Soft Animal* (SA): uncooked pork meat, fresh pig skin, and pig fat;
- *Soft Vegetal* (SV): raw sweet potato, carrots, grass.

Soft materials produce uneven patterns of small feather-terminated fractures. Working on soft animal material such as fresh skin, fat, or meat may result in extensive matte polishing and light edge rounding, while work on soft vegetal material mostly produced bright polish, if the tool was used long enough. Striations are not generally present.

- *Medium-soft Vegetal* (1M): fresh pine tree wood, fresh unknown tree wood;
- *Medium Animal* (MA): fish, dried pig skin;
- *Medium-hard Vegetal* (2M): dried hackberry tree wood, dried apple wood.

Scarring from woodworking typically exhibits medium-to-large sized scars and a "rolled-over" pattern on the working edge. The common scarring resulting from working on medium-soft wood material tends to be feather-terminated and poorly defined on the interior borders, whereas scarring from working on medium-hard wood material appears as hinged-terminated with distinct interior borders. Bright polish and medium-to-heavy rounding usually appear, but striations are rare. Abrasion of medium animal material includes rapid edge-rounding, matte polish, and occasional striations. Scarring of such material exhibits contiguous-to-clumped patterns of medium-sized feather or step termination.

- *Hard Animal* (1H): cooked pig limb bone, soaked cow limb bone, soaked red-tail deer antler;
- *Very Hard Animal* (2H): dried pig limb bone, dried cow limb bone, red-tail deer antler;
- *Hard Inorganic* (3H): soil with gravel, sandstone.

It should be noted that bone, antler, and hard, dry wood can occasionally produce similar wear (Odell and Odell-Verrecken 1980:101). The most diagnostic wear of this type is step-terminated and medium-to-large sized scarring, and roughened/crushed edges. The distinct features that distinguish hard animal and dried wood are: matte polish and contiguous patterns of large termination scars which normally appear when hard animal materials are worked; while bright polish and uneven patterns of large hinge fractures are produced when dried wood materials are worked. Hard inorganic materials produce heavy rounded edges, flattened projections and surfaces, diffused polish, and large step-terminated scars.

A retouched sickle blade (ES38) was used for cutting grass for 22 minutes (2500 strokes). It displayed edge rounding and tiny discontinuous feather scars on both sides which are identical to the wear on Odell's specimen EO131 (Photo 11; Odell 1996 Fig. 4.9d). Both specimens ES02 and TUA14 were used to saw hard animal material, and they exhibit similar scarring patterns (Photos

7 and 12). They display heavy rounding and matte polish in the middle of the used edge (Photo 8), accompanied by large hinge and feather fractures.

Tools worked on soft materials usually show a row of small feather scars as described above. Roughened edges, denticulate scarring, and bifacial grassy polish indicate meat cutting wear, as shown by specimen TUB137 (Photo 13). The ES16 flake was used for scraping a potato for a half-hour period (3500 strokes). Unifacial scarring on the working edge, along with several strips of polish perpendicular to the edge can be observed on the contact surface under high magnification (Photos 9 & 10).

An Onondaga chert flake (ES12) was used to drill a dried pig limb bone for 15 minutes. Heavy rounding of the tip with crushing scars on all ridges resulted (Photo 14). A few snap scars on the tip of a flake (ES33) are indicative of penetrating activity (Photo 15), whereas heavy rounding and bright polish on the contact tip of a graving tool (ES40) are also typical wear for this kind of activity (Photo 16). Polish is well developed on a wood chopping tool (ES41) after 15 minutes use and can be observed at 50x magnification (Photo 17).

The experiment also provides evidence suggesting that not all use wear types are useful in ascertaining tool motions and contact materials. Different raw materials, the length of use, and the topography of the flake tool surface, are all factors that affect the wear formations. As I briefly mentioned in the case cited at the end of the last section, two tools (ES02 and ES18) were used for a similar use-task, sawing a very hard animal material, their scarring patterns are quite different. While the former shows the typical wear type resulting from sawing hard material, the latter displays a scarring similar to that resulting from a scraping action. This is largely due to the differences in the shape of the working edge of the two tools. After close examination, however, it was noticed that these two tools have polish on both sides of the flake surface in a linear formation near the edge (Photos 6 & 8). This kind of polish formation differs from that which results from transverse motions such as scraping and shaving (Photo 10). In some cases where polish is poorly developed on the tool (ES19, ES36, TUB145), the direction of scar removals, scar distribution, and size and termination of scars provide helpful clues (Photo 18, 19, 20) as to the use-wear type.

A blind test was carried out under the supervision of Dr. Odell in early 1995, according to rules set out by him (Odell and Vereecken-Odell 1980). Twenty-two experimental pieces were picked by Odell from the TU collection and set aside before I started examining the collection. Four edge wear variables were tested: employed location, tool motions, relative hardness of the contact materials, and exact contact materials (Table 3.6). Out of 22 pieces, two pieces were misinterpreted for all categories. Use location and tool motion had a relatively high rate of correct identification, 77.8% and 81.8%, respectively. The relative resistance of the material had a 68.2% rate of accuracy, and the lowest correct assessment was the identification of the exact worked material: 59.1%.

This result was then compared with other use-wear blind tests (Table 3.7). My overall score is average, but I scored below average for identification of the exact material worked. The results of the blind tests indicate that the actual use-wear determinations are subject to error. This suggests that interpretations of tool use, or "tool function" as preferred by some lithic analysts, is problematic. With this in mind, the results of my use-wear analysis, presented in the following chapters, should be considered as a generalization of tool use patterning, rather than an absolute manifestation.

To a large degree, the erroneous identification of tool motions and contact materials, as shown in the blind test, were due to the short period of use of some of the experimental specimens (e.g., TUC8, which I misinterpreted was used for only 7 minutes or 350 strokes). This suggests that some used flaked tools may not exhibit use wear at all due to short period of use. However, the use-wear analysis applied in this study is based on a conservative assumption: if there is no use-wear detected on flaked tools, the tool is assumed not to have been used. Flake tools that show use-wear but where it was not possible to discriminate

Table 3.6: Use-Wear Blind Test.

N	SN	Location			Tool Motion			Relative WM			Exact WM		
		correction	test	score	correction	test	score	correction	test	score	correction	test	score
1	B59	2-3, 4, 6-7	2-3	1	scrape	scrape	1	2M	2M	1	wood	wood	1
2	C6	8-1	7	1	drill	drill	1	1M/2M	1M	1	maple	wood	1
3	B58	2-3	2-3	1	saw	carve/slice	1	H	H	1	soaked antler	bone	0
4	B68	6-7	6-7	1	saw	cut	1	SV	SV	1	potato	potato	1
5	B101	1-3, 6-8	2	0.5	scrape	scrape	1	VS	SV	1	hide	hide	1
6	B71	6-7	6-7	1	saw	saw	1	2M	2M	1	sea. wood	wood	1
7	A8	6-7	6-7	1	shave	shave	1	1H	SA	0	soaked antler	hide	0
8	B63	6-8	6-8	1	cut	saw	1	1M	H	0	bamboo	bone	0
9	D14	1	1	1	drill	drill	1	1M	1M	1	wood	wood	1
10	A17	8	8	1	bore	bore	1	SA	H	0	leather	bone	0
11	A10	1-2, 4-5	4-5	0	cut	grave	0	SV	SA	1	vegetable	hide	0
12	C8	7	6	0	bore	slice	0	1H	SA	0	bone	hide	0
13	B135	2-3	2-3	1	saw	scrape	0	1H	1H	1	bone	bone	1
14	B122	1-2	1-2	1	slice	slice	1	SA	SA	1	leather	leather	1
15	A4	6-7	6-7	1	cut	cut	1	1H	2M	0	bone	wood	0
16	B28	1-3	1-3	1	cut	slice	1	SA	SA	1	dear hide	hide	1
17	E3	7-8	8-1	1	chop	chop	1	2M	2M	1	plum log	wood	1
18	B134	5-7	6-7	0	slice	grave	0	SA	H	0	cow hide	antler	0
19	B136	8, 1-2	8, 1-2	0	slice	cut	1	SA	SA	1	cow hide	hide	1
20	B155	2-3	2-3	1	saw	carve	1	SA	SA	1	leather	meat/hide	1
21	B45	6-8	6-8	1	slice	slice	1	SA/1H	MA	1	butcher pig	meat/bone	1
22	B107	1-2, 6-7	2-3	0.5	saw	saw	1	1M	H	0	bamboo	bone	0
Total Correction				17			18			15			13
				77.30%			81.80%			68.20%			59.10%

Table 3.7: List of Comparative Use-Wear Blind Tests.

Sources	number tested	location of use	tool motion	worked material
Shen , 1995	22	77.80%	81.80%	68.2% (hardness) 59.1% (exact)
Shea (1986)	243	98%	91%	82%
Odell and Odell-Vereechen (1981)	31	79%	69%	61%(hardness) 38% (exact)
Keeley and Newcomer (1977)	16	88%	63%	75%
Vaughan (1985)	24	96%	79%	79%

the tool motions or contact material, were placed in an "indeterminate" category. These tools would be classified as "used" but with an unknown use-task.

Furthermore, this experimental study indicates that examination of use-wear should rely on an understanding of the configurations of microfracture and abrasion formations. The procedure used in my microscopic analysis is described below:

- The artifact was cleaned with liquid soap and clean water, and wiped with dry cotton cloth before examination. Cleaning was usually repeated during the examination in order to remove oily finger prints on the tool surface that can sometimes be confused with polish.
- The artifact was then scanned under a microscope with 13-20x magnification, to look for possible utilized locations. Edge damage and edge rounding were two main elements used to determine whether the piece was used or not.
- If the piece had possible use-wear, magnification was increased in order to detect polish and striations. I frequently used the range from 30x to 80x, but 80x plus was also often used for pieces having light use-traces (such as those found on pieces used on soft material). Once polish was confirmed, the employed unit and employed location were determined. Sometimes finding polish and/or striations was difficult and, in such cases, the pieces were either assigned to an undetermined category, or a "judgment call" was made if I was confident of diagnostic microfracture combinations and edge rounding.
- Once the employed location was determined, the piece was examined without the microscope. The morphology of the piece was inspected to determine possible activities in terms of edge shape and size, to observe possible holding or hafting positions and orientation, and to postulate how microfracture and abrasion might have been formed in relation to the piece's curvature.
- Then, at a proper range of magnification depending on the wear type, the artifact was examined for microfracture and abrasion patterns from one side of the surface to the other, from the edge inward. Scar size, termination, and distributions were sought and recorded, as well as patterns of rounding, polish, and striation.
- Finally, edge-wear variables were recorded, and the suggested tool motions and contact material were assigned.

Data Manipulation and Presentation

The metric variables are presented in this study as measures, including means, ranges, standard deviations, and interval histograms. The categorical variables are expressed in terms of relative frequencies of attribute states. Cross-tabulation of paired variables is used to examine the relationship between two correlated variables, e.g., platform facet vs. bulb shape, flake size vs. dorsal scar count, tool motion vs. contact material, and artifact type vs. tool motion. This method, a means of *paired attribute analysis*, is particularly useful for arriving at behavioral inferences, compared to the method relying on single variables (see above).

Statistical tests explore whether significant differences between archaeological samples at the intersite and intrasite levels exist. The t-test is employed to examine differences in means between two groups. The Mann-Whitney U test, as a nonparametric test, is employed for the discrete data (both nominal and ordinal) to evaluate the null hypothesis that two independent samples come from populations with identical distributions (Thomas 1986:261-344). The Mann-Whitney U test is used because it is the most sensitive nonparametric alternative to the t-test for independent samples. The level of statistical significance selected for these tests is set at 0.05. However, the level of statistical significance is not always coincident with archaeological meaning. The statistical results should be considered as a guide only. Interpretative or subjective approaches are still relevant to data presentation.

Lithic Assemblages and Sampling

Two of the three archaeological sites I investigated are single component occupations, while Grand Banks is a multi-component site (Table 3.8 and 3.9). Unfortunately only a portion of the lithic artifacts from Grand Banks could be selected for detailed study, due to the disturbed nature of Area C. The Late Archaic/Early Woodland lithic assemblage, recovered from Paleosol I, comes from only two square meters in Area A, and the sample is too small to be statistically significant.

Two sampling strategies were employed. All formed types and cores were selected for typo-technological and functional analysis because both are interpretatively significant and have reasonable sample sizes. However, gavin the large quantity of unretouched flakes, a sampling procedure was necessary. To achieve representative samples, a minimal sample size had to be determined. Minimal sample sizes were calculated based upon the formula as follows (Shennan 1988:308):

$$n = \frac{Z_a s^2}{d}$$

where n=sample size; Z_a=probability level; s=standard deviation; and d=tolerance. Here, Z_a may be set at 1.96 since we are interested in a 95% probability level (Shennan 1988:308). Because the standard deviation is normally not known prior to sampling, estimation of standard deviation (s) and specification of tolerance (d) are necessary. There are two ways of doing this. First, a small number of artifacts are selected for measurements as a pilot sample and the standard deviation is calculated. Or alternatively, it can be estimated by taking one-fourth of the range, according to the *Empirical Rule* (Mendenhall 1968:167). The latter approach has been adopted. The range in length of unretouched flakes is from 10.5 mm to 68 mm. A value of tolerance is then established based upon the precision desired. Then I have an estimate of s as 14.4 mm, and will be satisfied with a tolerance of +/- 3.0 mm. Therefore the minimal sample size for unretouched flakes is:

$$n = \frac{Z_a s^2}{d} = \frac{1.96(14.4)^2}{3.0} = 135.48$$

Because other metric attributes (width, thickness, quarterly width and thickness) exhibit a much smaller range than length, 136 artifacts are considered as a minimum sample for quantitative analysis of unretouched flakes for each assemblage or analytic unit (e.g., Area A vs. Area B). As far as qualitative attributes and edge wear attributes were concerned, sample sizes were arrived at intuitively, but were generally about 10 percent of the total unretouched flakes.

No samples of flaking debris was selected for variable observations, and no specific measurements of flaking debris were taken. Only the distribution of flaking debris frequency will be taken into account for comparisons between spatial units. However, in order to verify that this artifact class is a group of waste flakes, a small number of these artifacts were randomly selected for use-wear observations.

To establish the relevant intra-site or inter-site comparisons, I selected lithic artifacts only from cultural layers. Since cultural features at the sites are not substantial, only a few flaked stone tools were recovered from a limited number of pits. No

Table 3.8: Summary of Lithic Artifacts Recovered from the Grand Banks Site.

CLASS	Princess Point		Late Archaic/ Early Woodland (Area A)		Mixed Deposit (Area C)		Total	
	piece N	weight gram	piece N	weight gram	piece N	weight gram	piece N	weight gram
Core	47	829.1	1	7	56	1241.3	110	2436.1
Formed Type	146	449.5	2	4.6	110	729.6	269	1412.9
Unretouched Flake	6252	5304.2	75	3.5	4422	5034.7	10858	10577.9
Flaking Debris	11249	1417.9	111	5.2	4158	946.6	15533	2382
TOTAL	17694	8000.7	189	20.3	8746	7952.2	26770	16808.9

Table 3.9: Summary of Lithic Artifact Samples from Grand Banks, Lone Pine, and Young 1.

CLASS	GRAND BANKS		LONE PINE		YOUNG 1	
	piece N	weight gram	piece N	weight gram	piece N	weight gram
Core	47	829.1	41	849	1	18.6
Formed Type	146	449.5	122	320.9	4	3.7
Unretouched Flake	6252	5304.2	8301	6179	632	465.8
Flaking Debris	11249	1417.9	27254	2367.2	1606	157.3
TOTAL	17694	8000.7	35718	9716.1	2243	645.4

Table 3.10: Frequencies of Sampled Artifacts for Attribute Observations.

SITE	Lithic Class: Core	Formed Type	Unretouched Flake	Flaking Debris	*Utilized Flake**	Selected N by sites
Grand Banks						
Total Number	46	146	6252	11249	-	
Selected Number	46	146	657	52	-	901
Sampling Proportion	100%	100%	10.51%	0.46%	-	
Lone Pine						
Total Number	41	122	8301	27254	-	
Selected Number	41	122	829	0	-	992
Sampling Proportion	100%	100%	10.00%	0%	-	
Young 1						
Total Number	1	4	632	1606	-	
Selected Number	1	4	133	0	-	138
Sampling Proportion	100%	100%	21.04%	0%	-	
HH						
Total Number	-	-	-	-	335	
Selected Number	-	-	-	-	204	204
Sampling Proportion	-	-	-	-	60.90%	
Selected Number by Lithic Class	88	272	1617	56	204	**2237**

* *Utilized Flake* is the category that is defined by excavators of the HH site

clear household structures were found, although post-moulds were clearly identified. Thus those limited number of flaked artifacts recovered from features were not selected in the samples presently available. Surface collected artifacts from Lone Pine were also excluded for comparisons.

An additional sample from an earlier period was also studied for comparative purposes. The HH lithic assemblage and preliminary analytic data was generously provided by the Ontario Ministry of Transportation. First, it represents a late Middle Woodland culture. Its radiocarbon dates place it ca. A.D. 325 - 675, indicating the period immediately preceding the Princess Point Complex (Woodley 1996). Second, this site is located at the extreme western end of Lake Ontario, the area where the Princess Point Complex is well developed, but where so far few Middle Woodland sites have been found. Third, the excavation of the site was well controlled and lithic data have been systematically recorded and analyzed (Woodley 1996:59-110). Two hundred and four artifacts categorized as "utilized flakes" were selected for use-wear analysis only, in an effort to provide a picture of trends in the use-patterning of flaked stone tools from late Middle Woodland to early Late Woodland within the study region. No other attributes of HH artifacts were recorded for this study. The results of my use-wear analysis on the HH lithic materials are presented in Chapter 8.

Based on these sampling strategies, the typo-technological and use-wear analyses were done on 2,237 artifacts four lithic assemblages (Table 3.10). The following four chapters present analyses for each class of lithic artifacts. Presentations of analytical results will focus on Grand Banks, considering its importance and relevance to the overall study. Data from the Lone Pine assemblage are mainly presented in tabular form. However, data drawn from all lithic assemblages were interpreted for comparative purposes.

CHAPTER 4

ANALYSIS OF CORES

Cores comprise only a small percentage -- less then one percent -- of the total lithic assemblages from both Grand Banks and Lone Pine (Table 4.1). The percentages are greatly affected by the inclusion in the artifact counts of abundant small flaking debris. By weight (Table 4.2), cores account for about 10 percent of the assemblages at Grand Banks and Lone Pine. Since only one core was recovered from Young 1, it will be excluded from this discussion. A total of 88 cores (47 from Grand Banks and 41 from Lone Pine) were analyzed for both typo-technological and use-wear aspects.

Typo-Technological Analysis

Typo-technological analysis of cores allows for: (1) reconstructing raw material procurement and selection; (2) representing core reduction modes at the sites through observations of core morphology, core size, and core facet patterns; and (3) reconstructing tool blank production through the examination of core facets. The results of core analysis suggest that: (1) the Princess Point people acquired local Onondaga cherts, and probably brought only small-sized raw material to the occupation site for further tool production; (2) the Princess Point people employed a "transformed core reduction" strategy in order to maximize flake production; and (3) most tool blanks were likely removed by a prepared core reduction mode.

Raw Material Procurement

About 85 percent of cores from Grand Banks are on Onondaga chert (Table 4.3). Grand Banks cores also include some other local materials such as Haldimand (6.4%) and Ancaster (4.3%). Only a few (N=2) are on non-local cherts.

As I have briefly outlined in Chapter 2, Onondaga is a kind of high-quality chert suitable for tool-making and it is locally accessible (Fig. 2.6). The closest quarry sites of Onondaga chert to Grand Banks are less than 10 km distant. Most sources are on the Lake Erie shore, between 15 and 30 km from Grand Banks. Based upon a study of the chert chemistry using Instrumental Neutron Activation Analysis, Quin (1966) confirmed that the Onondaga chert from these local quarry sites along the north shore of Lake Erie was employed for tool-making. The sources of Blois Blanc Formation chert (conventionally called Haldimand), on the other hand, are also close to the sites (10-15 km), but they were utilized less in tool manufacture. This is likely in large part due to the poor quality of the Blois Blanc Formation (Eley and von Bitter 1989:19). The short distance to chert sources provided an opportunity for the Grand Banks inhabitants to frequently visit quarry sites and to select suitably-sized raw material or to reduce core nodules to manageable sizes to be brought back to habitation sites for further reduction.

It is generally assumed that extensively used cores may not present the original cortex. Thus, cortex presence on small sized cores may possibly be indicative of the small size of core nodules brought back to the sites. At Grand Banks, cores with cortex covering 30% to 50% of the surface account for about 12 percent of the total core assemblage, and cores with 1% to 30% cortex coverage account for nearly 40 percent (Table 4.4). Such cores are exceptionally rare at Lone Pine, where there are very few cores with over 30% cortical coverage (4.9%). This differentiation probably reflects different raw material procurement strategies employed by the occupants

Table 4.1: Distributions of Lithic Classes at the Three Sites by Count.

	GRAND BANKS		LONE PINE		YOUNG 1	
CLASS	N	%	N	%	N	%
Core	47	0.27%	41	0.11%	1	0.04%
Formed Type	146	0.83%	122	0.34%	4	0.18%
Unretouched Flake	6252	35.33%	8301	23.24%	632	28.18%
Flaking Debris	11249	63.58%	27254	76.30%	1606	71.60%
TOTAL	17647	99.73%	35677	99.89%	2242	99.96%

Table 4.2: Distributions of Lithic Classes at the Three Sites by Weight.

	GRAND BANKS		LONE PINE		YOUNG 1	
CLASS	weight (g)	%	weight (g)	%	weight (g)	%
Core	829.1	10.36%	849	8.74%	18.6	2.88%
Formed Type	449.5	5.62%	320.9	3.30%	3.7	0.57%
Unretouched Flake	5304.2	66.30%	6179	63.60%	465.8	72.17%
Flaking Debris	1417.9	17.72%	2367.2	24.36%	157.3	24.37%
TOTAL	7171.6	90%	8867.1	91%	626.8	97%

Table 4.3: Distribution of Raw Material Presented in the Core Assemblages.

Chert Type	GRAND BANKS		LONE PINE	
	N	%	N	%
Onondaga	40	85.11	36	87.80
Haldimand	3	6.38	0	0.00
Ancaster	2	4.26	3	7.32
Others	2	4.26	2	4.88
Total	47	100	41	100

Table 4.4: Distribution of Cortex Coverage in the Core Assemblages.

Cortex	GRAND BANKS		LONE PINE	
	N	%	N	%
no coverage	23	48.94	26	63.41
1%-30% coverage	18	38.30	13	31.71
30%-50% coverage	6	12.77	2	4.88
Combined	47	100.00	41	100.00

Table 4.5: Distributions of Heat Treatment Presented in the Core Assemblages.

Heat Treatment	GRAND BANKS		LONE PINE	
	N	%	N	%
not indicated	42	89.36	33	80.49
indicated	5	10.64	8	19.51
Combined	47	100.00	41	100.00

of the two sites, but to what extent is unclear at present. As expected, no cores with over 50% cortical coverage were found in either assemblages, Cores that were brought back to sites were extensively utilized for tool production. Some cores with original cortex are relatively small, and at or near exhaustion (see below for data), suggesting the possibility that small core nodules were selected from quarry sites.

Other evidence supporting the statement that the Princess Point people likely transported small core nodules to the site can be derived from the fact that most raw materials at Onondaga quarries are present in small to medium size. During my field surveys of Onondaga chert sources, I observed that common core nodules are about 10-20 centimeters in three dimensions (length, width, and thickness). Core nodules over 30 x 30 x 30 cubic centimeters are present, but not common. Eley and von Bitter (1989:17) also state that the approximate thickness of such chert blocks is 2-8 cm on average. This by no means indicates that this occurrence reflects the prehistoric situation at the quarries. However, this characteristic of Onondaga chert may have led to the pattern of utilization of raw materials found at Princess Point sites.

Thermally altered cores are not common at either sites (Table 4.5). About 90% of Grand Banks cores manifest no indication of thermal damage on the surface. Lone Pine presents a relatively higher frequency of heated cores than Grand Banks. However, given the fact that most cores were recovered from the excavation areas where there are hearths, the relatively higher frequency of thermally damaged cores (19.5%) at Lone Pine may be due to the accidental heating. Furthermore, most thermally damaged cores are identified in the category of core fragments. This suggests that thermal damage was caused by accidental burning after core reduction. In other words, there is little, if any, evidence for intentional heat treatment as a process of core reduction modes at the Princess Point site.

Core Types and Reduction Modes

Over half of the core assemblages from both sites are core fragments: small and irregular blocks whose platforms are unidentifiable and whose core facets are incomplete (Table 4.6). By *core facet* I refer to the latest and complete negative flake scar on the core surfaces, which indicates an intentional removal from either a prepared or unprepared core platform (Fig. 4.1). The presence of numerous core fragments is indicative of reduction activities that took place on site too, because these blocky pieces are the result of blowoffs during flintknapping. It is impossible to determine whether core fragments were the result of biface reduction or flake core reduction since they may well be the by-products of both. These core fragments have a small to medium size range, with a mean of 33.9 mm by 22.5 mm by 14.4 mm.

Morphologically, three types of cores are recog-

Table 4.6: Core Typology.

Core Type	GRAND BANKS			LONE PINE			YOUNG 1		
		N	%		N	%		N	%
Prepared Core		8	17.02		4	9.76		1	100.00
single platform core	5			3			1		
opposed platform core	1			1			-		
ninety-degree platform core	2			-			-		
Amorphous Core		8	17.02		7	17.07		0	
flake core	8			7			-		
blade core	-			1			-		
Bipolar Core		5	10.64		5	12.20		0	
flake core	3			4			-		
blade core	2			1			-		
Core Fragment		26	55.32		24	58.54		0	
Grand Total		47	100.00		41	97.56		1	100.00

nized from both sites (Table 4.6). At Grand Banks, amorphous and prepared cores are equally common while bipolar cores have the lowest frequency. At Lone Pine, amorphous cores are the most common, followed by bipolar cores, and finally, prepared cores.

The prepared core mode of reduction in the Northeast Woodlands is not a specialized strategy, though it is usually regarded as such in the Old World archaeological context (e.g., Levallois cores). It, relatively speaking, reflects a skillful process used to obtain desired blanks (e.g., certain sizes of flakes or blades), as compared to the amorphous core reduction. During prepared core reduction, flake byproducts are removed in designated directions through careful core platform preparation. Techniques of the prepared core reduction mode have been recognized in terms of platform appearance. At Grand Banks, the predominant technique is the single platform core reduction, although the two other techniques are also present: opposed platform core and ninety-degree platform core reduction. Single platform and opposed platform cores are usually in the form of pyramidal and cylinder shapes, while ninety-degree platform cores are relatively square in plain view (Fig. 4.2a-b). The prepared core mode of reduction was usually abandoned when desirable blanks were obtained or when no more desirable blanks could be obtained. As a result, prepared cores are typically larger than other core

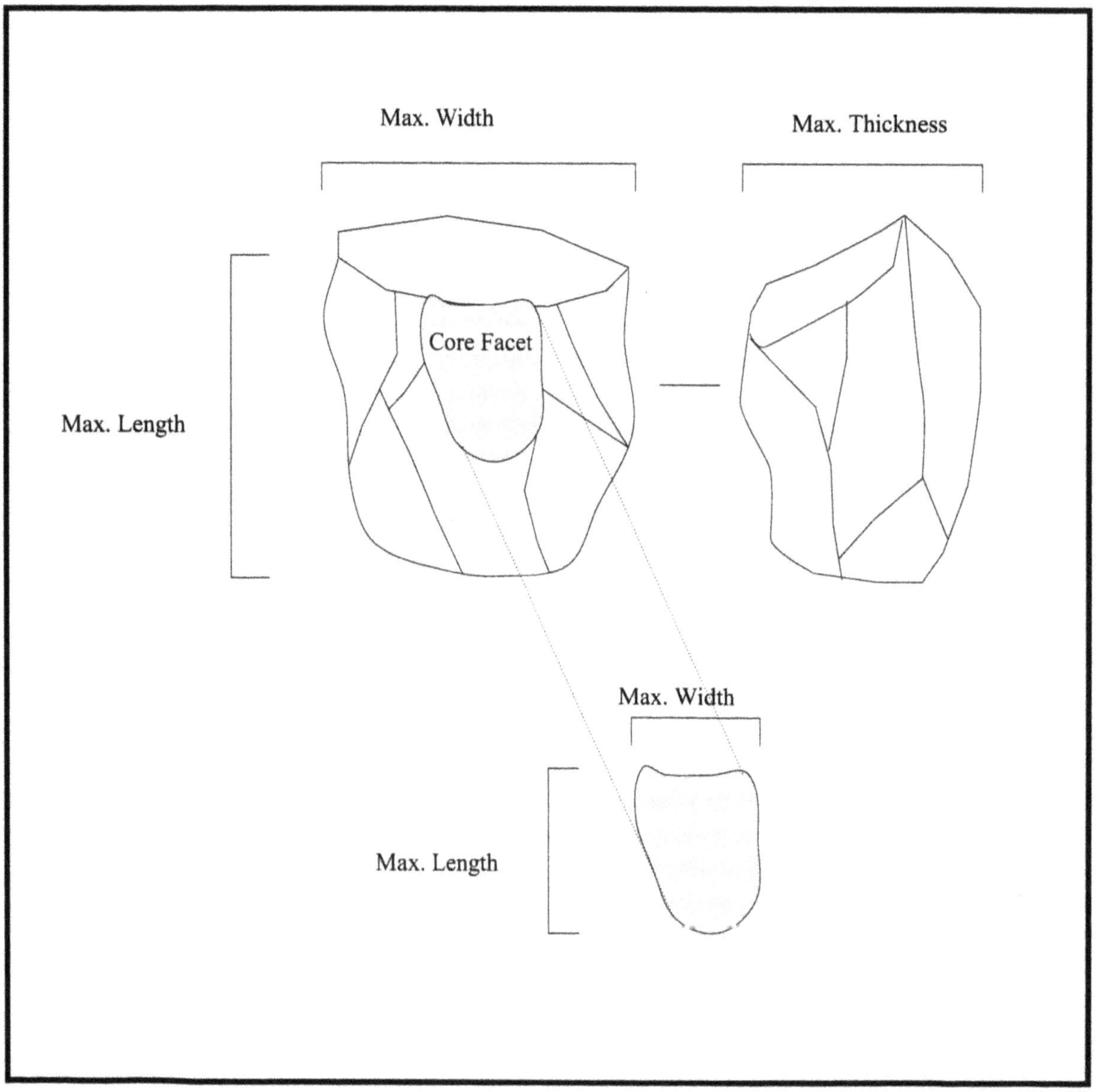

Fig. 4.1: Diagram of Metric Attributes of Cores and Core Facets.

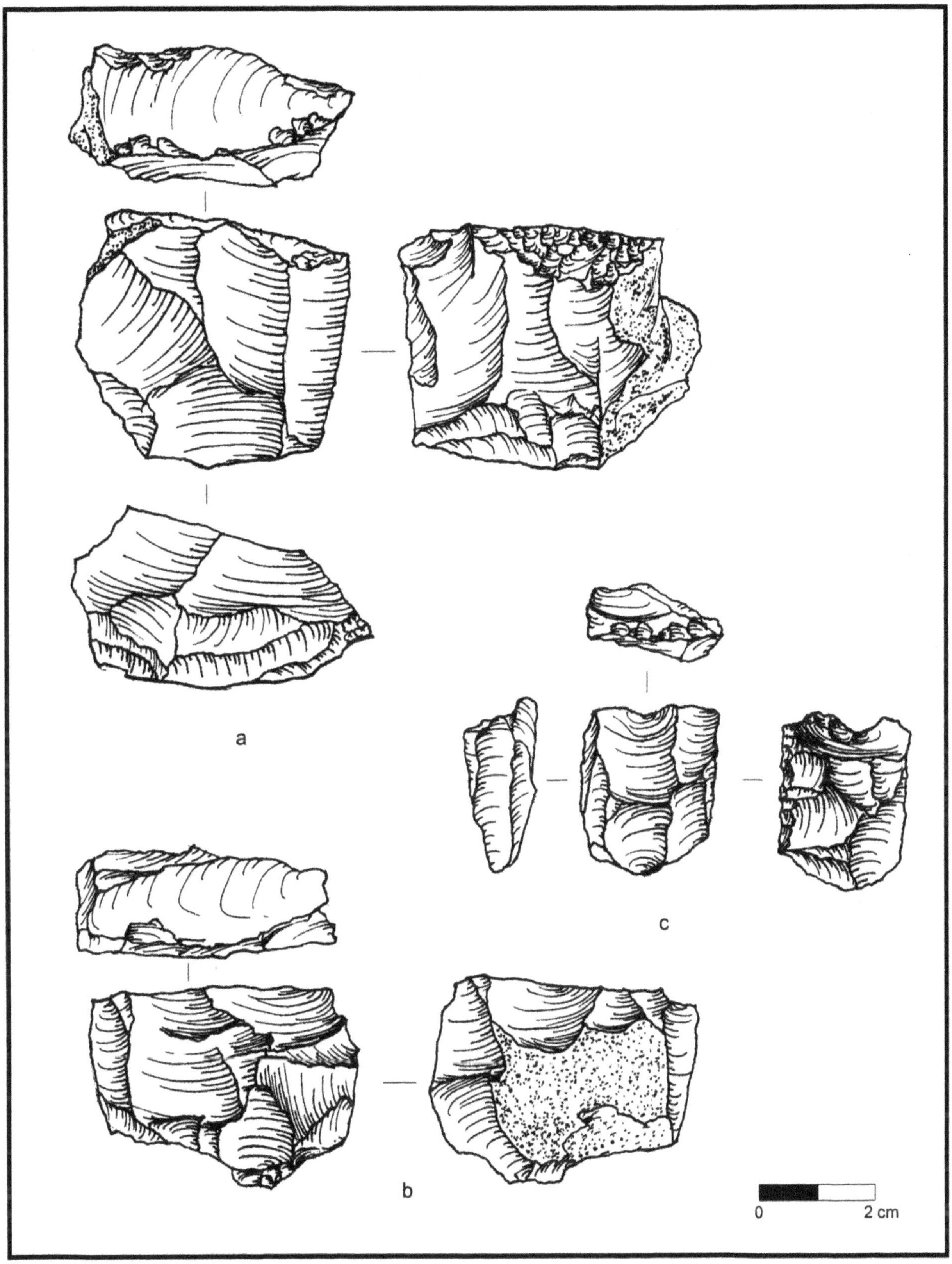

Fig. 4.2: Illustrations of Cores: a-b: prepared core; c: bipolar core.

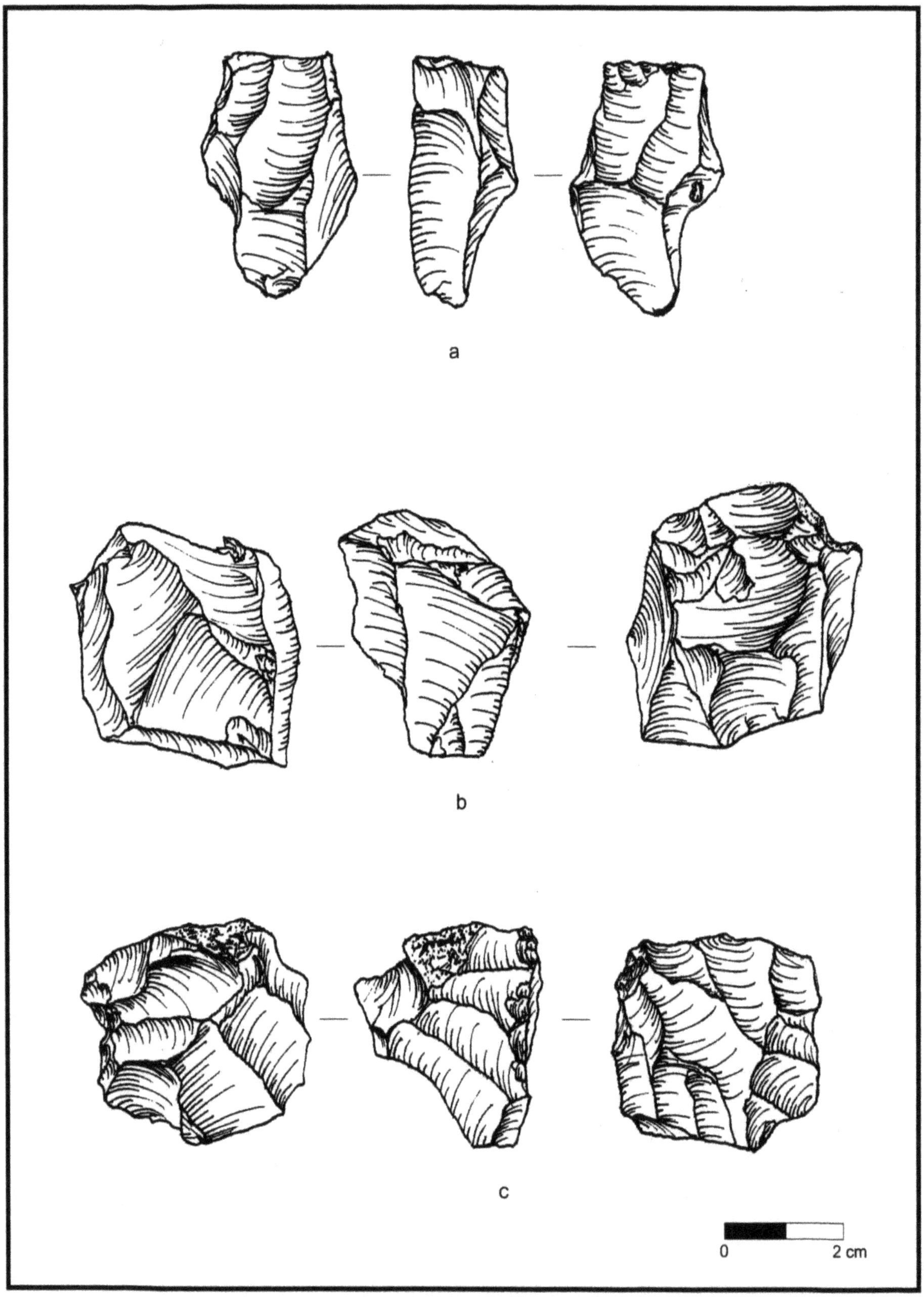

Fig. 4.3: Illustrations of Cores: a: bipolar core; b-c: amorphous core.

types (Table 4.7).

The amorphous core mode of reduction has been described as "expedient technology" by others (e.g., Johnson 1986; Johnson and Morrow 1987). The authors suggest that this reduction mode allows one to quickly remove flakes. With this mode of reduction, flakes are removed in multiple directions with no regularity, as long as edges or surfaces provide conditions allowing for flake removal (Fig. 4.3b-c). It was argued that the amorphous core reduction mode represents a wasteful strategy in terms of raw material conservation, and that application of such strategy is in large part due to an increase in sedentism and the availability of local raw material (Parry and Kelly 1987; Johnson 1987). Although the Lower Grand River Valley during the Princess Point period possibly presents increased sedentism and abundant local raw materials, another possibility for applying such a reduction mode also exists. Patterson (1987) pointed out that amorphous cores may be residual byproducts, produced by a variety of reduction modes. For instance, the amorphous core mode of reduction may reflect the last stages of the prepared core reduction mode where an attempt is made to maximize flake production. This mode of reduction could be applied in the following situations, as summarized by Patterson (1987:51), (1) at habitations where large flakes were not needed, (2) where specialized stone tools were not used, (3) where there was abundant raw materials, or (4) where sizes of raw materials were small.

Bipolar core reduction is common in most eastern North American sites. This technique could conserve raw material and maximize flake production because it removes flakes from relatively small sized cores. Accordingly, bipolar cores are usually small and elongated, and their core facets are also relatively small as well (Fig. 4.2c and Fig. 4.3a). However, bipolar cores are rare at both sites.

Therefore, I hypothesize that the amorphous core reduction mode applied at Grand Banks represents an economizing strategy for maximizing raw materials. Following Patterson (1987), I suggest that with a "transformed core reduction" strategy, the Grand Banks people seemed to apply the prepared core reduction mode first in order to obtain the designated flakes or blade-like flakes for tool blanks. When cores were no longer suitable for obtaining such blanks, amorphous core or bipolar core reduction modes were applied to "residual" cores in order to maximize flake production.

To support this hypothesis, we would expect lithic data demonstrating the following patterns: (1) overall, the prepared cores should be larger than both amorphous cores and bipolar cores; (2) the size of core facets from the prepared cores should be larger than those from the other two types of cores; (3) the length of core facets from prepared cores should fit into tool blank dimensions to a greater extent than those from the other two cores; and (4) the core facet pattern from the prepared cores should possess more regular flakes than those from the amorphous core and bipolar cores. The core analysis presented below will, therefore, focus on core size, core facet size, and core facet patterning.

Core Size and Core Facet Size

Metric data on cores is presented in Tables 4.7 and 4.8. Although the average length of the three core types is statistically similar, the range in length of prepared cores (*sd*=13.4) is much greater than the range of the other two types (*sd*=8.7 for amorphous core length and *sd*=3.7 for bipolar core length). Amorphous cores display similarity in length and width, having a mean length of 39.5 mm and a mean width of 31.0 mm. This is in contrast to the observations concerning the other two types, and was likely an effect of the technique of multi-directional flake removal.

It appears that at Grand Banks there are differences in both the mean width and the mean thickness between prepared and of amorphous cores. The bipolar reduction mode at Grand Banks resulted in much thinner and more elongated cores than the other two reduction modes. Overall, the mean core size of prepared cores at Grand Banks are about 45.9 x 34.5 x 16.1 mm^3, relatively larger than the mean core size of amorphous cores (39.5 x 31.0 x 19.5 mm^3) and that of bipolar cores (33.8 x 22.7 x 13.4 mm^3). Overall sizes of prepared cores are also larger than those of the other two kinds of cores at Lone Pine

Table 4.7: Core Size Data.

a: Grand Banks	Max. Length (mm)					Max. Width (mm)					Max. Thickness (mm)				
Core Type	Mean	S. D.	Min.	Max.	N	Mean	S. D.	Min.	Max.	N	Mean	S. D.	Min.	Max.	N
Prepared Core	45.93	13.39	31.82	66.78	8	34.49	9.06	26.42	52.01	8	16.06	2.27	13.10	20.06	8
Amorphous Core	39.51	8.65	27.27	56.79	8	31.00	5.03	24.68	37.95	8	19.45	3.99	14.81	25.22	8
Bipolar Core	33.82	3.70	30.04	38.38	5	22.66	1.66	20.30	24.31	5	13.35	2.57	10.04	15.52	5
Combined	40.60	10.72	27.27	66.78	21	30.34	7.74	20.30	52.01	21	16.70	3.83	10.04	25.22	21

b: Lone Pine	Max. Length (mm)					Max. Width (mm)					Max. Thickness (mm)				
Core Type	Mean	S. D.	Min.	Max.	N	Mean	S. D.	Min.	Max.	N	Mean	S. D.	Min.	Max.	N
Prepared Core	58.47	17.75	45.52	83.81	4	36.66	5.10	29.05	40.05	4	21.19	3.99	16.90	26.53	4
Amorphous Core	38.07	7.93	29.23	53.62	8	30.97	5.95	21.88	38.24	8	21.31	3.70	15.35	26.45	8
Bipolar Core	35.29	13.89	21.40	58.60	5	26.40	7.81	18.18	38.54	5	11.99	5.29	6.11	20.31	5
Combined	42.05	14.97	21.40	83.81	17	30.96	7.09	18.18	40.05	17	18.54	5.91	6.11	26.53	17

Table 4.8: Core Facet Size Data.

a: Grand Banks	Core Facet Length (mm)					Core Facet Width (mm)					Core Facet L/W Ratio				
Core Type	Mean	S. D.	Min.	Max.	N	Mean	S. D.	Min.	Max.	N	Mean	S. D.	Min.	Max.	N
Prepared Core	24.14	7.98	13.66	31.97	8	19.87	12.75	11.31	49.54	8	1.79	0.91	0.50	2.97	8
Amorphous Core	20.02	3.88	14.62	27.75	8	13.96	6.98	20.75	3.88	8	1.55	0.43	0.96	2.12	8
Bipolar Core	22.08	8.47	12.55	31.50	5	12.34	3.62	8.55	18.11	5	1.77	0.41	1.46	2.49	5
Combined	22.08	6.73	12.55	31.97	21	15.83	8.85	6.98	49.54	21	1.58	0.50	2.68	0.59	21

b: Lone Pine	Core Facet Length (mm)					Core Facet Width (mm)					Core Facet L/W Ratio				
Core Type	Mean	S. D.	Min.	Max.	N	Mean	S. D.	Min.	Max.	N	Mean	S. D.	Min.	Max.	N
Prepared Core	25.91	10.03	11.40	34.00	4	19.18	9.92	6.02	29.83	4	1.94	1.79	0.61	4.57	4
Amorphous Core	20.57	4.00	12.89	26.06	8	14.70	9.21	8.42	36.71	8	1.62	0.48	0.71	2.15	8
Bipolar Core	17.39	7.27	7.97	27.13	5	14.47	5.45	31.15	10.36	5	1.39	0.39	0.87	1.90	5
Combined	20.89	7.08	7.97	34.00	17	15.69	9.30	5.45	36.71	17	1.63	0.88	0.61	4.57	17

(Table 4.7).

Differences in the dimensions of core facets are also affected by core reduction modes (Table 4.8). The amorphous core reduction mode featured removals of much smaller flakes during the final stage of core reduction than the other two types did. The average length of amorphous core facets is 20.0 mm, the shortest of the three types. The majority of amorphous core facets fall in the 10 mm to 20 mm length interval (Fig. 4.4a). On the other hand, over 60 percent of prepared core facets are between 20 mm and 40 mm long (Fig. 4.4b). The average width of prepared core facets is also much greater than that of amorphous and bipolar core facets (19.9 mm vs 13.9 mm and 12.3 mm, respectively). The diagrams of width intervals show that most amorphous core facets fall within the 5-15 mm wide intervals, while the majority of prepared core facets fall within the 10-20 mm wide intervals (Fig. 4.4a-b). Furthermore, with a core facet L/W ratio mean of 1.7, the bipolar core reduction mode produced narrower and more elongated flakes than the other two reduction modes. In sum, the sizes of the core facets of prepared cores are longer and wider than that of amorphous and bipolar cores.

Core Facet Pattern and Tool Blank Production

Having presented the dimensional differentiation of overall core sizes and core facet sizes, the question now is: do flakes produced through the

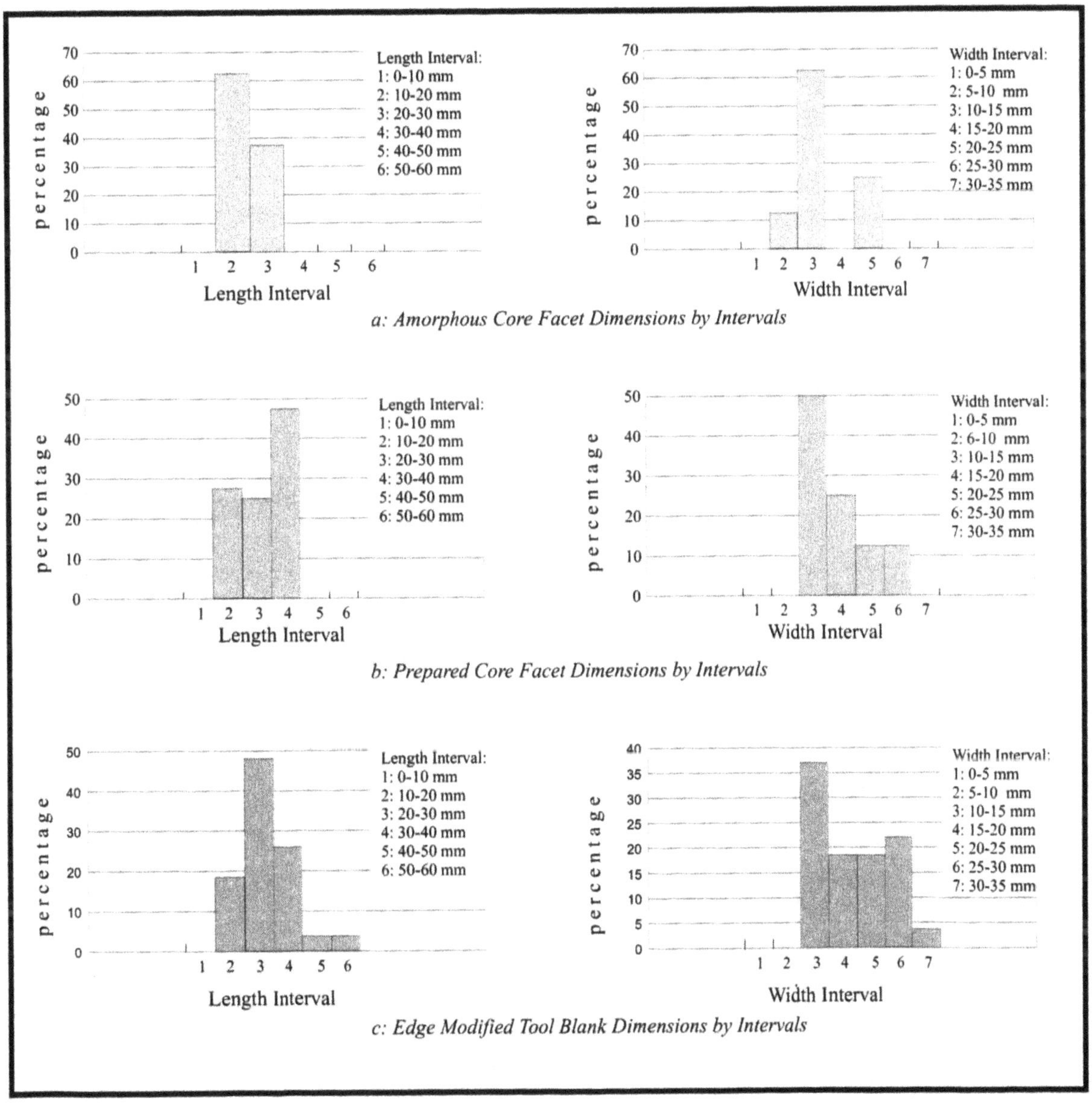

Fig. 4.4: Comparison of Length and Width Intervals by Selected Core Facets and Tool Blanks.

prepared core reduction mode have a special purpose, specifically, as blanks for tool production? To answer this question, it is first necessary to compare the dimensions of core facets of both prepared cores and amorphous cores with the dimensions of flake tool blanks drawn from scrapers and retouched pieces (the two most common flake tool types in the assemblages). Bipolar core facets were excluded from the comparison to avoid statistical bias, as they represent only a small portion of the assemblage. The majority of tool blanks as drawn from scrapers and retouched pieces fall between 20 and 30 mm in length and 10 and 20 mm in width (Fig. 4.4c).

A comparison of the length and width intervals (Fig. 4.4) shows that the majority of the flakes removed from prepared cores in the last stage fall within the dimensions which obtain in the range of tool blank sizes. However, this does not apply to the latest flakes removed from amorphous cores. In other words, this may imply that flakes, which were produced using the prepared core reduction mode immediately before the core was discarded, could have been selected for flake tool blanks. On the other hand, smaller flakes removed at the last stage from amorphous cores would not fit within tool blank dimensions. It should be kept in mind that earlier in the reduction sequence the amorphous core reduction may have produced larger flakes that could have been used as tool blanks as well.

In order to support the hypothesis that the prepared core reduction mode was possibly used for blank production at Grand Banks, I examined the relationship between core types and core facet coverage (Table 4.9). Most prepared cores have fewer removal scars than do amorphous and bipolar cores. In contrast, all but one of the amorphous cores have full coverage of flake removals on the surface, indicating continued use of these cores until exhaustion in order to maximize flake production. All bipolar cores have 50% to 75% core facet coverage. I believe this can be attributed to the fact that bipolar reduction at Grand Banks featured concentration on two sides of the core tablets for removal of bipolar flakes. These differences in core facet coverage, which correlated with core type, strongly suggest that prepared cores might have been discarded earlier in the reduction sequence than amorphous and bipolar cores.

Core facet types and core types are cross-examined to support the proposed blank production strategy at Grand Banks (Table 4.10). Half of the core facets from prepared cores are blade facets, while all those from amorphous cores are flakes. Bipolar core reduction also produced some blade

Table 4.9: Core Type and Core Facet Coverage from the Grand Banks Core Assemblage.

Core Type	Core Facet Coverage				
	1%-25%	25%-50%	50%-75%	75%-100%	Total
Prepared Core	0	3	3	2	8
Amorphous Core	0	0	1	7	8
Bipolar Core	0	0	5	0	5
Combined	0	3	9	9	21

Table 4.10: Core Type and Core Facet Type from the Grand Banks Core Assemblage.

Core Type	Core Facet Type		
	Flake	Blade	Total
Prepared Core	4	4	8
Amorphous Core	8	0	8
Bipolar Core	4	1	5
Combined	16	5	21

facets, but flakes are the predominant products of bipolar core reduction. From those observations, I suggest that most blades at Grand Banks could possibly have been produced using the prepared core reduction mode.

Use-wear Analysis

Grand Banks

The possibility that cores were utilized tools in eastern North America has not been discussed until recently. Usewear analysis of lithics from some Illinois Valley assemblages suggests that cores have been used in a wide variety of activities, including longitudinal, transverse, and rotation motions (Odell 1996:155, also his Table 6.9 and Table 10.10). A similar use-wear study of the 88 cores from Grand Banks/Lone Pine has underscored this property of cores. It indicates that representative proportions of the core assemblages have been utilized at these sites, accounting for 15 percent at both Grand Banks and Lone Pine.

All seven used cores from Grand Banks show a single employed unit per piece. Tool motions at Grand Banks are limited to four types: cutting/sawing, planing/whittling, drilling, and chopping (Table 4.11). Contact materials commonly include woods and soft to hard resistance animal substances (Table 4.12). From a cross-tabulation of tool motion and contact material (Table 4.13), the use-tasks associated with core tools include meat-cutting (n=1); fresh bone/soaked antler cutting/sawing (n=1); planing/whittling fresh wood (n=2); planing/whittling dried or hard wood (n=1); drilling dried wood (n=1); and chopping hard wood (n=1).

Table 4.11: Tool Motion in the Core Assemblages.

Tool Motion	GRAND BANKS		LONE PINE	
	N	%	N	%
indeterminate	0	0.0	0	0.0
cut/saw	2	28.6	0	0.0
slice/carve	0	0.0	0	0.0
scrape/shave	0	0.0	4	50.0
plane/whittle	3	42.9	1	12.5
projection	0	0.0	0	0.0
bore/drill	1	14.3	1	12.5
grave	0	0.0	1	12.5
chop	1	14.3	0	0.0
hoe/dig	0	0.0	1	12.5
pound/grind	0	0.0	0	0.0
wedge	0	0.0	0	0.0
pick	0	0.0	0	0.0
total employed units	7	100.0	8	100.0

Table 4.12: Contact Material in the Core Assemblages.

Contact Material	GRAND BANKS		LONE PINE	
	N	%	N	%
Indeterminate	0	0.0	1	12.5
Soft animal (SA)	1	14.3	0	0.0
Soft vegetal (SV)	0	0.0	0	0.0
Medium-soft vegetal (1M)	2	28.6	2	25.0
Medium animal (MA)	0	0.0	0	0.0
Medium-hard vegetal (2M)	3	42.9	0	0.0
Hard animal (1H)	1	14.3	3	37.5
Very hard animal (2H)	0	0.0	1	12.5
Hard inorganic (3H)	0	0.0	1	12.5
Total employed units	7	100.0	8	100.0

The evidence of meat cutting on a prepared core is indicated by a row of small to medium feather scars unevenly distributed along a thin edge (A) (Fig. 4.5a; Photo 25), a pattern similar to those seen on experimental specimens used for cutting meat. Incipient polish is weakly developed on both sides of the surface, but can still be observed at about 50x magnification. A blunt edge with small crushed scars indicates that location B may have been used by a forefinger as the holding position (Photo 26).

A chopping tool on a medium sized core fragment has a heavily rounded edge and a few crushed medium to large hinged scars at location A where there was contact with hard wood (Fig. 4.5b). These bifacially distributed scars, perpendicular to the edge, are characteristic of chopping activities, as demonstrated experimentally. A snap scar with a rounded edge is observed on a projection (B), and I suspect that it could be used for various tasks such as drilling or picking, assisting in the chopping process. I placed this use-wear case, however, into the indeterminate category due to a lack of clear polish. Another small fragment was employed in wood planing (Fig. 4.5c). The working edge of this tool consists of a thin, notched edge (location A) where there is a density of unifacially distributed feather and

Table 4.13: Tool Motion by Contact Material in the Grand Banks Core Assemblage.

Tool Motion	Contact Material									
	?	SA	SV	1M	MA	2M	1H	2H	3H	
indeterminate	-	-	-	-	-	-	-	-	-	0
cut/saw	-	1	-	-	-	-	1	-	-	2
slice/carve	-	-	-	-	-	-	-	-	-	0
scrape/shave	-	-	-	-	-	-	-	-	-	0
plane/whittle	-	-	-	2	-	1	-	-	-	3
projection	-	-	-	-	-	-	-	-	-	0
bore/drill	-	-	-	-	-	1	-	-	-	1
grave	-	-	-	-	-	-	-	-	-	0
chop	-	-	-	-	-	1	-	-	-	1
hoe/dig	-	-	-	-	-	-	-	-	-	0
pound/grind	-	-	-	-	-	-	-	-	-	0
wedge	-	-	-	-	-	-	-	-	-	0
pick	-	-	-	-	-	-	-	-	-	0
total employed units	0	1	0	2	0	3	1	0	0	7

Table 4.14: Tool Motion by Contact Material in the Lone Pine Core Assemblage.

Tool Motion	Contact Material									
	?	SA	SV	1M	MA	2M	1H	2H	3H	
indeterminate	-	-	-	-	-	-	-	-	-	0
cut/saw	-	-	-	-	-	-	-	-	-	0
slice/carve	-	-	-	-	-	-	-	-	-	0
scrape/shave	-	-	-	-	-	-	3	1	-	4
plane/whittle	-	-	-	1	-	-	-	-	-	1
projection	-	-	-	-	-	-	-	-	-	0
bore/drill	-	-	-	1	-	-	-	-	-	1
grave	1	-	-	-	-	-	-	-	-	1
chop	-	-	-	-	-	-	-	-	-	0
hoe/dig	-	-	-	-	-	-	-	-	1	1
pound/grind	-	-	-	-	-	-	-	-	-	0
wedge	-	-	-	-	-	-	-	-	-	0
pick	-	-	-	-	-	-	-	-	-	0
total employed units	1	0	0	2	0	0	3	1	1	8

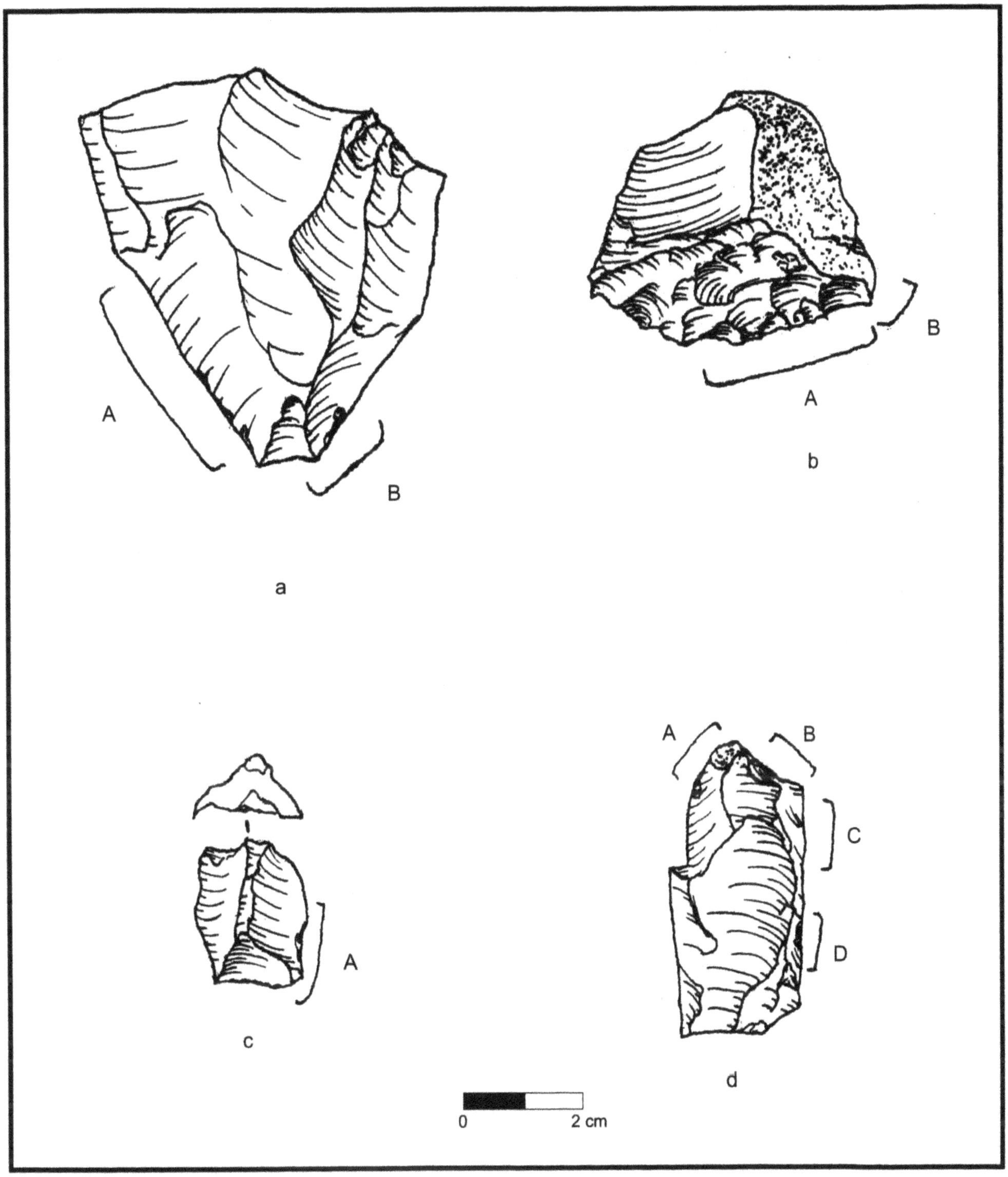

Fig. 4.5: Illustrations of Utilized Cores: a: AfGx-3:5963; b: AfGx-3:6132; c: AfGx-3:6321; d. AfGx-113:3999.

stepped scars and light polish (Photo 27).

All prehensile wear on core tools belong to the hand-holding category. One can reasonably imagine that core tools are much easier to use hand-held than hafted.

Lone Pine

Eight employed units identified from Lone Pine cores are involved in the five types of tool motions and in the four kinds of material worked (Table 4.11 and 4.12). Four employed units are engaged with scraping bone or antler, a higher rate than with other use-tasks (Table 4.14). The two tasks involved planing and drilling fresh or seasonal wood. One core graving tool was identified but the contact material is unknown. A discarded single platform core was also used as a digging tool.

A medium-sized core fragment (Fig. 4.5d) would seem to have been used for multiple use-tasks. Three employed units display similar wear traces identified as hardwood scraping (A, B, and C). Unifacially distributed small to medium feather scars are associated with heavy rounding on all the worked edges, and a rolled-over scarring pattern is indicative of woodworking. Light polish developed on all three units, and is found on all contact surfaces, which is an indicator of transverse tool motions. Heavy rounding and a few crushed scars indicate that this piece was either used intensively or used on material of hard-resistance. Intensive use of this piece is indicated by a clear and heavily rounded edge at location D where prehensile wear has been identified. As at Grand Banks, no core tools at Lone Pine were hafted during use.

Discussion and Summary

Core analysis clearly demonstrates the utilization of local Onondaga chert at the sites. Given the overall small size of cores and the presence of original cortex on the surface, it is thought that raw materials (or cores reduced at quarry sites) which were brought to the sites for tool production were small. Evidence for heat treatment to aid in flintknapping is minimal.

Results from core analysis suggest that three basic core reduction modes were used at the Princess Point site, Grand Banks, and also at the later period site, Lone Pine. Observations of core types, core sizes, core facet size, and core facet patterns yielded fruitful results for reconstructing the modes of core reduction on-sites. The analysis demonstrates that a unique reduction sequence was likely employed at Grand Banks: the "transformed" core reduction. It suggests that most desired blanks (flakes or blades) were probably removed first from the prepared cores by careful platform preparation. It is possible that amorphous core or bipolar core reduction modes could have been applied to discarded prepared cores for the sake of the conservation of raw material. People at Grand Banks may have produced amorphous flakes from small amorphous cores, regardless of the shape and the size of the flakes they could obtain until core exhaustion.

The special core reduction sequence should be verified by re-fitting techniques, which are beyond the scope of this study. The supportive evidences for this scenario are from Grand Banks as follows. The overall sizes of prepared cores are larger than those of both amorphous cores and bipolar cores, which indicates that prepared cores were discarded at an early stage after no more suitable blanks could be removed through the prepared core reduction mode. The core facet sizes of amorphous cores are smaller than those of prepared cores, indicating that the final stages of flake removal were probably associated with the amorphous core reduction mode rather than prepared core reduction. Most lengths of prepared core facets fit within the dimensional ranges of flake tool blanks (scrapers and retouched pieces), while lengths of amorphous core facets are smaller than those of tool blanks, suggesting that most flakes removed in the final stages of prepared core reduction could be directly selected for tool modification. Finally core facet patterns exhibit more blade negative scars on prepared cores than on amorphous cores, suggesting that blades were often removed through the prepared core reduction, presumably as desired blanks.

Therefore, the three reduction modes may represent different stages of core reduction processes. The frequency distributions of core types are not necessarily indicative of the particular reduction mode which was predominant. Moreover, in contrast to conventional understanding, the amorphous core reduction mode represents an economizing strategy, since it was applied to already-reduced cores in order to maximize flake production. However, the economizing behavior involved in raw material procurement should not be judged by core morphology and/or reduction modes alone. Instead, it must be considered from the perspective of overall patterns of lithic production. Thus, I will discuss this issue in some detail in Chapter 6.

Use-wear analysis of core tools from Grand Banks showed that they were predominantly used in wood-working. By contrast, analysis of cores from Lone Pine demonstrated that these tools were used on hard material like bone or antler. Overall, this analysis demonstrates that cores were extensively used as tools at Princess Point sites.

The piece from Lone Pine discussed above, with multiple employed units, was used in a consistent manner: for scraping hard wood. It might have been used for a single task on different occasions or on the same occasion but using different parts of the tool. This phenomena may be interpreted as evidence of intensive use of individual tools. This intensified utilization of tools is also evidenced by use-wear on unmodified flake tools, as will be illustrated in Chapter 6.

CHAPTER 5

ANALYSIS OF FORMED TYPES

A total of 272 artifacts recovered from Grand Banks, Lone Pine, and Young 1, were categorized as "formed types:" pieces showing intentional modification. Not surprisingly, each formed type assemblage accounts for less than one percent of the total flaked stone assemblage (Table 4.1). While Grand Banks has a relatively higher proportion of formed types (0.8%), Lone Pine and Young 1 occur fewer shaped artifacts, 0.3% and 0.2%, respectively.

Typo-Technological Analysis

Typo-technological analysis of formed types concentrates on the following topics: (1) typological descriptions; (2) tool blank production and selection; and (3) retouch techniques.

Typological Descriptions

The Young 1 assemblage yielded only two projectile points (fragments), one simple endscraper on a flake, and one retouched piece. Because this sample is not representative, I have excluded Young 1 formed types from the discussion. Typologically, the Grand Banks and Lone Pine assemblages present high frequencies of projectile points and retouched pieces, followed by bifaces and scrapers. Detailed morphological types are listed in Table 5.1. Different distributions of formed types exist between Grand Banks and Lone Pine. The Grand Banks assemblage is dominated by projectile points (42%), although over half of them (34 out of 62 pieces) are fragmentary. At Lone Pine projectile points comprise only 21% of the formed types and are the second most common group of the formed type component. Retouched pieces account for slightly more than one-third of the formed types at Lone Pine, but less than one quarter at Grand Banks. Both sites have bifaces in roughly equal percentages, 16% at Grand Banks and 18% at Lone Pine. Scrapers, a common artifact type in eastern North America, rank fourth in frequency at both sites, although the Grand Banks assemblage contains fewer. Other types including unifaces and drills, which are found at both sites, have less than 10% of total formed types. Burins and perforators are present only at Lone Pine, and microliths and truncations are found only at Grand Banks (Fig. 5.1).

At the Grand Banks site, formed type artifacts from the two excavation areas (Areas A and B) exhibit similar distribution trends. There are, however, two notable differences: Area B has a much higher frequency of bifaces and its variety of types is more limited than Area A (Fig. 5.2).

Projectile Points

Twenty-two of 62 projectile points found at Grand Bank are triangular in shape. These complete points, together with 6 triangular fragmentary bases, comprise nearly half of the projectile point component at Grand Banks (45.2%). Morphologically these points are similar to the Madison point (Ritchie 1971), but in general they fall within Justice's (1987:224-230) Late Woodland-Mississippian Triangular Cluster. As compared to Justice's data, the triangular points found at Grand Banks are relatively longer and narrower in shape, having a mean length to width ratio of 1.43 (Justice's ratio: 0.60). Their average length is 31.4 mm (sd=6.6), width 21.9 mm (sd=3.9), and thickness 6.0 mm (sd=1.6). These measurements are similar to those of the triangular projectile points found at Grand Banks by Stothers (1974) (Length=37.9 mm, Width=25.7 mm, and Thickness=5.9 mm). Most of these triangular points exhibit straight or nearly straight

Table 5.1: Formed Type Typology.

ARTIFACT TYPES	GRAND BANKS			LONE PINE			YOUNG 1		
		N	%		N	%		N	%
Projectile Points		62	42.47		26	21.31		2	50.00
side-notched point	2			1			-		
stemmed point	1			-			-		
nonstemmed, non-fluted point	-			2			-		
triangular point	22			4			-		
type-unidentifiable point	3			1			-		
fragment, side-notched base	-			1			-		
fragmentary, stemmed base	1			-			-		
fragmentary, nonstemmed base	-			1			-		
fragment, triangular base	6			1			-		
fragment, point tip	25			13			2		
fragment, medium section	2			2			-		
Retouched Pieces		35	23.97		45	36.89		1	25.00
unilateral-obverse retouch	5			11			-		
unilateral-inverse retouch	4			8			1		
bilateral-obverse retouch	1			2			-		
bilateral-inverse retouch	1			0			-		
alternate retouch	3			7			-		
partial retouch	4			4			-		
retouch fragment	17			13			-		
Bifaces		24	16.44		22	18.03		0	0.00
Stage 1 biface	3			-			-		
Stage 2 biface	6			1			-		
Stage 3 biface	1			2			-		
fragment, stage 1 biface	1			2			-		
fragment, stage 2 biface	10			9			-		
fragment, stage 3 biface	3			8			-		
Scrapers		11	7.53		20	16.39		1	25.00
simple endscraper	4			4			1		
endscraper without lateral retouch	2			6			-		
thumbnail endscraper	2			1			-		
shouldered/nosed endscraper	-			1			-		
sidescraper, straight	-			1			-		
sidescraper, convex	1			2			-		
double sidescraper	-			2			-		
endscraper bit	1			2			-		
endscraper fragment	1			1			-		
Unifaces		5	3.42		1	0.82		0	0.00
Drills		4	2.74		3	2.46		0	0.00
Square-based, biface	2			-			-		
drill tip	-			2			-		
drill fragment	2			1			-		
Notches		2	1.37		2	1.64		0	0.00
single notch without lateral retouch	2			-			-		
notch fragment	-			2			-		
Microliths		2	1.37		0	0.00		0	0.00
Truncations		1	0.68		0	0.00		0	0.00
Burins		0	0.00		2	1.64		0	0.00
angle burin on snap or old surface	-			1			-		
angle burin on straight truncation, oblique	-			1			-		
Perforators/gravers		0	0.00		1	0.82		0	0.00
Grand Total		**146**	**100.00**		**122**	**100.00**		**4**	**100.00**

base shapes, but concave and convex bases are also present in relatively low frequencies (Fig. 5.3a, b). In addition to the triangular points, one side-notched and one stemmed point are found at Grand Banks. All the triangular points were made of local Onondaga chert, while the stemmed one was made of Haldimand chert.

The Lone Pine projectile points show comparatively more variety. Triangular points, including one typical Levanna point (Fig. 5.3d), account for only 15.4% of the projectile point. Other types include a side-notched point (Fig. 5.3e) and a non-stemmed point with a non-fluted base. Both were well made on non-local raw materials.

Previous study suggests that the Levanna- or Madison-like triangular projectile point is an "index-fossil" of the Princess Point lithic assemblage (Stothers 1977; Fox 1990). Although a relatively high frequency of triangular points was found in the assemblage from Grand Banks, in this study I do not apply the principle that projectile points may be considered cultural markers. It has been argued that placing a particular type of projectile point into a temporal sequence may provide too narrow an interpretation of cultural history (Michlovic 1976). Experimentation with replicating a variety of projectile points suggests that morphological projectile point typologies are

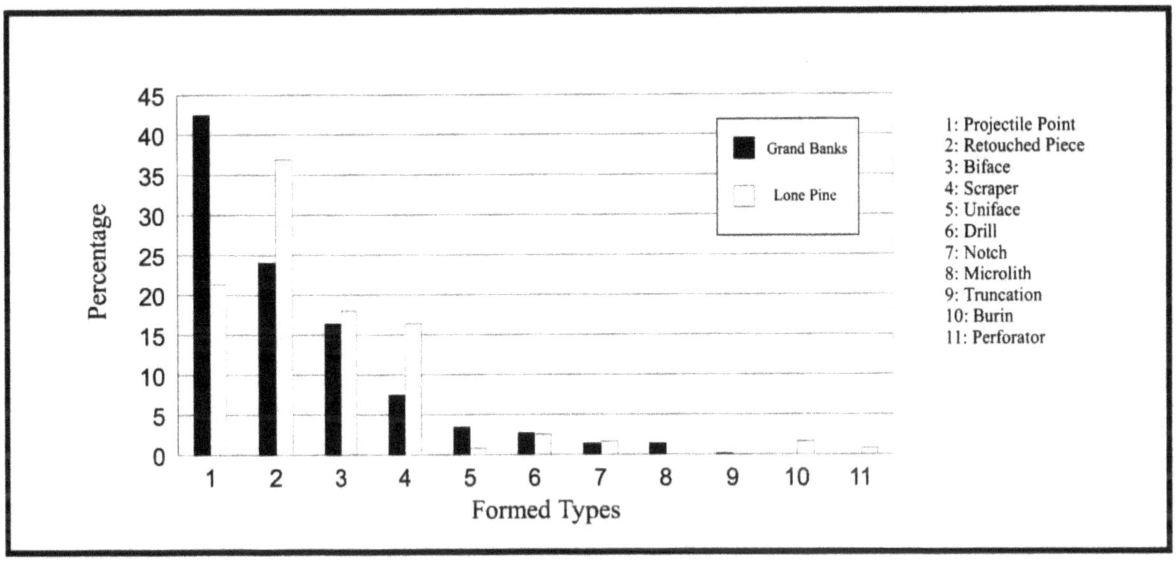

Fig. 5.1: Distribution of Formed Types from Grand Banks and Lone Pine.

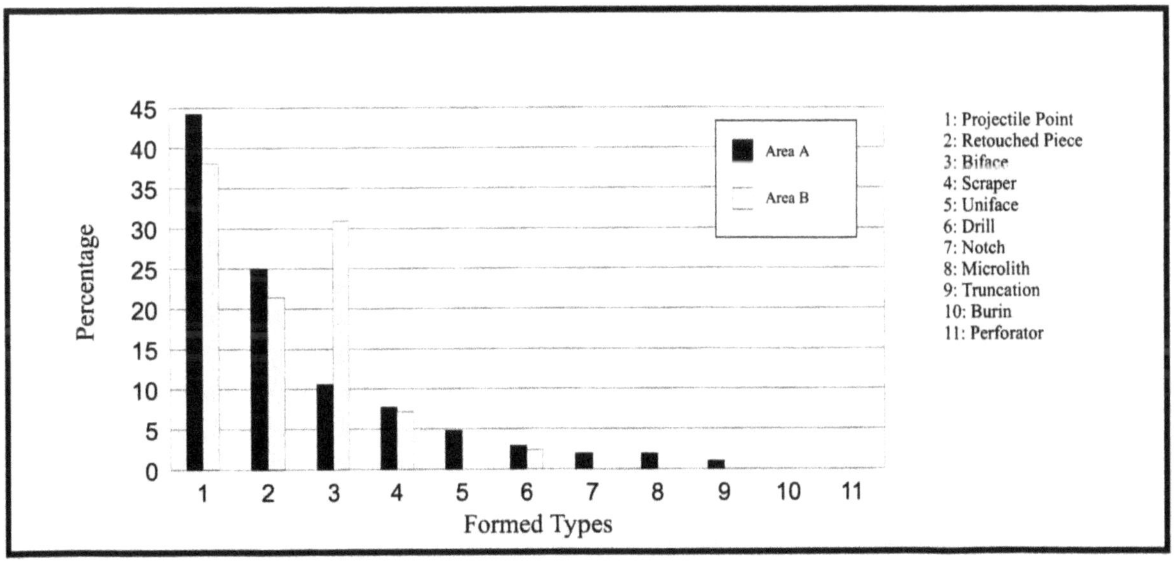

Fig. 5.2: Distribution of Formed Types from Area A and Area B at Grand Banks.

not consistently reliable temporal or cultural markers, but may present different stages of manufacture or use (Flenniken and Raymond 1986). On the other hand, triangular Levanna and Madison-like point types have a widespread distribution in the Eastern Woodlands from A.D. 500 onward, and especially during the Late Woodland period (e.g., Jelks 1993; Railey 1992; Borse 1994:89-110).

Qualitative measurements of the quality of point making have been made in an attempt to distinguish "well-made" from "roughly-made" points. Although the criteria for determination of the quality of point making is subjective, distinctions are based on the following factors: the stage of thinning, final retouch, and symmetry. Of the 62 points from Grand Banks examined, less than one third (29%) fall into the category of well-made points. Over half (54.8%) are of moderate quality, while 16.1% are roughly made. Roughly the same distribution of point making quality is present at Lone Pine, with 34.6% of points being well-made. Most of these well-made points at both sites are fashioned on non-local raw materials. Seven pieces (5 from Grand Banks and 2 from Lone Pine) are made on flake blanks with only unifacial retouch and edge modification (Fig. 5.3h).

Among the point fragments from Grand Banks, tip fragments are the most common (Fig. 5.3c, f). The angle of these tips in plane view was measured (Fig. 5.4). It was found that most of the tips (40%) are between 40° and 50° with a mean of 43°. Complete triangular points exhibit a predominance of narrow angles: between 35° and 40° in plane view (63.6%), with a mean angle of 39.3°. It is clear that the fragmentary tips have wider angles than that of complete triangular points (Fig. 5.5). The difference between the two groups is also reflected by tip thickness, which was measured at about 10 mm below the tip. T-test results suggest that there is a significant difference at 0.05 level (*t-value*=3.6922; *df*=61, *p-value*=0.000477). The mean thickness of complete triangular points (6.07 mm) is much greater than that of tip fragments (4.68 mm). The results may imply that the fragmentary tips are not broken parts of the triangular points which dominate the Grand Banks assemblage. They could be unfinished or unsuccessful sections from various point types, or broken portions of others.

Bifaces

The Grand Banks assemblage contains 10 complete bifaces and 14 fragments, while Lone Pine has 3 and 19, respectively. Most bifaces are made of local Onondaga chert, with only one made of local Haldimand chert. From observations of the reduction trajectory, it is clear that most of the bifaces were thinned from small core tablets. Failure ratios are relatively high at Grand Banks, as is suggested by the relatively higher frequencies of Stage 1 and Stage 2 bifaces (including fragments) present at Grand Banks (83.3%). It is hard to determine the morphological types of these artifacts because there are few complete Stage 3 bifaces. The only complete Stage 3 biface at Grand Banks has a rectangular-ended shape with a pointed tip (Fig. 5.6a). It was fabricated through thinning a core tablet, and its lateral edges were well modified into straight working edges. One stage 3 biface from Lone Pine presents a similar production technique, but it is oval in shape (Fig. 5.6c). The other Lone Pine stage 3 biface is also oval in form but was roughly made on a flake blank. One of the stage 2 bifaces from Lone Pine (Fig. 5.6d), used in slicing animal-hide or fat, seems to have met with failure in the flake-thinning process. The convex base of the object was sharpened through marginal retouch for use as a cutting edge, and a retouched notch on the lateral edge of this tool likely provided a blunt edge to be used when holding the tool in the hand.

Unifaces

This formed type is distinguished by invasive retouch on one side of the lithic object. Five of six unifaces are from Grand Banks. Unifaces from Grand Banks are made on flake blanks, including one on a large thick primary flake (Fig. 5.6b, Fig. 5.7f). One Grand Banks uniface is made on a thin plain flake with parallel lateral edges. Overall the unifaces have the greatest thickness (mean 13 mm) of all formed types.

Scrapers

Simple endscrapers on flakes dominate the scraper class at both sites (Fig. 5.7c, e). Other

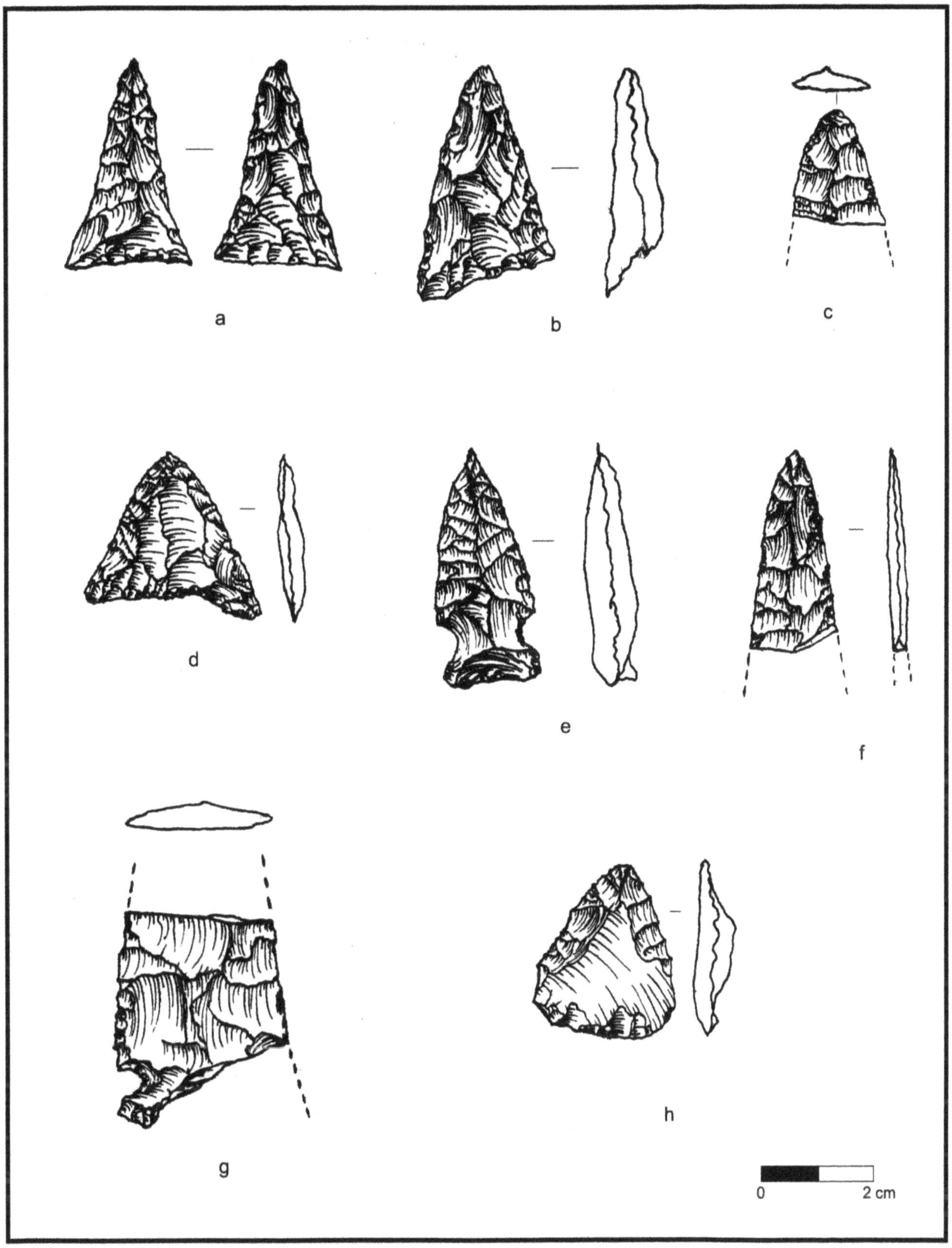

Fig. 5.3: Illustration of Formed Types: projectile points

types of scrapers are present, but in low frequencies. Thumbnail endscrapers appear at both sites (Fig. 5.7g). A shoulder/nosed endscraper was identified only at Lone Pine (Fig. 5.7d). It was characterized by steep lateral retouch adjacent to the "scraping" bit which gives the bit an isolated or "nosed" appearance in plane view. Sidescrapers are more common in the Lone Pine assemblage; one concave sidescraper was recovered from Grand Banks. The bits of broken endscrapers also occur consistently, although in low numbers. Without exception, all of these scrapers are made using edge retouch from local chert.

Drills

A few drills were found at both sites, and all com-

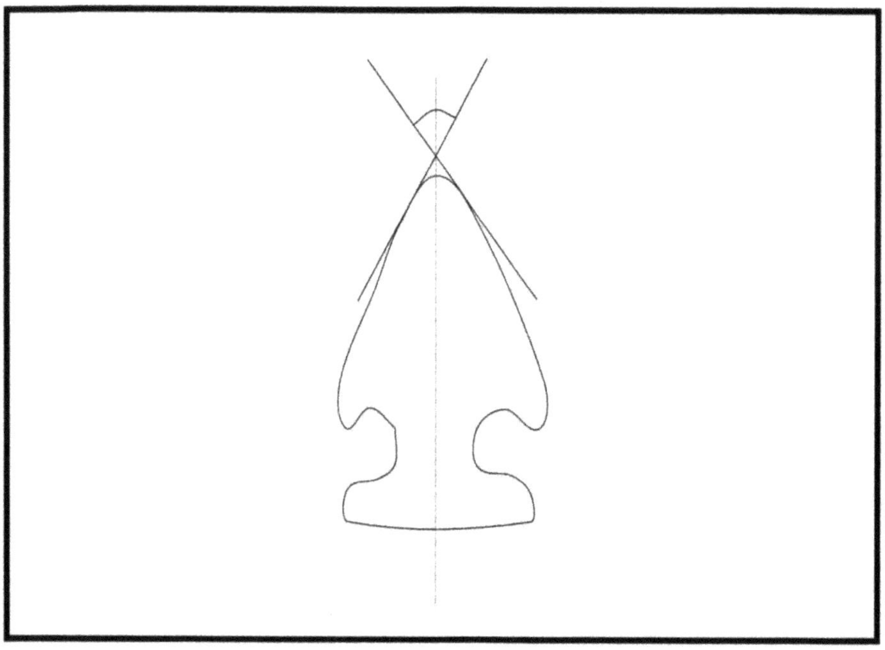

Fig. 5.4: Measurement of Projectile Point Tip-angle in Plane View.

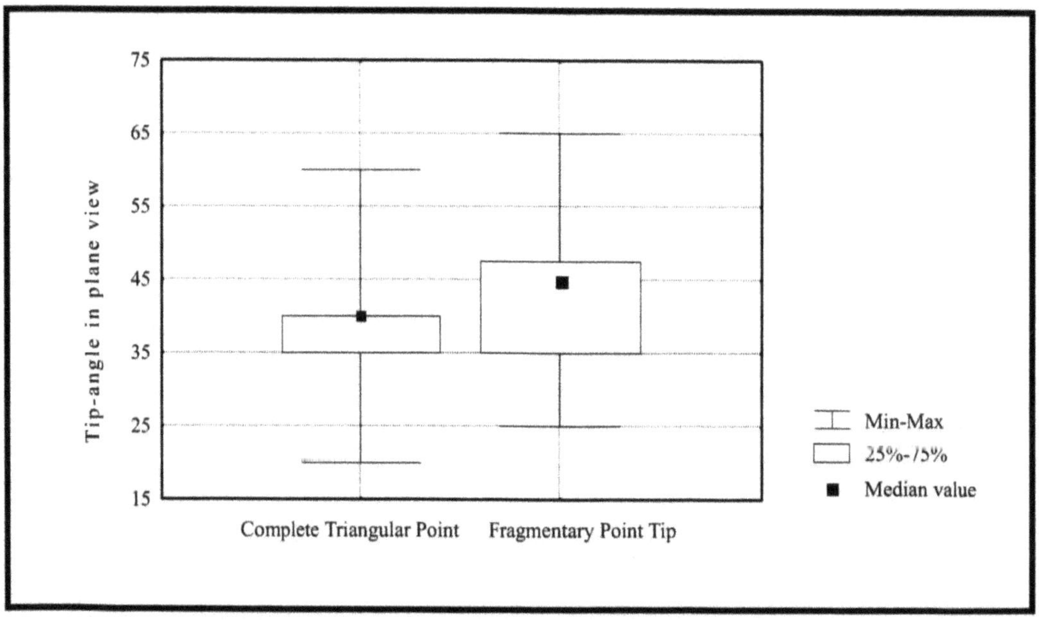

Fig. 5.5: Box & Whisker Plot Comparison of The Tip-angle of Projectile Points from Grand Banks.

plete ones have a square-shaped base (Fig. 5.7a, b). Technologically most drills, as is commonly the case in eastern North America, are made through bifacial retouch. However, three unifacially retouched drills were also found (one from Grand Banks and the other two from Lone Pine). One Grand Banks drill has been bifacially retouched around the tip, leaving the rest of the object unmodified on a thin flake. Morphologically, the drills exhibit no uniform types. One complete drill from Grand Banks has two bits. Sizes of drill blanks vary a great deal, ranging from 32 mm to 45.8 mm in length.

Notches

Notches identified from both sites comprise a very small proportion of the formed type assemblages. Two incomplete notches from Lone Pine display a single notch on the lateral side, and both are made on relatively thick plain flakes removed from Onondaga cores (Fig. 5.8f). The two complete notches from Grand Banks are classified as a single notch type without lateral retouch. One of the two is made on a bipolar flake of non-local chert. (Fig. 5.8d, e) Although metric measurements were not taken for the two broken notches from Lone Pine, it is visually apparent that blanks for making notches from Lone Pine are much larger than those from Grand Banks.

Burins

Two burins are identified from Lone Pine only. Both pieces present a single burin blow on the lateral edge. The one made on a small and thin flake is classified as an angle burin on a snap surface (Fig. 5.8j); the other one, made on a large and thick flake, is an angle burin on an old platform (Fig. 5.8k). However, the use-wear analysis showed that this large burin was first used as a wedge tool, then resharpened into a burin form by striking off one burin blow from the old platform. The small burin is 30.6 mm by 13.4 mm, with a thickness of 5.1 mm. The other one is 46.2 mm long and 21.7 mm wide, with a thickness of 9.8 mm.

Perforators

One perforator on a flake, without lateral retouch, was found at Lone Pine (Fig. 5.8l). Its tip was deliberately modified into a sharp point as a working bit. The tip shows usewear though drilling hard resistant material such as fresh bone or antler. The size of this perforator is about 37.1 mm long, 28.5 mm wide, and 8.3 mm thick.

Microliths

It must be noted first that the presence of microlithcs in the assembalges doesn not necessary mean that microlithic technique was involved in the production at the Princess Point site. The identification of this type is suggested according to the morphological form only. Two microlith-like formed types are identified from Grand Banks. They are characterized as bladelets with micro-retouch modification. One is formed as a geometric-rectangular shape, by deliberate retouch on both lateral edges and truncation of the distal section (Fig. 5.8a). The second microlith is a non-geometric type with a pointed shape. Its lateral sides near the distal section have been alternately retouched to form a pointed tip (Fig. 5.8b). Both pieces, made on local Onondaga chert, have a mean length to width ratio of 2.11.

Truncations

Only one truncation was found from Grand Banks. This piece, in the form of oblique truncation without lateral retouch, was made on a blade of local chert. It presents abrupt retouch across the distal section (Fig. 5.8c). This piece is 20.66 mm in length, 10.46 mm in width, and 4.28 mm in thickness.

Retouched Pieces

This group of modified artifacts has not been substantially treated in the lithic analysis conducted from the study region. Traditionally, pieces with edges irregularly retouched are likely catalogued as "utilized flakes." However, these pieces, with intentional modifications, comprise a relatively large portion of the formed types in the assemblages. Retouched piece fragments are normally present in high numbers. Excluding fragmentary pieces, unilateral obverse retouched pieces dominate at both sites (Fig. 5.8i). While Grand Banks has roughly equal numbers of unilateral inverse retouched pieces (Fig. 5.8g) and partially retouched pieces, Lone Pine seems to have a relatively high frequency of alternate retouched pieces (Fig. 5.8h).

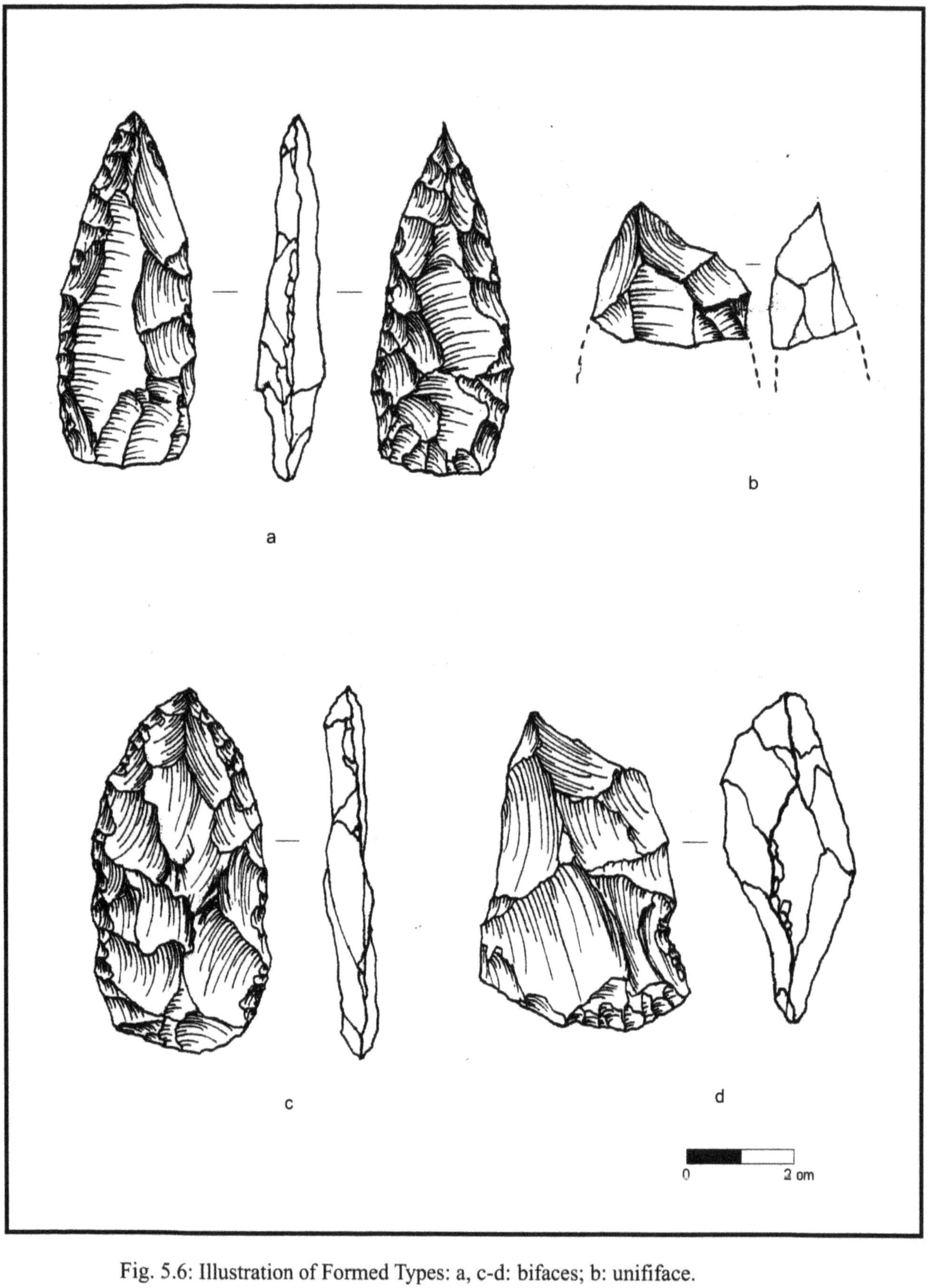

Fig. 5.6: Illustration of Formed Types: a, c-d: bifaces; b: unififace.

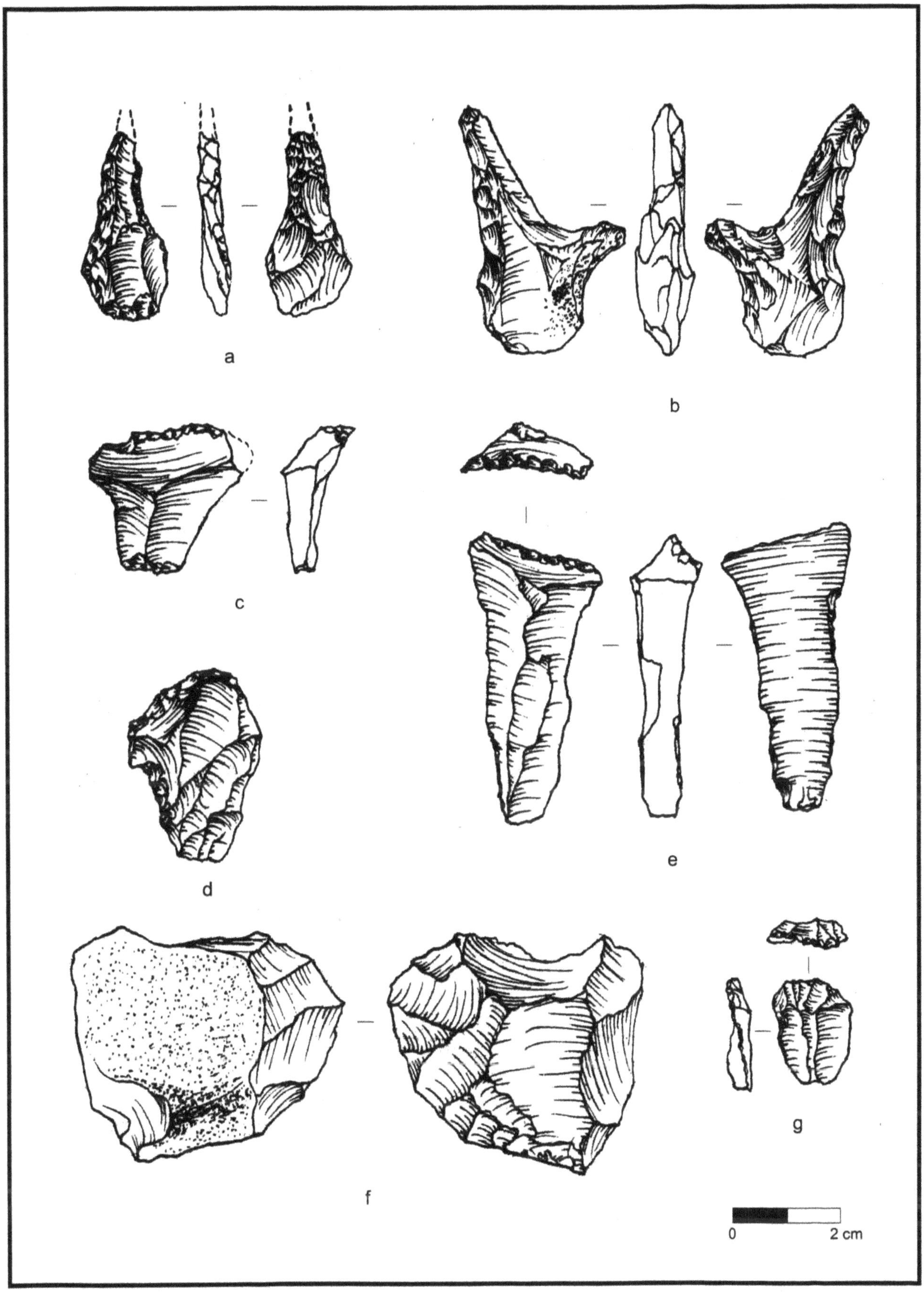

Fig. 5.7: Illustration of Formed Types: a-b: drills; c, e: simple endscraper on flakes; d: shouldered endscraper; f: uniface; g: thumbnail endscraper.

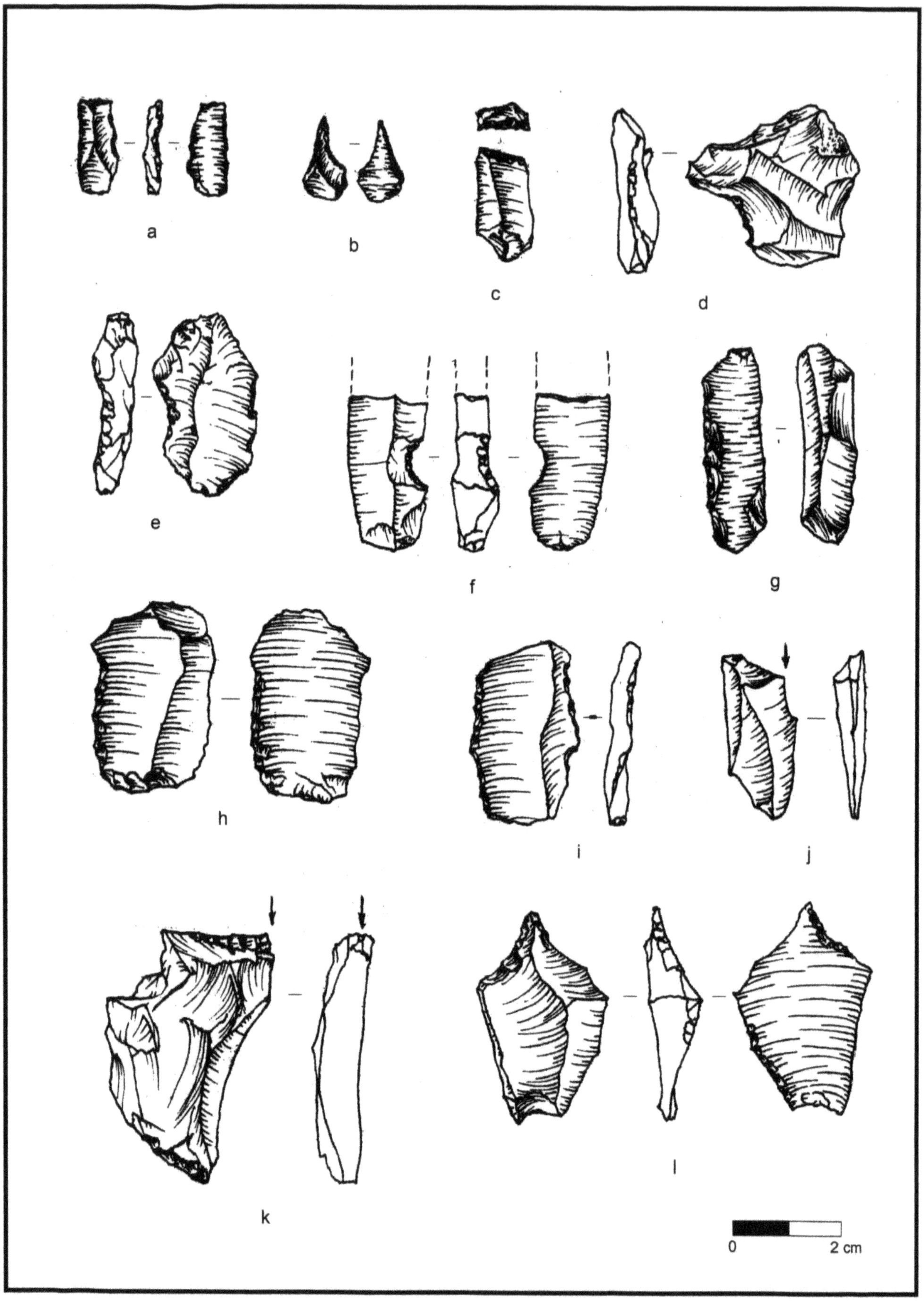

Fig. 5.8: Illustration of Formed Types: a-b: microliths; c: truncation; d-f: notches; g-i: retouched pieces; j-k: burins; l: perforator.

Tool Blank Production and Selection

Having described the formed types morphologically, the next questions which arise are: what kinds of blanks were chosen for modification into the above formed type artifacts? and were different blanks selected for modification into different formed types? Apparently, two types of blanks were used for modification: core tablets and unretouched flakes. As noted above, most bifaces and unifaces are made on core tablets (blanks made on flattened cores), although some of these artifacts were also fashioned on flakes. It is difficult to determine the types of blanks on which projectile points were made, due to the extensive bifacial thinning process of projectile points. It was possible to determine, however, that at least seven projectile points were made on flakes. Combined with the blanks used to make the rest of the formed types, it is clear that unretouched flakes were the major source of tool blanks.

Metric data for all the groups of formed types are presented in Table 5.2. In general, blanks for making bifaces and unifaces are much longer and wider than blanks for making the other formed types. The mean length to width (L/W) ratios of bifaces and unifaces are 1.2 and 1.3, respectively, at Grand Banks. Although scrapers and notches have the same L/W ratio as bifaces and unifaces, their overall sizes are much smaller. As far as overall sizes of formed type blanks are concerned, the t-tests show that there are significant differences in length, width, and thickness, between the surface modified formed types (projectile point, bifaces, and unifaces) and the edge modified formed types (e.g., scrapers, notches, retouched pieces) (Table 5.3). The majority of tool blanks used for making the surface modified formed types are 29-36 mm long, 19-27 mm wide, and 5-7 mm thick, whereas the predominant blanks used to make edge modified formed types are smaller in all dimensions (Fig. 5.9). Within the group of edge modified formed types, blanks for manufacturing retouched pieces are more elongated and narrower than scraper blanks (L/W ratio comparison: 1.5 vs. 1.2, respectively). In addition, three retouched pieces and one scraper are made on blades. I found that there are no significant differences in the sizes of blanks sizes between Grand Banks Areas A and B (Table 5.4).

At the intersite level, I discovered that there is only a slight difference in the blank sizes between the two sites examined. Statistically, there is only a difference in the L/W ratios of blanks used to make the surface modified formed types between the two assemblages (Table 5.5). For the edge modified formed types, there are no statistical differences between the two sites in blank sizes (Table 5.6). According to these quantitative parameters, at Grand Banks, blanks selected for producing surface modified formed types are relatively shorter and broader (mean L/W ratios at 1.37), as compared to those from Lone Pine (mean L/W ratios at 1.72). However, Grand Banks blanks are slightly thicker (mean 6.6 mm) than ones from Lone Pine (mean 6.0 mm).

As far as utilization of raw materials is concerned, 94.5% of formed types at Grand Banks were made on local Onondaga chert. The others were made on other local cherts such as Haldimand and Ancaster (Table 5.7). This finding is consistent with the pattern of raw material utilization found in the core analysis (Chapter 4). Three types of unknown chert were imported to the Grand Banks site and fabricated into projectile points and notches.

Three tools at Grand Banks were made on blanks from the category of primary flakes with cortex over 50% of the dorsal surface (Table 5.8). All of these have been modified into either bifaces or unifaces. Bifaces also embrace a few blanks with 30% to 50% cortex on the dorsal surface. Together this usage of blanks may be indicative of the bifacial reduction sequence. It is clearly suggestive of a strategy of on-site biface production. In contrast, the Lone Pine assemblage does not present the same pattern in the use of blanks with cortex. All five blanks bearing over 30% cortex on the dorsal surface are found on retouched pieces. At Grand Banks only one retouched piece has been made from a blank with cortex.

About 10% of formed types at Grand Banks show thermal damage on their blank surface (Table 5.9). Again, this resembles the distribution thermal damage on cores (Table 4.5). Among the pieces showing thermal damage, most (n=8 out of

Table 5.2: Formed Type Metric Data.

a: Grand Banks	Max. Length (mm)			Max. Width (mm)			Max. Thickness (mm)			L/W Ratio		
Formed Type	Mean	S. D.	N	Mean	S. D.	N	Mean	S. D.	N	Mean	S. D.	N
Projectile Point	30.58	6.26	28	21.92	3.66	28	5.39	1.43	62	1.41	0.22	28
Biface	38.28	11.32	10	30.38	10.71	10	11.88	5.31	10	1.32	0.30	10
Uniface	37.72	7.26	3	32.58	13.31	3	13.00	3.82	3	1.24	0.37	3
Scraper	28.06	10.64	9	23.00	5.94	9	5.25	2.34	9	1.24	0.40	9
Retouched Piece	26.24	7.35	18	18.05	5.82	18	4.39	0.93	18	1.54	0.55	18
Drill	38.91	9.74	2	22.18	7.95	2	8.96	2.62	2	1.79	0.20	2
Notch	29.27	2.75	2	24.46	7.42	2	6.55	0.09	2	1.27	0.50	2
Truncation	20.66	0.00	1	10.46	0.00	1	4.28	0.00	1	1.98	0.00	1
Microlith	16.06	1.71	2	7.65	0.64	2	1.89	0.05	2	2.12	0.40	2
Combined	30.22	9.03	75	22.22	7.89	75	6.03	3.18	109	1.43	0.40	75

b: Lone Pine												
Formed Type	Mean	S. D.	N	Mean	S. D.	N	Mean	S. D.	N	Mean	S. D.	N
Projectile Point	34.49	7.35	8	22.53	6.95	8	5.50	1.99	22	1.74	0.82	8
Biface	47.38	16.80	3	27.89	4.84	3	9.75	4.79	3	1.67	0.29	3
Scraper	27.51	7.74	16	19.11	4.75	16	4.01	1.13	16	1.45	0.27	16
Retouched Piece	33.09	16.67	31	20.25	7.26	31	5.31	3.89	31	1.68	0.64	31
Combined	32.48	14.07	58	20.65	6.67	58	5.26	3.11	72	1.62	0.58	58

15) are found within the projectile point group. A few bifaces (n=3) and retouched pieces (n=4) are also made on blanks with thermal damage. At Lone Pine, a higher percentage of heated blanks are observed, accounting for 25.4% (N=31) of the formed type assemblage. This different distribution in thermal damage flakes may be biased by the sampling strategy carried out at the two sites, because most artifacts from Lone Pine were recovered from areas near hearths. However, the pattern of use of heated blanks is similar to that found at Grand Banks; most of those pieces showing thermal damage were modified into projectile points (n=11), bifaces (n=8), and retouched pieces.

Five fracture attributes (Fig. 5.10) were further observed on both scraper and retouched piece

Table 5.3: T-test Results of Blank Sizes between Surfacially Modified and Edge Modified Formed Types at Grand Banks.

Group 1: Surfacially modified formed types　Group 2: Edge modified formed types

Statistics t-test	Mean Group 1	Mean Group 2	S. D. Group 1	S. D. Group 2	t-value	df	p-value	Valid N Group 1	Valid N Group 2
*Max. Length	32.98	26.88	8.40	8.74	3.07128	73	0.002994	41	34
*Max. Width	24.77	19.15	7.87	6.85	3.262106	73	0.001683	41	34
*Max. Thickness	6.56	4.86	3.49	1.97	2.65289	107	0.009196	75	34
L/W Ratio	1.37	1.51	0.25	0.51	-1.48651	73	0.14145	41	34

* indicates significant difference between the two groups

Table 5.4: T-test Results of Blank Sizes between the Area A and Area B at Grand Banks.

Group 1: Area A　Group 2: Area B

Statistics t-test	Mean Group 1	Mean Group 2	S. D. Group 1	S. D. Group 2	t-value	df	p-value	Valid N Group 1	Valid N Group 2
Max. Length	30.57	29.30	8.96	9.37	0.548711	73	0.584878	54	21
Max. Width	22.53	21.42	8.28	6.89	0.539492	73	0.591189	54	21
Max. Thickness	5.98	6.15	3.06	3.53	-0.25052	107	0.802668	79	30
L/W Ratio	1.43	1.43	0.36	0.49	0.014225	73	0.988689	54	21

* indicates significant difference between the two groups

Table 5.5: T-test Results of Blank Sizes of the Surfacially Modified Formed Types between the Two Formed Types Assemblages.

Group 1: Grand Banks　Group 2: Lone Pine

Statistics t-test	Mean Group 1	Mean Group 2	S. D. Group 1	S. D. Group 2	t-value	df	p-value	Valid N Group 1	Valid N Group 2
Max. Length	32.98	38.01	8.04	11.43	1.628297	50	0.109748	41	11
Max. Width	24.76	23.99	7.86	6.69	-0.29685	50	0.767808	41	11
Max. Thickness	6.56	6.01	3.48	2.71	-0.72364	98	0.471009	75	25
*L/W Ratio	1.37	1.71	0.25	0.70	2.621419	50	0.011571	41	11

* indicates significant difference between the two groups

Table 5.6: T-test Results of Blank Sizes of the Edge Modified Formed Types between the Two Formed Type Assemblages.

Group 1: Grand Banks　Group 2: Lone Pine

Statistics t-test	Mean Group 1	Mean Group 2	S. D. Group 1	S. D. Group 2	t-value	df	p-value	Valid N Group 1	Valid N Group 2
Max. Length	31.42	26.88	14.14	8.74	1.666045	82	0.09952	50	34
Max. Width	20.26	19.15	6.92	6.84	0.722225	82	0.472196	50	34
Max. Thickness	5.04	4.86	3.27	1.97	0.284266	82	0.776923	50	34
L/W Ratio	1.60	1.51	0.56	0.51	0.764193	82	0.444946	50	34

* indicates significant difference between the two groups

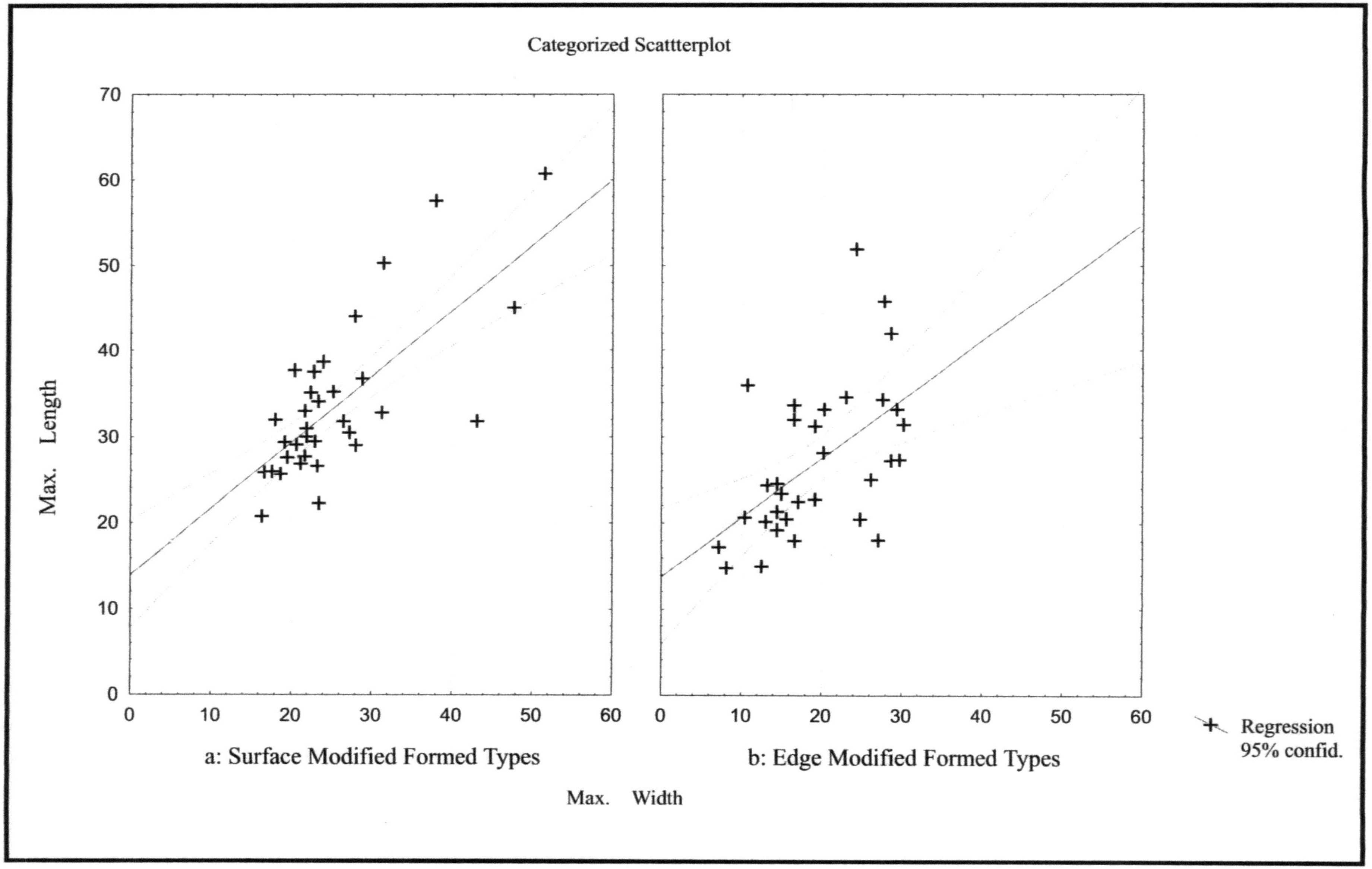

Fig. 5.9: Scatterplot Comparison of Blank Sizes Between Surface Modified and Edge Modified Formed Types from Grand Banks

blanks. The distributions of these attributes are found in Table 5.10. The platform facets on blanks are predominately single facet, followed by crushed platform. In the Grand Banks assemblage, lipped bulbs are the most common, followed by unlipped bulbs. This contrasts with the Lone Pine assemblage in which there are roughly equal numbers of blanks with lipped and unlipped bulbs. The cross-tabulation of platform facet and bulb shape (Table 5.11) shows that most blanks (42.4%) at Grand Banks are characterized by a single platform with a lipped bulb, suggesting that these pieces have been removed by soft-hammer percussion. Blanks with crushed platforms and unlipped bulbs, which may suggest hard-hammer percussion, are also present but in low percentages (18.2% at Grand Banks and 15.5% at Lone Pine).The lateral edge alignment of flakes from both sites is predominately expanding or parallel. This is a characteristic that has been used to identify "bullet flakes" removed through soft hammer percussion (Hayden and

Table 5.7: Raw Material Presented in the Formed Type Assemblages.

	GRAND BANKS		LONE PINE	
Chert Types	N	%	N	%
Onondaga	138	94.52	111	90.98
Haldimand	1	0.68	1	0.82
Ancaster	2	1.37	4	3.28
Others	5	3.42	6	4.91
Total	146	99.99	122	99.99

Table 5.8: Formed Type by Cortical Coverage in the Formed Type Assemblages.

a: GRAND BANKS

Formed Type	no cortex	1%-30% cortex	30%-50% cortex	50-99% cortex	100% cortex	
Projectile Point	62	-	-	-	-	
Biface	17	5	2	-	-	
Uniface	3	1	-	-	1	
Scraper	11	-	-	-	-	
Retouched Piece	34	1	-	-	-	
Drill	4	-	-	-	-	
Notch	2	-	-	-	-	
Microlith	2	-	-	-	-	
Truncation	1	-	-	-	-	
Combined Number	136	7	2	0	1	N=146
Combined Percentage	93.15	4.79	1.37	0.00	0.68	%=100

b: LONE PINE

Formed Type	no cortex	1%-30% cortex	30%-50% cortex	50-99% cortex	100% cortex	
Projectile Point	26	-	-	-	-	
Biface	22	-	-	-	-	
Uniface	1	-	-	-	-	
Scraper	20	-	-	-	-	
Retouched Piece	40	2	3	-	-	
Drill	3	-	-	-	-	
Notch	2	-	-	-	-	
Burin	2	-	-	-	-	
Perforator	1	-	-	-	-	
Combined Number	117	2	3	0	0	N=122
Combined Percentage	95.90	1.64	2.46	0.00	0.00	%=100

Hutchings 1989).

The tool blanks were further examined for attributes which would indicate whether chosen blanks were removed during the early stages or later stages of core reduction. The dorsal scar count and scar pattern are informative. Overall, the three intervals of scar counts (0-2, 3-5, >5) show roughly equal frequencies (Table 5.10). Since small flake size affects dorsal scar counts as demonstrated experimentally, I divided the blanks into two size grades: small (length or width < 25 mm) and medium to large (length or width > 25 mm). With the size grade taken into account a clear pattern emerges (Table 5.12): for medium to large size tools, the majority of blanks display higher dorsal scar counts. As for small-sized blanks, 80% of them have moderate to high dorsal scar counts at Grand Banks.

At Lone Pine, although just about half of the small blanks exhibit few scar counts, there are still 42.1% of small blanks with dorsal scar counts over 3 removals (Table 5.12). This leads me to believe that the blanks selected for modification into flake tools, such as scrapers and retouched pieces, were those removed during later reduction stages. If this is indeed the case, then one must ask whether they were intentionally produced as tool blanks? The distribution of scar patterns (Table 5.10) suggests that over half of the selected blanks exhibit parallel and parallel-opposed patterns of dorsal scarring. Parallel scar patterns may indicate prepared core reduction. Very few blanks, if any, display non-parallel scar patterning. Bearing in mind that the majority of blanks have a high dorsal scar count (Table 5.13), this could indicate that these blanks were carefully removed, through core preparation, during the later stages of core reduction. Therefore, it appears that the selection of tool blanks may have been an important procedure in tool production at Grand Banks.

Retouch Technique

The morphological appearance of retouch is indicative of the retouch technique which Princess Point people chose to modify tool blanks. Four retouch variables have been chosen for further observations: *retouch type*, *retouch orientation*, *retouch shape*, and *retouch pattern* (Fig. 5.11). Their frequency distributions are presented in Table 5.14. It is interesting to ask whether these populations have the same statistical distributions at the intersite and intrasite levels. The Mann-Whitney U nonparametric test proved that there are no significant differences between the two areas at Grand Banks (Table 5.15) and between the two sites (Table 5.16). On this basis, our data manipulations are mainly drawn from Grand Banks.

Table 5.9: Formed Type by Thermal Damage Presented. in the Formed Type Assemblages.

Formed Types	GRAND BANKS thermal damage not indicated	GRAND BANKS thermal damage indicated		LONE PINE thermal damage not indicated	LONE PINE thermal damage indicated	
Projectile Point	54	8		15	11	
Biface	21	3		14	8	
Uniface	5	-		1	-	
Scraper	11	-		18	2	
Retouched Piece	31	4		36	9	
Drill	4	-		2	1	
Notch	2	-		2	-	
Microlith	2	-		x	x	
Truncation	1	-		x	x	
Burin	x	x		2	-	
Perforator	x	x		1	-	
Combined	131	15	N=146	91	31	N=122
Percentage	89.73	10.27	%=100	74.59	25.41	%=1-00

x: indicated that such types do not exist in the formed type assemblage

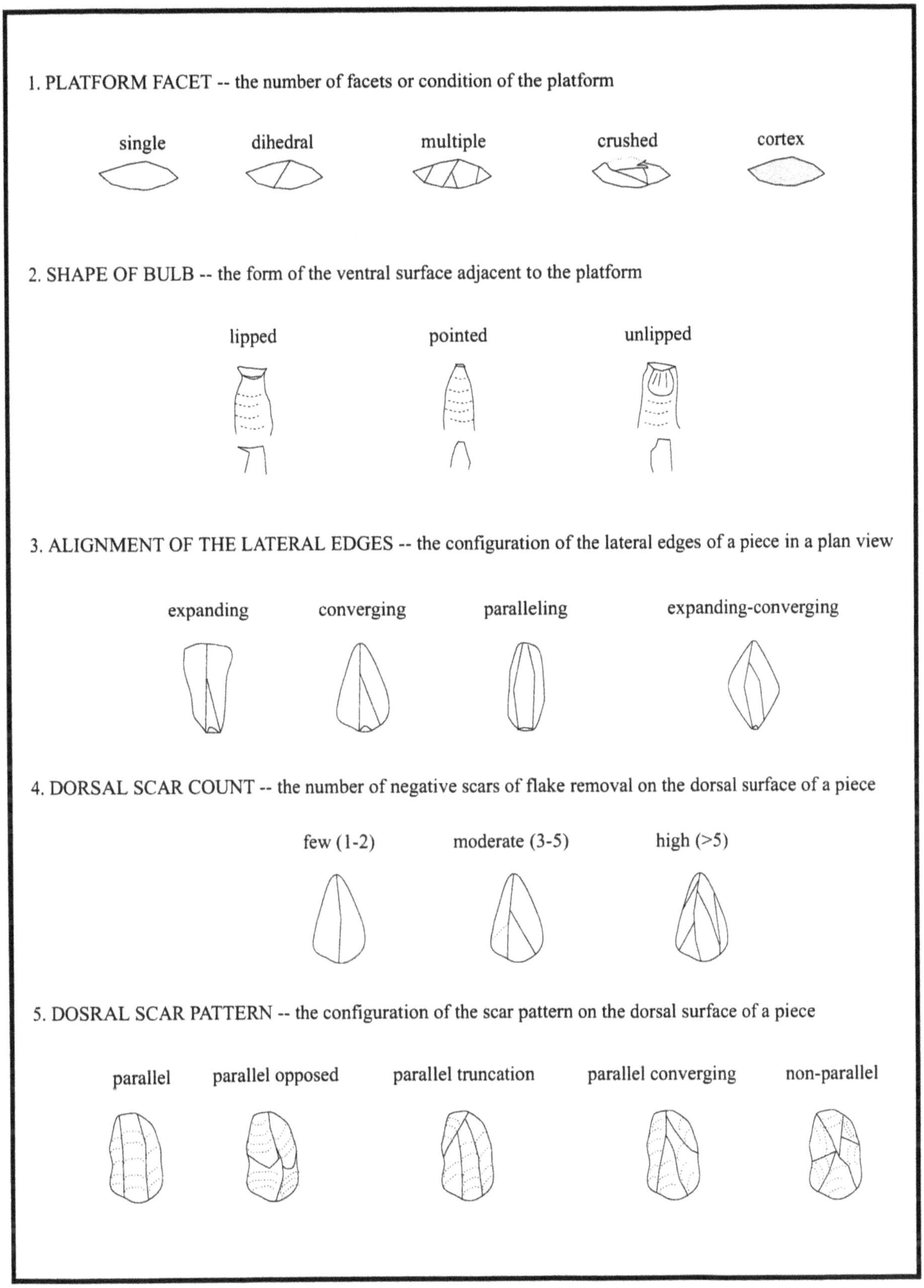

Fig. 5.10: Diagrams of Fracture Attributes.

I have illustrated five types of retouch methods (Fig. 5.11:A); they are defined mainly on the basis of angle and extent of retouch (Inizan et al. 1992:67-76). I consider them to be indicative of stylistic preferences because all of the retouch methods aim at edge modification with fewer technological inferences. The results show that *normal abrupt retouch* is more common in the assemblage (62.5%). This is followed by *obverse* or *inverse semisteep retouch* (20.8%) and *micro-edging retouch* (11.5%). *Alternating retouch* and *backing retouch* types are present, but in very low frequencies at Grand Banks. Clearly, the normal abrupt retouch technique was the method of retouch preferred at Grand Banks for further tool production.

The distribution of retouch scar appearance also

Table 5.10: Distributions of Five Fracture Attributes Observed on Edge Modified Formed Types.

	GRAND BANKS		LONE PINE	
	N	%	N	%
1. PLATFORM FACET				
indeterminate	5	15.15	11	18.97
single	17	51.52	24	41.38
dihedral	3	9.09	6	10.34
multiple	1	3.03	4	6.90
crushed	6	18.18	11	18.97
cortex	1	3.03	2	3.45
Observed Number	33	100.00	58	100.00
2. SHAPE OF BLUB				
indeterminate	4	12.12	11	18.97
lipped	17	51.52	22	37.93
pointed	2	6.06	4	6.90
unlipped	10	30.30	21	36.21
Observed Number	33	100.00	58	100.00
3. ALIGNMENT OF LATERAL EDGE				
indeterminate	1	3.03	10	17.24
expanding	14	42.42	18	31.03
converging	1	3.03	6	10.34
paralleling	16	48.48	17	29.31
expanding-converging	1	3.03	7	12.07
Observed Number	33	100.00	58	100.00
4. DORSAL SCAR COUNT				
indeterminate	0	0.00	3	5.45
few (0-2 scar facets)	6	19.35	16	29.09
moderate (3-5 scar facets)	16	51.61	21	38.18
high (>5 scar facets)	9	29.03	15	27.27
Observed Number	31	100.00	55	100.00
5. DORSAL SCAR PATTERN				
indeterminate	0	0.00	7	19.44
parallel	9	42.86	16	44.44
parallel opposed	6	28.57	3	8.33
parallel truncation	2	9.52	7	19.44
parallel converging	2	9.52	3	8.33
non-parallel	2	9.52	0	0.00
Observed Number	21	100.00	36	100.00

indicates that the normal abrupt retouch was preferred. Most retouch units have parallel retouch scar patterning (50%), followed by sub-parallel and scalar patterns. Parallel retouch scars are likely the result of normal-abrupt retouching.

Retouch units occur most often on the dorsal side of the blanks (60%), while retouch on the ventral side and alternating retouch are present in equally low percentages. Retouch shape is dominated by straight retouch (43.33%), followed by convex (23.33%) and notched (10%) retouch. All these retouch attributes point to a single dominance (Fig. 5.12), possibly suggestive of stylistic preferences, that is, normal abrupt straight parallel retouch on the dorsal side of the blanks. This analytic result is interesting and should be used in the comparison of this material with other Princess Point lithic assemblages in future studies.

Although I have stressed the stylistic aspects of retouch techniques, this does not necessarily mean that retouch techniques are unaffected by technological and functional considerations. In fact, in addition to producing a general form for tools, there are other reasons for retouching: (a) to shape specific portions for use; (b) to shape particular parts for insertion; and (c) to blunt

Table 5.11: Cross-Tabulation of Platform Facet and Shape of Bulb of the Formed Type Blanks.

a: GRAND BANKS

Platform Facet	Shape of Bulb				
	indeterminate	lipped	pointed	unlipped	
indeterminate	4	-	1	-	5
single	-	14	1	2	17
dihedral	-	2	-	1	3
multiple	-	-	-	1	1
crushed	-	-	-	6	6
cortex	-	1	-	-	1
Combined	4	17	2	10	33

b: LONE PINE

Platform Facet	Shape of Bulb				
	indeterminate	lipped	pointed	unlipped	
indeterminate	10	-	-	1	11
single	-	12	4	8	24
dihedral	-	5	-	1	6
multiple	-	3	-	1	4
crushed	-	2	-	9	11
cortex	1	-	-	1	2
Combined	11	22	4	21	58

Table 5.12: Distributions of Dorsal Scar Count by Blank Size Grades in the Formed Type Assemblages.

	GRAND BANKS				LONE PINE			
Dorsal Scar Count	Small		Medium to Large		Small		Medium to Large	
	N	%	N	%	N	%	N	%
indeterminate	0	0	0	0.00	1	5.26	2	5.56
few	2	20	4	19.05	10	52.63	6	16.67
moderate	5	50	11	52.38	4	21.05	17	47.22
high	3	30	6	28.57	4	21.05	11	30.56
Observed Number	10	100	21	100	19	100	36	100

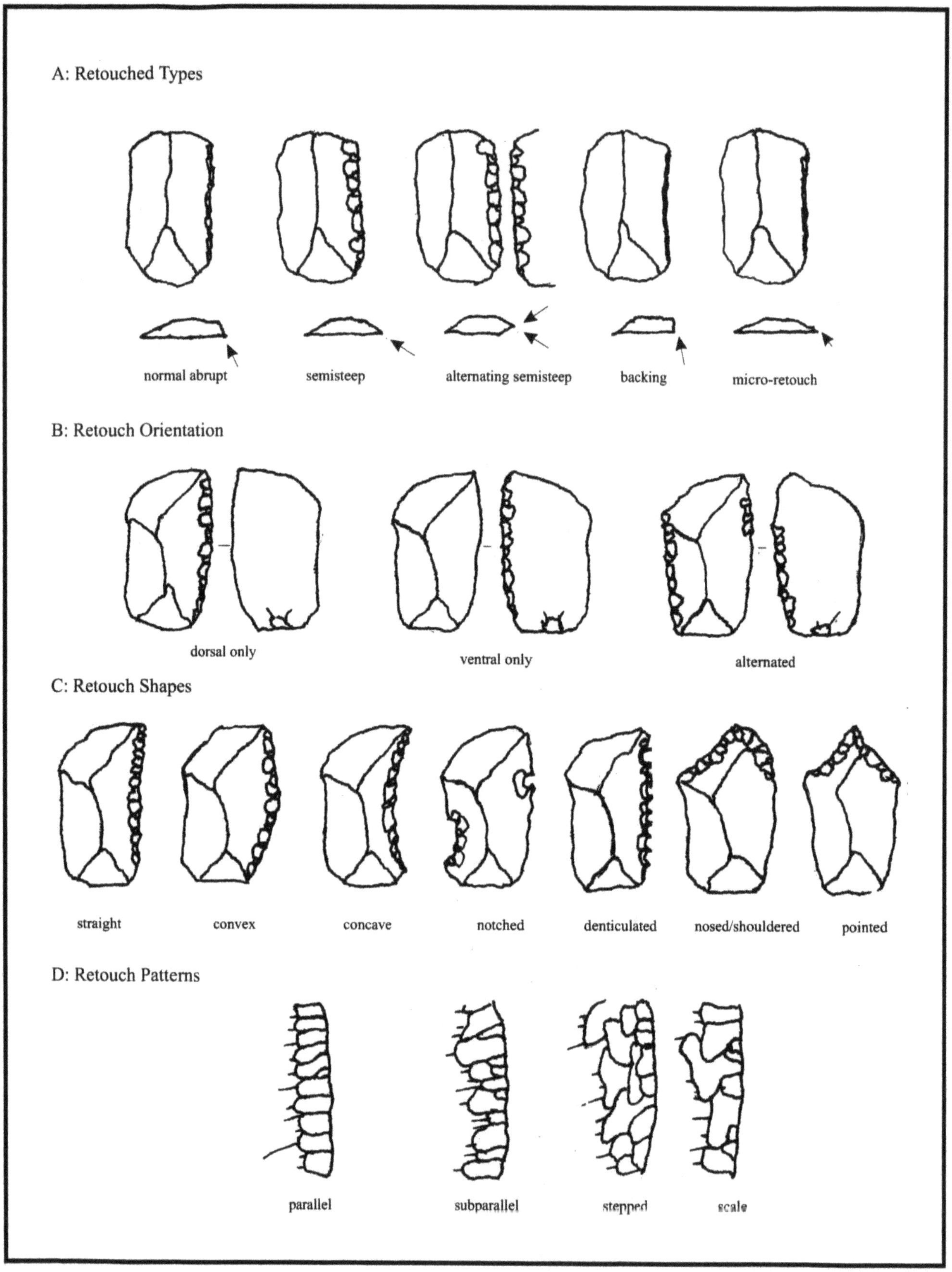

Fig. 5.11: Diagram of Retouch Attributes.

Table 5.13: Cross-Tabulation of Dorsal Scar Count and Scar Pattern in the Formed Type Assemblages.

a: GRAND BANKS

Scar Count	Indeterminate	Parallel	Par. opposed	Par. truncation	Par. converging	non-parallel	
indeterminate	-	-	-	-	-	-	0
few	-	4	-	-	-	2	6
moderate	-	3	7	3	2	1	16
high	-	-	6	-	3	-	9
Combined	0	7	13	3	5	3	31

b: LONE PINE

Scar Count	Indeterminate	Parallel	Par. opposed	Par. truncation	Par. converging	non-parallel	
indeterminate	4	-	-	-	-	-	4
few	3	9	2	2	-	-	16
moderate	2	6	6	6	2	-	22
high	4	9	1	1	1	-	16
Combined	13	24	9	9	3	0	58

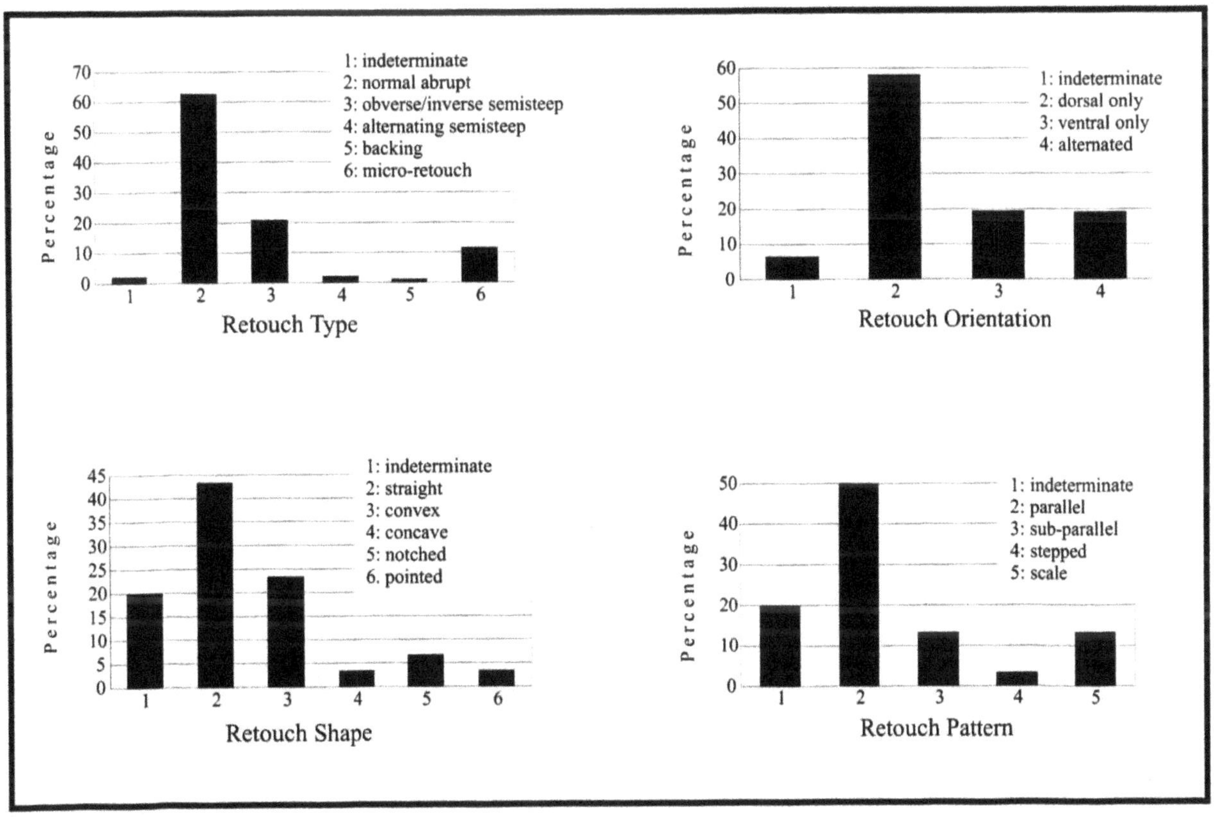

Fig. 5.12: Distributions of Retouch Attributes of the Grand Banks Formed Types.

sharp protruding edges to facilitate grasping in the hand (Odell 1994b:77). To illustrate this, I have selected all retouched pieces as an example for examination of both retouch and use-employed locations. Among 35 retouched pieces from Grand Banks, 14 pieces display use-wear. Within this group, 7 pieces have the same location of both modification and utilization. It is observed that the edges on three such pieces have been modified for cutting or scraping, and that edges of the other four have been either altered for hafting devices or blunted for hand-holding. It is worth noting, however, that modifications for the purpose of use appear to be rare at Grand Banks. Nevertheless, this still represents variability in the application of retouch techniques.

Table 5.14: Distributions of Four Retouch Attributes of Formed Type.

	GRAND BANKS		LONE PINE	
	N	%	N	%
1. Retouch Type				
indeterminate	2	2.08	5	6.10
normal abrupt	60	62.50	37	45.12
obverse/inverse semisteep	20	20.83	32	39.02
alternating semisteep	2	2.08	5	6.10
back	1	1.04	2	2.44
micro-retouch	11	11.46	1	1.22
Observed Number	96	100.00	82	100.00
2. Retouch Orientation				
indeterminate	2	6.45	2	3.08
dorsal only	18	58.06	46	70.77
ventral only	6	19.35	7	10.77
alternated	5	16.13	10	15.38
Observed Number	31	100.00	65	100.00
3. Retouch Shape				
indeterminate	6	20.00	6	9.38
straight	13	43.33	20	31.25
convex	7	23.33	20	31.25
concave	1	3.33	6	9.38
notched	2	6.67	7	10.94
denticulated	0	0.00	2	3.13
shoulder/nosed	0	0.00	1	1.56
pointed	1	3.33	0	0.00
Observed Number	30	100.00	64	96.88
4. Retouch Pattern				
indeterminate	6	20.00	4	6.35
parallel	15	50.00	32	50.79
sub-parallel	4	13.33	9	14.29
stepped	1	3.33	6	9.52
scale	4	13.33	12	19.05
Observed Number	30	100.00	63	100.00

Table 5.15: The Mann-Whitney U Test Results of Retouch Attributes between Area A and Area B at Grand Banks.

Group 1: Area A	Group 2: Area B								
Statistics the Mann-Whitney	Rank Sum Group 1	Rank Sum Group 2	U	Z	p-level	Z adjusted	p-level	Valid N Group 1	Valid N Group 2
retouch type	7517.5	3213.5	2057.5	-0.54687	0.584474	-0.58050	0.561579	104	42
retouch orientation	361	135	90	-0.39167	0.695301	-0.43974	0.660130	22	9
retouch shape	317	148	86	-0.38468	0.700478	-0.40596	0.684772	21	9
retouch pattern	324	141	93	-0.06788	0.945878	-0.07307	0.941754	21	9

* indicates significant difference between the two groups

Table 5.16: The Mann-Whitney U Test Results of Retouch Attributes between the Two Sites.

Group 1: Grand Banks	Group 2: Lone Pine								
Statistics the Mann-Whitney	Rank Sum Group 2	Rank Sum Group 1	U	Z	p-level	Z adjusted	p-level	Valid N Group 2	Valid N Group 1
retouch type	19326.5	16184.5	8595.5	-0.26347	0.792189	-0.27588	0.782643	146	120
retouch orientation	1675.5	2980.5	835.5	-1.34771	0.177762	-1.46079	0.144082	31	65
*retouch shape	1165.5	3299.5	700.5	-2.10482	0.035315	-2.18381	0.028983	30	64
retouch pattern	1215	3156	750	-1.60262	0.109029	-1.72661	0.084247	30	63

* indicates significant difference between the two groups

Use-wear Analysis

Grand Banks

Microscopic examination of all 146 formed artifacts from Grand Banks indicates that only one-third (N=44) show evidence of use. Table 5.17 indicates that morphologically shaped scrapers, notches, drills, and retouched pieces, were efficiently utilized; however projectile points and bifaces were not. Of the 62 projectile points from Grand Banks, only 12 (19.35%) display use traces and/or prehensile wear. A lower degree of utilization of bifaces is also indicated.

The forty-four utilized pieces yield a total of 70 employed units, including 47 utilized wear units and 23 prehensile wear units. Scraping/shaving motions dominate the assemblage, accounting for 38.3% of the total (Table 5.18). When scraping/shaving motions are combined with planing/whittling, transverse motions account for nearly half the employed units. Longitudinal motions, i.e., cutting/sawing, are well represented (19.2%) and projecting or penetrating motions are present in small but fairly representative numbers (10.6%). Other tool motions such as graving, drilling, digging, and wedging are indicated in small numbers.

If formed types are cross-tabulated with tool motions (Table 5.19), some variability in the use of typed artifacts can be noted. Not surprisingly, retouched pieces are the only group of artifacts associated with a wide variety of activities. Half the retouched pieces, however, have been used for scraping. Scrapers and drills at Grand Banks are functionally consistent with their morphological names; both tool types were used for transverse (scraping) and rotation (drilling) motions, respectively. Three employed units observed on unifaces are associated with scraping (2) and graving (1), as their morphological shapes may indicate. Two bifaces were used for cutting and scraping.

An unexpected discovery is that most utilized projectile points were not used for projecting or penetrating but rather for cutting and digging (7 out of 13 EdUs). Only three projectile points show impact wear on, or near, their tips. The two notches employed at Grand Banks were used in transverse motions: shaving and planing.

The usewear analysis of formed type artifacts indicates that a variety of materials was worked. Evidence of working on medium-soft vegetal (1M) is slightly more common than other material (Table 5.20). When this category is combined with the hard wood category (2M), woodworking may be inferred to have been a relatively common tool use. Other substances are also represented, and there appears to have been an approximately equal exploitation of hard material (1H, 2H), medium animal material (MA), and soft vegetal material (SV). Evidence of use on inorganic substances was present on only three employed units.

When artifact types were compared with contact materials (Table 5.21), I found that both projectile points and retouched pieces were used on every resistance-grade of material. Three projectile points were contacted with inorganic material such as soil and stone gravel. One possible interpretation of this is that a small portion of projec-

Table 5.17: Utilization Rates of Formed Types.

Formed Types	GRAND BANKS			LONE PINE		
	total N	used	use rate (%)	total N	used	use rate (%)
Projectile Point	62	12	19.4	26	9	34.6
Biface	24	2	8.3	22	8	36.4
Uniface	5	2	40.0	1	0	0.0
Scraper	11	7	63.6	20	14	70.0
Retouched Piece	35	17	48.6	45	18	40.0
Drill	4	2	50.0	3	2	66.7
Notch	2	2	100.0	2	2	100.0
Microlith	2	0	0.0	x	x	
Truncation	1	0	0.0	x	x	
Burin	x	x		2	1	50.0
Perforator	x	x		1	1	100.0
Combined	146	44	30.1	122	55	45.1

x: indicates that such types do not exist in the assemblage

Table 5.18: Tool Motion in the Formed Type Assemblages.

Tool Motion	GRAND BANKS		LONE PINE	
	N	%	N	%
indeterminate	5	10.6	9	14.8
cut/saw	9	19.1	14	23.0
slice/carve	0	0.0	2	3.3
scrape/shave	18	38.3	21	34.4
plane/whittle	3	6.4	3	4.9
projection	5	10.6	2	3.3
bore/drill	2	4.3	5	8.2
grave	1	2.1	0	0.0
chop	0	0.0	1	1.6
hoe/dig	3	6.4	1	1.6
pound/grind	0	0.0	1	1.6
wedge	1	2.1	1	1.6
pick	0	0.0	1	1.6
Total Employed Units	47	100.0	61	100.0

tile points was, accidentally or expediently, used for digging soil or the like. Bifaces and unifaces were used on relatively soft material, such as fresh wood (1M) and plants or grass (SV), with the exception of one uniface (identified as a graving tool) which was used on fresh bone or antler (1H). One third of the scrapers were used for woodworking (1M). Another third of the scrapers were utilized on animal substances (SA, 1H, 2H) suggesting that they may have been used in hide-working or bone/antler-working. The notches, bearing steeply retouched indentations on the lateral edges, show evidence of having been used in working hard materials like bone and antler, whereas drills appear to have been utilized only for woodworking.

Another way to arrive at inferences of use-tasks is to examine associations between tool motions and contact materials. Table 5.22 suggests that most scraping/shaving activities were carried out on soft to medium resistance materials like

Table 5.19: Formed Type by Tool Motion in the Grand Banks Assemblage.

Tool Motion	Formed Type							
	Projectile Point	Biface	Uniface	Scraper	Retouched piece	Drill	Notch	Combined
indeterminate	3	-	-	2	-	-	-	5
cut/saw	4	1	-	-	4	-	-	9
slice/carve	-	-	-	-	-	-	-	0
scrape/shave	-	-	2	7	8	-	1	18
plane/whittle	-	1	-	-	1	-	1	3
projection	3	-	-	-	2	-	-	5
bore/drill	-	-	-	-	-	2	-	2
grave	-	-	1	-	-	-	-	1
chop	-	-	-	-	-	-	-	0
hoe/dig	3	-	-	-	-	-	-	3
pound/grind	-	-	-	-	-	-	-	0
wedge	-	-	-	-	1	-	-	1
pick	-	-	-	-	-	-	-	0
Total Employed Units	13	2	3	9	16	2	2	47

Table 5.20: Contact Material in the Formed Type Assemblages.

Contact Material	GRAND BANKS		LONE PINE	
	N	%	N	%
indeterminate	5	10.6	12	19.7
soft animal (SA)	6	12.8	7	11.5
soft vegetal (SV)	1	2.1	3	4.9
medium-soft vegetal (1M)	12	25.5	12	19.7
medium animal (MA)	7	14.9	7	11.5
medium-hard vegetal (2M)	5	10.6	12	19.7
hard animal (1H)	7	14.9	7	11.5
very hard animal (2H)	1	2.1	0	0.0
hard inorganic (3H)	3	6.4	1	1.6
Total Employed Units	47	100.0	61	100.0

Table 5.21: Formed Type by Contact Material in the Grand Banks Assemblage.

Contact Material	Projectile Point	Biface	Uniface	Scraper	Retouched piece	Drill	Notch	Combined
	Formed Types							
indeterminate	1	-	-	3	-	-	1	5
soft animal (SA)	1	-	-	1	-	-	4	6
soft vegetal (SV)	-	1	-	-	-	-	-	1
medium-soft vegetal (1M)	2	1	2	3	1	-	3	12
medium animal (MA)	3	-	-	-	-	-	4	7
medium-hard vegetal (2M)	2	-	-	-	1	-	2	5
hard animal (1H)	1	-	1	1	-	2	2	7
very hard animal (2H)	-	-	-	1	-	-	-	1
hard inorganic (3H)	3	-	-	-	-	-	-	3
Total Employed Units	13	2	3	9	2	2	16	47

Table 5.22: Tool Motion by Contact Material in the Grand Banks Formed Type Assemblage.

Tool Motion	Contact Material									
	?	SA	SV	1M	MA	2M	1H	2H	3H	Comb.
indeterminate	3	-	-	-	-	2	-	-	-	5
cut/saw	1	1	1	3	1	-	2	-	-	9
slice/carve	-	-	-	-	-	-	-	-	-	0
scrape/shave	1	5	-	5	1	1	4	1	-	18
plane/whittle	-	-	-	2	-	-	1	-	-	3
projection	-	-	-	-	5	-	-	-	-	5
bore/drill	-	-	-	1	-	1	-	-	-	2
grave	-	-	-	1	-	-	-	-	-	1
chop	-	-	-	-	-	-	-	-	-	0
hoe/dig	-	-	-	-	-	-	-	-	3	3
pound/grind	-	-	-	-	-	-	-	-	-	0
wedge	-	-	-	-	-	1	-	-	-	1
pick	-	-	-	-	-	-	-	-	-	0
Total Employed Units	5	6	1	12	7	5	7	1	3	47

Table 5.23: Prehensile Wear Type in the Formed Type Assemblages.

Prehensile Wear Type	GRAND BANKS		LONE PINE	
	N	%	N	%
general prehensile wear	3	13.04	3	21.43
hand-held wear	5	21.74	6	42.86
haft wear	15	65.22	5	35.71
Total Employ Units	23	100.00	14	100.00

woods or plants and softer bones or antlers. Since transverse motions comprise most of the EdUs from the Grand Banks formed type assemblage, it is not surprising that scraping, shaving, and planing register their manifestations on almost all kinds of contact materials. Projecting use-tasks are primarily associated with medium resistance materials, possibly indicative of penetrating animal skins and contacting bone. Cutting/sawing are also concentrated on soft materials, both animal and vegetal. Due to the relatively small sample of utilized formed types, other associations between activities and contact materials are poorly represented.

Twenty-three out of 44 utilized formed types exhibit clear prehensile wear, either hand-held or hafted (Table 5.23). Tools with hafted devices account for 65.2% of the prehensile wear. Three pieces have wear which cannot be determined to be either a result of hand held use or hafting, so they have been placed in a general prehensile wear category (13.04%). As expected, hafting wear is primarily associated with projectile points (60%), but it is also found on scrapers and retouched pieces (Table 5.24). When the relationship between prehensile wear and tool motions is examined (Table 5.25), it was found, not surprisingly, that one-third of hafted tools were used for projecting use-tasks. A quarter of them were utilized on inorganic materials, whereas another quarter of hafted tools were associated with scraping activities.

In considering tool utilization patterns between Areas A and B, I observed that although Area B has fewer formed types than Area A, there is no statistical difference in the distributional frequencies of tool utilization between the two areas (*Chi-square*=0.09, *df*=1, *p-value*=0.7636). The Mann-Whitney U test also indicates that there are no significant differences for employing tool motions and for contact materials between the two areas (for tool motion: *U*=2319.00, *Z*=-

Table 5.24: Formed Type by Prehensile Wear Type in the Grand Banks Assemblage.

Formed Type	Prehensile Wear Type			
	General	hand	haft	Combined
projectile point	1	-	9	10
scraper	-	1	3	4
drill	-	1	-	1
retouched piece	2	3	3	8
Total Employed Units	3	5	15	23

Table 5.25: Tool Motion by Prehensile Wear Type in the Grand Banks Assemblage.

Tool Motion	Prehensile Wear Type			
	general	hand	haft	Combined
indeterminate	-	-	2	2
cut/saw	1	1	-	2
scraper/shave	2	1	3	5
plane/whittle	-	1	-	1
projection	-	-	4	4
drill	-	1	-	1
hoe/dig	-	-	3	3
Total Utilized Pieces	3	4	12	19

1.26043, *p-value*=0.207525; for contact material: *U*=1253.00, *Z*=-1.13270, *p-value*=0.25349).

Lone Pine

Use-wear data generated for the Lone Pine formed types are presented in a similar manner to that of the Grand Banks data. I will not give a detailed discussion of the results since the tabulations (Tables 5.17-4.18, 5.20, 5.26-5.28) speak for themselves. However, differences in tool utilization from this late-period site should be highlighted. Statistical analysis shows that there are no significant differences between the two sites in the application of tool motions and the occurrences of contact materials. However, use-tasks at Lone Pine were more diverse than those at Grand Banks. We see an approximately equal emphasis on both transverse and longitudinal motions at Lone Pine (Table 5.18). Several use-tasks are found in low numbers at Lone Pine, but are com-

Table 5.26: Formed Type by Tool Motion in the Lone Pine Assemblage.

Tool Motion	Formed Type									
	projectile Point	biface	uniface	scraper	retouched piece	drill	notch	burin	perforator	Combined
indeterminate	2	1	-	3	2	-	1	-	-	9
cut/saw	2	4	-	3	5	-	-	-	-	14
slice/carve	2	-	-	-	-	-	-	-	-	2
scrape/shave	-	1	-	11	9	-	-	-	-	21
plane/whittle	-	1	-	-	1	-	1	-	-	3
projection	-	1	-	-	1	-	-	-	-	2
bore/drill	2	-	-	-	-	2	-	-	1	5
grave	-	-	-	-	-	-	-	-	-	0
chop	-	-	-	-	1	-	-	-	-	1
hoe/dig	-	-	-	-	1	-	-	-	-	1
pound/grind	-	1	-	-	-	-	-	-	-	1
wedge	-	-	-	-	-	-	-	1	-	1
pick	-	-	-	-	1	-	-	-	-	1
Total Employed Units	8	9	0	17	21	2	2	1	1	61

Table 5.27: Formed Type by Contact Material in the Lone Pine Assemblage.

Contact Material	Formed Types									
	projectile point	biface	uniface	scraper	retouched piece	drill	notch	burin	perforator	Combined
indeterminate	3	1	-	3	4	-	1	-	-	12
soft animal (SA)	1	-	-	4	2	-	-	-	-	7
soft vegetal (SV)	1	-	-	1	1	-	-	-	-	3
medium-soft vegetal (1M)	-	2	-	1	9	-	-	-	-	12
medium animal (MA)	1	3	-	2	1	-	-	-	-	7
medium-hard vegetal (2M)	2	2	-	4	2	-	1	1	-	12
hard animal (1H)	-	1	-	2	1	2	-	-	1	7
very hard animal (2H)	-	-	-	-	-	-	-	-	-	0
hard inorganic (3H)	-	-	-	-	1	-	-	-	-	1
Total Employed Units	8	9	0	17	21	2	2	1	1	61

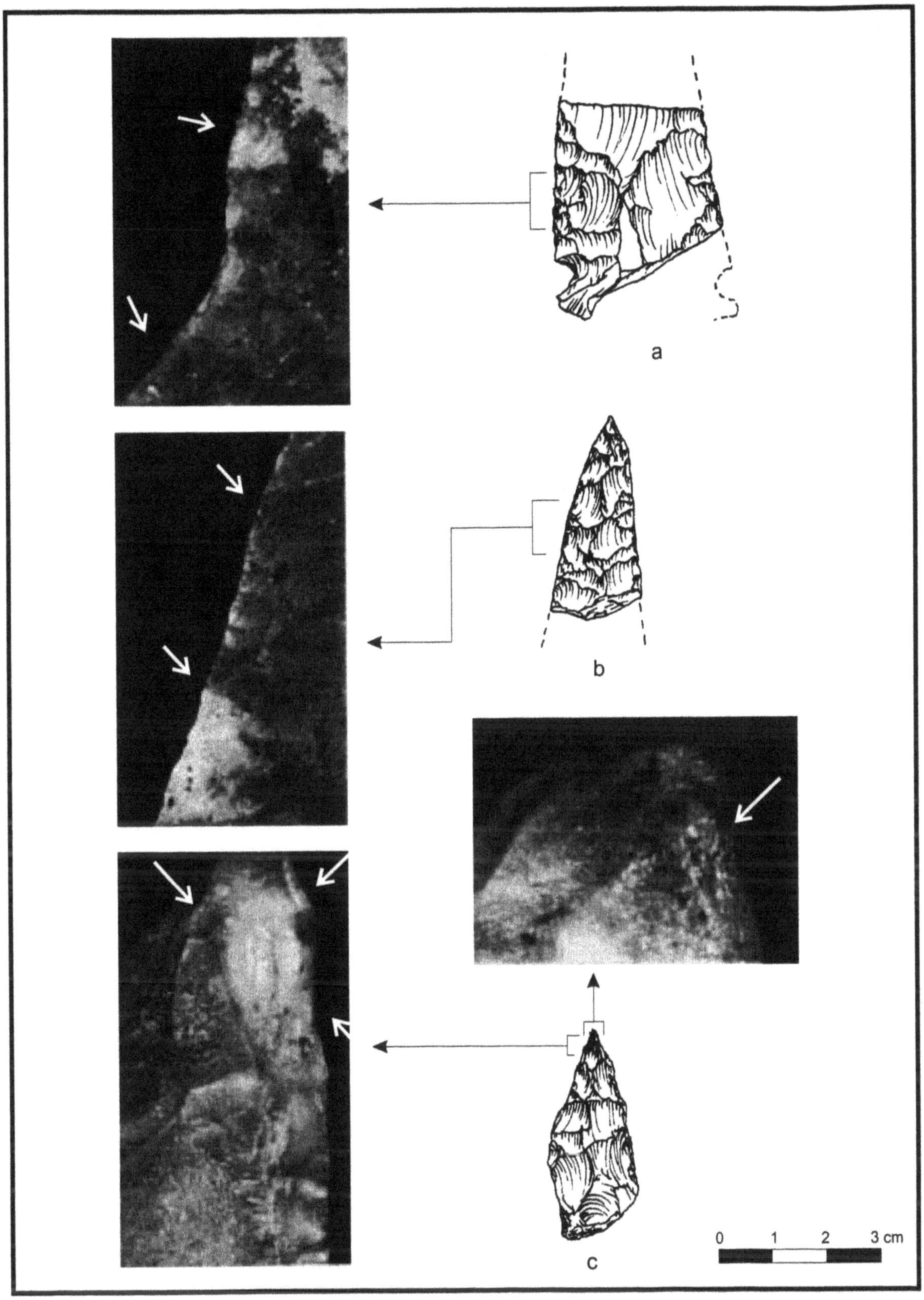

Fig. 5.13: Lone Pine Projectile Points with Use-Wear Images.

a: point (#614) used as wood-cutting tool, noting medium sized feather and hinged scars, and heavy rounded edge. 17

b: point (#372) used as meat-cuting or butching tool, noting a row of small feather scars and light polish. 13x

c: point (#962) used as wood-drilling tool, noting strong polish, heavy rounded edge, and crushed scar patterns, 10x (left) and 28x (top) pletely absent from Grand Banks. These include chopping, slicing/carving, pounding/grinding, and picking activities at Lone Pine. Also in contrast to the Grand Banks assemblage, contact materials from Lone Pine are equally distributed within the eight resistance-grade categories (Table 5.20). Woodworking (e.g., scraping, cutting/sawing, drilling, chopping), bone-working (e.g., scraping, drilling, graving), butchering (e.g., slicing/carving, penetrating), and soil digging, are all well represented at Lone Pine (Table 5.28).

This general pattern in tool usage at Lone Pine can also be observed in cross-tabulations between formed types and tool motions/contact materials (Tables 5.26 & 5.27). The utilization of projectile points for a number of use-tasks is again witnessed at Lone Pine (Fig. 5.13). Bifaces have relatively high rates of utilization, and were mostly used as knives for cutting/sawing (Table 5.26). Scrapers at Lone Pine have been used substantially on animal hide or skin as well as bone or antler. Retouched pieces at Lone Pine are mainly associated with woodworking (Table 5.27).

Very little prehensile wear was detected in the Lone Pine formed type assemblage. Among 14 prehensile units identified, hafting wear accounts for only slightly over one-third (35.71%) of total prehensile wear; a feature that distinguishes this assemblage from the Grand Banks assemblage (Table 5.23). The five hafted tools from Lone Pine are of various morphological types.

Discussion and Summary

Formed types are defined in this study as objects with intentional modifications. These pieces could be either surface modified or edge modified, although the latter may not greatly change the form of blanks. The study of the morphology of formed types in the Princess Point Complex can be used as a basis for characterizing the typology of the Princess Point lithic assemblage. Typologically, surface modified artifacts dominate the formed type assemblage. Among these, projectile points are the most common and distinctive formed types. As previously observed (Stothers 1977; Fox 1990; Bursey 1995), the Grand Banks projectile points consist primarily of the triangular-shaped type, some of which were made on flake blanks. Bifaces are produced in moderate frequencies on-site. Edge modified formed types include mainly scrapers and retouched pieces. All of these forms were made on flakes removed from Onondaga chert cores. Scrapers and retouched pieces comprise nearly one-third of the formed type assemblage. Among them, simple endscrapers and unilateral-obverse retouched pieces are predominant types. Square-based drills and single notches also form part of Grand Banks flaked stone toolkits. Although

Table 5.28: Tool Motion by Contact Material in the Lone Pine Formed Type Assemblage.

Tool Motion	Contact Material									
	?	SA	SV	1M	MA	2M	1H	2H	3H	Comb.
indeterminate	8	-	-	1	-	-	-	-	-	9
cut/saw	1	4	2	2	3	2	-	-	-	14
slice/carve	-	1	-	-	1	-	-	-	-	2
scrape/shave	1	2	1	7	2	5	3	-	-	21
plane/whittle	-	-	-	1	-	2	-	-	-	3
projection	1	-	-	-	1	-	-	-	-	2
bore/drill	1	-	-	-	-	1	3	-	-	5
grave	-	-	-	-	-	-	-	-	-	0
chop	-	-	-	-	-	1	-	-	-	1
hoe/dig	-	-	-	-	-	-	-	-	1	1
pound/grind	-	-	-	-	-	-	1	-	-	1
wedge	-	-	-	-	-	1	-	-	-	1
pick	-	-	-	1	-	-	-	-	-	1
Total Employed Units	12	7	3	12	7	12	7	0	1	61

occurring in fewer numbers, microliths and truncations are present within the assemblage and appear to be distinguishable types of the Grand Banks lithic typology.

Overall, the Grand Banks lithic typology demonstrates that the Princess Point lithic industry represents a lesser degree of diversity and complexity of tool types as compared to tool types of Paleo-Indian and Archaic lithic industries. Diversity denotes the number of distinct tool types or classes (Shott 1986:19), whereas complexity denotes the different morphological forms of tool types or classes. It is obvious that flaked stone tools at Grand Banks have not been intentionally modified into various forms to a great degree. Observations of retouched flakes suggest that edge modifications made to the flake tools were primarily for the purpose of facilitating the use of the tools: providing suitable edges for direct working, hafting, or hand-holding.

Some have authors argued that less diversity and complexity of tool types may have occurred in response to increased mobility (e.g., Shott 1986; Torrence 1983). "The response to constraints imposed by settlement mobility should involve a limit on the size of the tool inventory or even a reduction in its size if increasing mobility reduces the overall transport capacity of the group" (Shott 1986:20). However, it is unlikely that the tools at Grand Banks were modified to enhance their portability since evidence of highly mobile settlement organization was not found at the site. The lesser diversity and complexity of tool types at Grand Banks, as I argue here, is one of the characteristics of generalized stone tool production which represents the Princess Point lithic industry (details in Chapter 9).

The manufacture of formed types is characterized by several distinguishing features at the Grand Banks site. Core tablets and unretouched flakes were selected for making these formed artifacts. Production of both surface modified and edge modified tools required the selection of different sizes and forms of tool blanks, as well as different retouch techniques. Bifacial tool production, which I suggest was an on-site activity at Grand Banks, produced a high percentage of triangular projectile points, most of which do not bear evidence of use. The relatively high number of roughly made products and unfinished bifacial tools at Grand Banks may be interpreted as evidence of the early stages of bifacial tool production. Finished bifacial products might have been taken away from habitation sites for use during the hunting seasons, and then discarded or lost off-site. If this assumption is correct, we can infer that Grand Banks must have functioned, in part, as a location where tool manufacturing took place, and where tools were produced for special events. Those projectile points or bifaces which remained at the site might have been employed for a variety of use-tasks, as shown in the use-wear analysis.

However, an alternative explanation for the large quantity of unused projectile points on site should be considered. I suggest that these unused projectile points could have been produced for purposes other than hunting or something similar. This opens room for discussion of the social relations that attain in tool production. Given the fact that most unused points are often extremely well-fashioned on exotic raw materials, and that the points used for cutting, scraping, and boring were made more casually from local raw materials, it is hypothesized that once people were settled in farming villages they continued to make arrowheads that may have been produced as ornamental objects, for trade, or to pass on stone tool-making technology. This speculative hypothesis cannot be tested with the data currently available, but it should be taken into account when discussing social dynamics in terms of lithic production.

I do not imply that the use of arrowheads in hunting activities was not important after the introduction of agriculture. In contrast, ethnohistoric data (Chapter 1) suggests that proto- or historic Iroquian people depended to a large degree on deer hunting with arrowheads. However, objects produced on exotic raw material through a standardized and skillful procedure could also serve for purposes other than hunting. Recently, Neusius (1996) re-analyzed a large sample of projectile points collected by Arthur Parker in 1906, and suggested that technological input of these projectile points may likely point to "the creation of somewhat stylized points which were never used and perhaps never intended for use." He further hypothesizes that those points found in cemeteries probably served as grave goods.

The manufacture of edge modified tools involved two important stages: producing or selecting blanks and secondary retouching. The results of formed type analysis suggest that blanks for flake tools were carefully removed using soft-hammer percussion during later stages of core reduction, perhaps using prepared core reduction techniques. A few blanks were also selected from those produced using bipolar core reduction. Normal abrupt retouch techniques were the most prevalent method of modification at Grand Banks. Most object modification was done through straightening edges and retouching dorsal sides.

Utilization of formed types is relatively low in the Grand Banks assemblage. It has been assumed that formed types should bear a relatively high occurrence of use-wear because these pieces were modified with great interest, design, and energy. However, this low rate of formed type utilization does not seem inappropriate when the large quantity of unused projectile points are taken into account. Other special formed tools such as scrapers, drills, and notches, have much higher use rates, and they present a functional consistence with their type names in this study. Retouched pieces, on the other hand, were employed in a variety of use-tasks.

In general, formed type tool use at Grand Banks represents a relatively specialized model of use-tasks. Tools with evidence of having been hafted are primarily associated with either projecting, digging, or scraping. Transverse motions on wood and other soft substances are common at Grand Banks, and are indicative of limited use-tasks. Formed type tool use at Lone Pine, in contrast, represents a greater range of inhabitant use-tasks.

Not surprisingly, use-wear evidences detected from projectile points suggest that most of those items were possibly used as general tools other than as hunting weapons. At Grand Banks, only three projectile points show tip impact evidence, suggesting that they were used as projectiles. However, the use-patterning of projectile points for digging, drilling, cutting, or sawing, does not mean that they were intended to be made for such activities. These items, worked well as supplementary tools, could have been occasionally selected for multi-purpose by inhabitants at the sites. This use-wear evidence from projectile points also implies that Grand Banks inhabitants were engaged in a great variety of activities on site.

CHAPTER 6

ANALYSIS OF UNRETOUCHED FLAKES

The objectives of this chapter are: (1) to describe the proportional occurrences of flake types; (2) to reconstruct on-site core reduction strategies; (3) to explore technological variability; and (4) to interpret the utilization of unretouched flakes.

Unretouched flakes comprise about one-third of the total lithic assemblages at the three sites (Table 4.1). However, by weight, unretouched flakes make up over two-thirds of each assemblage (Table 4.2). A total of 1,617 pieces were selected for the analysis from the three lithic assemblages (see chapter 3). Only the complete flakes selected were subject to both metric and categorical measurements, while both complete and incomplete flakes were examined for edge wear variables. The incomplete unretouched flakes sampled comprise 24.51% of the total sampled assemblage from Grand Banks, 38.48% from Lone Pine, and 33.08% from Young 1. All imcomplete unretouch flakes were examined microscopically for potential use traces.

Typo-Technological Analysis

Proportional Occurrences of Unretouched flakes

Eight types of unretouched flakes were identified, as shown in Table 6.1. The detailed descriptions for these eight types are given in Chapter 3. Of all the byproducts from core reductions, plain flakes are the most common. The occurrence of plain flakes is slightly higher at Grand Banks (81.2%), compared to the other two sites. Biface thinning flakes are the next most common type of unretouched flakes in all assemblages. Together, plain flakes and biface thinning flakes comprise over 95 percent of the unretouched flakes at all three sites.

The ratio of plain flakes to cores is very high: at Grand Banks there are 108 plain flakes per core. It is possible that cores might be extensively utilized on-site during flake production. Despite this, the following explanations could also account for the high flake to core ratio: (1) the ratio is skewed by the inclusion of a very large number of fragmentary flakes (Table 6.1); (2) totally reduced cores are unrecognizable as such; (3) reduced cores were re-deposited elsewhere; and (4) cores were fashioned into bifacial formed types after flakes were removed

The Grand Banks assemblage presents a relatively high occurrence of blade/bladelets (3.0%) compared to the Lone Pine (0.9%) and Young 1 (0.6%) assemblages. Overall, the low frequencies of blade/bladelet could represent the accidental production of blades, rather than specialized production. However, at Grand Banks the ratio of blades/bladelets to prepared cores is relatively high: 23:1. Although we can not assume that all blades/bladelets were produced from the prepared cores recovered at the site, it is reasonable to state that the presence of blades or blade-like flakes at Grand Banks probably represents a designated flake removal through the prepared core reduction mode for blank production (see below). However, a true blade production industry is not evidenced at the Grand Banks site, because (1) blades do not present a substantial proportion of unretouched flakes and (2) there are no true blade cores which indicate that a series of blades were removed consecutively.

Primary flakes and core trimming flakes are present at all sites in very low frequencies. Their existence, however, indicates that initial core reduction processes took place on site. Core trimming flakes, as they are defined, are by-products of core preparation. At Grand Banks, the ratios of primary flakes to cores is 1.3 and core trimming flakes to the prepared cores is 4.5. Most primary flakes and core trimming flakes are complete (Table 6.1).

Bipolar flakes represent less than 0.2% of the assemblages from both Grand Banks and Lone Pine. This reflects the fact that bipolar core reduction produces few classic bipolar flakes, as demonstrated by Casey's (1993) experiment. According to Casey, most flake byproducts of bipolar core reduction exhibit hinge or snap distal termination (Casey 1993: 223). I examined the distal termination (Fig. 6.1) of plain flakes, and

Table 6.1 Unretouched Flake Types.

Unretouched Flake Type	GRAND BANKS			LONE PINE			YOUNG 1		
		N	%		N	%		N	%
Primary Flake		63	1.01		13	0.16		1	0.16
complete	55			12			1		
fragmentary	8			1			0		
Core Trimming Flake		36	0.58		81	0.98		2	0.32
complete	36			69			2		
fragmentary	0			12			0		
Bifacial Thinning Flake		877	14.03		2690	32.41		30	4.75
complete	733			1519			21		
fragmentary	144			1171			9		
Bipolar Flake		9	0.14		15	0.18		0	0.00
complete	9			15			0		
fragmentary	0			0			0		
Plain Flake		5078	81.22		5418	65.27		595	94.15
complete	2006			1974			127		
fragmentary	3083			3505			468		
Blade/Bladelet		186	2.98		78	0.94		4	0.63
complete	128			44			2		
fragmentary	58			34			2		
Burin Spall		3	0.05		5	0.06		0	0.00
Microburin		0	0.00		1	0.01		0	0.00
Total		6252	100.00		8301	100.00		632	100.00

Table 6.2: Distribution of Distal Termination of Plain Flakes.

Distal Termination	GRAND BANKS		LONE PINE	
	N	%	N	%
feather	223	80.51	273	72.41
hinge	42	15.17	67	17.77
step	6	2.16	7	1.86
snap	6	2.16	30	7.96
Total	277	100	377	100

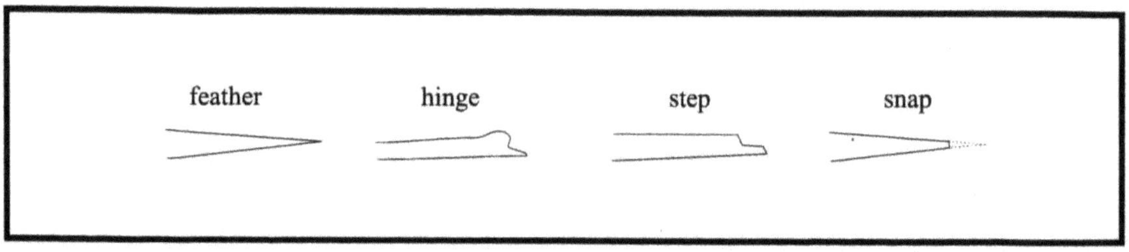

Fig. 6.1: Diagram of Distal Termination of Unretouched-Flakes.

discovered about 20% of plain flakes having hinged or snap terminations (Table 6.2). This may imply that a small proportion of plain flakes were probably byproducts of bipolar core reduction. The low proportions of bipolar flakes in the unretouched flakes assemblages, however, are consistent with the low occurrences of the bipolar cores in the core class.

Finally the unretouched flake assemblages from Grand Banks and Lone Pine include two types not normally found in Woodland assemblages from Ontario: burins spalls and microburins. A few burin spalls appear at Grand Banks and Lone Pine, indicating the production of burins. A single microburin from Lone Pine does not necessarily indicate the use of the microburin technique to manufacture microliths, it may simply be fortuitous.

Core Reduction Strategies

Proportional occurrences of unretouched flakes indicate that the on-site core reduction strategies are dominated by flake core reduction, followed by bifacial reduction. As we have seen from the results of the core analysis, flake core reduction consists of amorphous flake production, bipolar flake production, and possibly blade production (or precisely blade-like flake production through prepared cores). Although the morphological characteristics of unretouched flakes and cores can provide preliminary evidence on these on-site core reduction strategies, verification from other attribute observations is necessary.

Odell's experiment (1989a) demonstrates that appropriate attributes for discriminating flake core reduction from bifacial reduction include the dorsal perimeter scarring, the platform facet, and the midsection thickness. The dorsal perimeter scarring measures the coverage of scarring damage to the dorsal surface just below its juncture with the striking platform (Fig. 6.2). It is reported by Odell (1989a) that flakes removed during bifacial reduction have a greater degree of dorsal perimeter scarring than the ones removed during flake core reduction. This is because scarring damage might have been caused by platform preparation or edge abrasion. Four intervals of this variable are used, and their distributions are presented in Table 6.3. Overall, flakes with scarring over half of the perimeter are present in very low frequencies at all sites. In comparison, there is a lower degree of such scarring on flakes at Grand Banks than at Lone Pine. This pattern corresponds to the lower frequency of bifacial thinning flakes at Grand Banks. It by no means suggests that the flakes with a greater coverage of dorsal perimeter scarring must have been produced through bifacial reduction. However, the distribution pattern of this attribute does support the observation that the predominant core reduction strategy at Grand Banks was flake core reduction, not bifacial reduction.

The similar distribution pattern of platform modification types (Table 6.4) also supports the above statement. Bifacial reduction is more likely to produce flakes with multiple faceted platforms. These flakes from the unretouched flake assemblages comprise a similar small proportion of the analyzed samples, around 15% at both Grand Banks and Lone Pine. About half of the samples have a single platform facet.

Midsection thickness was measured on plain flakes and biface thinning flakes, and measurements from both flake types were statistically compared (Table 6.5). At Lone Pine there is a significant difference between the two groups in the mean of midsection thickness at Lone Pine. This may be a result of the two different reduction strategies. Although the midsection thickness of

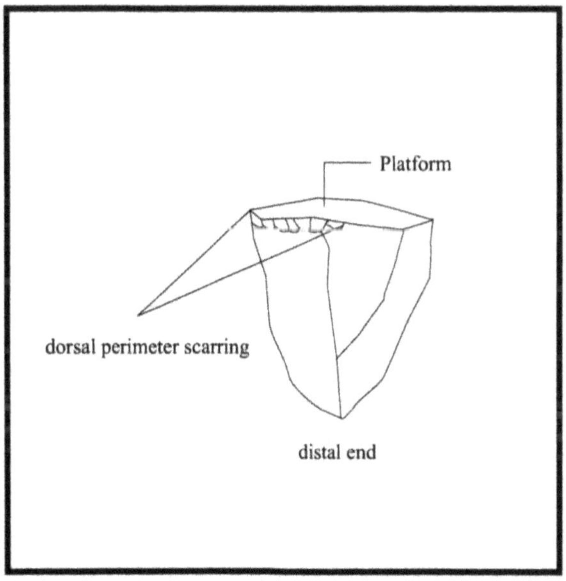

Fig. 6.2: Diagram of Dorsal Perimeter Scarring of Unretouched-Flakes.

Table 6.3: Dorsal Perimeter Scarring in the Unretouched Flake Assemblages.

Dorsal Perimeter Scarring	GRAND BANKS		LONE PINE		YOUNG 1	
	N	%	N	%	N	%
indeterminate	19	4.18	28	7.43	2	2.35
absent	356	78.24	254	67.37	71	83.53
less than half	44	9.67	45	11.94	3	3.53
more than half	26	5.71	33	8.75	4	4.71
entirely	10	2.20	17	4.51	5	5.88
Combined	455	100.00	377	100.00	85	100.00

Table 6.4: Platform Facet Presented in the Unretouched Flake Assemblages

Platform Facet	GRAND BANKS		LONE PINE		YOUNG 1	
	N	%	N	%	N	%
indeterminate	27	7.03	34	12.14	0	0.00
single facet	183	47.66	125	44.64	9	56.25
dihedral facet	22	5.73	27	9.64	4	25.00
multiple facet	63	16.41	44	15.71	1	6.25
crushed facet	51	13.28	39	13.93	2	12.50
cortical facet	38	9.90	11	3.93	0	0.00
Combined	384	100.00	280	100.00	16	100.00

Table 6.5: T-test Results of Midsection Thickness of Bifacial Thinning Flakes and Plain Flakes from the Grand Banks and Lone Pine Assemblages.

Group 1: Biface Thinning Flake Group 2: Plain Flake

Statistics t-test	Mean Group 1	Mean Group 2	S. D. Group 1	S. D. Group 2	t-value	df	p-value	Valid N Group 1	Valid N Group 2
Grand Banks Midsection Thickness	2.89	3.60	1.53	1.83	-1.96441	229	0.050693	28	203
*Lone Pine Midsection Thickness	2.70	3.75	1.69	2.06	-3.792	302	0.00018	65	239

* indicates significant difference between two groups

Table 6.6: T-test Results of Platform Dimensions of Bifacial Thinning Flakes and Plain Flakes from the Grand Banks and Lone Pine Assemblages.

Group 1: Biface Thinning Flake Group 2: Plain Flake

Statistics t-test	Mean Group 1	Mean Group 2	S. D. Group 1	S. D. Group 2	t-value	df	p-value	Valid N Group 1	Valid N Group 2
Grand Banks platform length	6.67	7.89	3.21	3.83	-1.76287	246	0.079164	34	214
Grand Banks platform width	2.64	2.63	2.24	1.40	0.020297	306	0.98382	37	271
* Lone Pine platform length	6.32	9.46	2.88	5.97	-3.93277	256	0.000108	60	198
* Lone Pine platform width	1.98	2.92	1.33	2.12	-3.36589	336	0.000851	64	274

* indicates significant difference between two groups

plain flakes and biface thinning flakes from Grand Banks are not significantly different at 0.05 level, there is some indication of a slight difference in midsection thickness as indicated by the low *p-value* (0.050693).

There are also significant differences in platform sizes of biface thinning flakes and plain flakes from Lone Pine (Table 6.6). As expected, flakes produced during bifacial reduction have smaller platforms. At Grand Banks, there is no clear difference between the two flake types in platform sizes. These t-test results may suggest that bifacial reduction was probably more prevalent at Lone Pine than at Grand Banks.

Despite the above indications that bifacial reduction took place at Grand Banks and Lone Pine, flake core reduction undoubtedly was the predominant strategy at both sites. As we have seen, there are three different byproducts of flake core reduction: plain flakes, bipolar flakes, and blades/bladelets. In order to determine whether these three products were produced separately from the different flake core reductions, metric data from three flake types are considered and compared.

Twelve metric measurements (Fig. 6.3) were taken of the three flakes types and compared (Tables 6.7-6.9). At Grand Banks most plain flakes, presumably produced through amorphous flake production, are relatively short and broad in shape, with a mean L/W ratio of 1.41. Blades/bladelets are more elongated and narrower in form, with a mean L/W ratio of 2.48 for blades and at 2.67 for bladelets. Only four bipolar flakes were selected for the analysis; their mean L/W ratio is moderate at 1.61. The dimensional differences between plain flakes and blades may be a reflection of their respective definitions: blades are those flakes at least two times

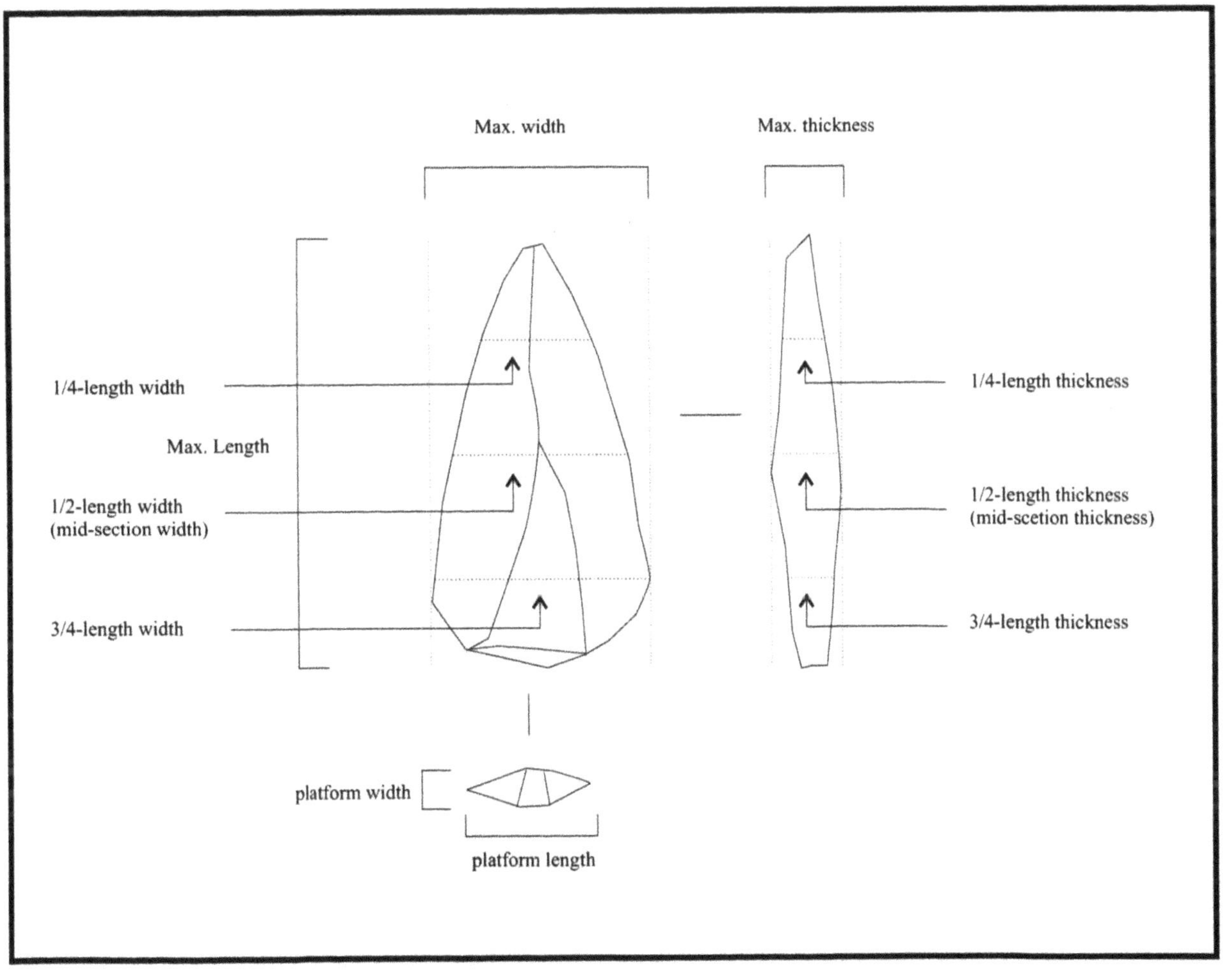

Fig. 6.3: Diagram of Metric Variables of Unretouched-Flakes.

Table 6.7: Metric Data on Unretouched Flakes from the Grand Banks Assemblage by Flake Type.

Metric Variables	plain flake			blade			bipolar flake			biface thinning flake			core trimming flake			primary flake		
	N	*x*	*sd*	*N*	*x*	*sd*	*N*	*x*	*sd*	*N*	*x*	*sd*	*N*	*x*	*sd*	*N*	*x*	*sd*
Max. Length	276	22.12	7.57	59	28.20	5.59	4	24.68	3.41	65	18.73	7.83	29	28.79	14.47	36	22.09	9.28
Max. Width	276	17.15	6.54	59	11.48	2.47	4	16.20	5.41	65	13.92	7.24	29	17.70	7.67	36	17.10	7.28
Max. Thickness	276	4.21	2.00	59	3.66	1.17	4	6.98	3.65	65	3.05	2.17	29	8.07	3.21	36	5.95	2.91
L/W Ratio	276	1.42	0.74	59	2.48	0.31	4	1.61	0.37	65	1.46	0.49	29	1.87	1.01	36	1.45	0.63
1/4-length Width	204	14.49	6.07	58	9.38	2.07	3	13.60	5.78	28	13.74	6.58	24	12.38	6.19	31	12.76	7.33
1/4-length Thickness	203	2.96	1.67	58	2.59	0.95	3	3.99	1.91	28	2.35	1.04	24	6.20	2.81	31	3.68	2.01
1/2-length Width	203	15.65	5.77	58	10.40	2.43	3	14.80	4.99	28	14.71	6.70	24	13.59	5.25	29	14.27	5.91
1/2-length Thickness	203	3.60	1.83	58	2.87	0.93	3	5.47	1.81	28	2.89	1.53	24	6.96	3.17	29	4.78	2.15
3/4-length Width	203	13.70	4.43	58	9.76	2.30	3	13.40	3.25	28	12.26	5.44	24	12.86	5.45	29	13.30	5.41
3/4-length Thickness	203	3.70	1.66	58	2.98	1.05	3	6.67	3.43	29	3.11	2.39	24	7.24	3.50	29	5.34	2.54
Platform Length	214	7.89	3.83	57	6.03	1.89	3	8.38	4.34	34	6.67	3.21	27	8.81	5.13	33	7.13	3.41
Platform Width	271	2.63	1.40	58	2.20	0.96	4	4.18	3.96	37	2.64	2.24	28	4.47	2.31	36	3.47	1.57

Table 6.8: Metric Data on Unretouched Flakes from the Lone Pine Assemblage by Flake Type.

Metric Variables	plain flake			blade			bipolar flake			biface thinning flake			core trimming flake			primary flake		
	N	*x*	*sd*	*N*	*x*	*sd*	*N*	*x*	*sd*	*N*	*x*	*sd*	*N*	*x*	*sd*	*N*	*x*	*sd*
Max. Length	288	22.80	8.33	19	29.81	8.45	4	24.74	5.52	115	17.50	3.90	33	27.26	8.58	9	24.68	8.13
Max. Width	287	19.19	7.59	18	13.93	4.45	4	13.78	8.26	114	12.35	4.80	34	19.04	9.12	10	22.77	4.27
Max. Thickness	287	4.42	2.54	18	3.76	1.29	4	3.94	1.76	114	2.63	1.26	34	7.82	3.44	10	7.47	3.63
L/W Ratio	286	1.28	0.45	18	2.42	0.78	4	2.28	1.26	114	1.65	1.35	33	1.74	0.90	9	1.10	0.25
1/4-length Width	239	16.15	6.86	18	11.56	4.53	4	11.50	6.09	65	11.52	4.97	33	14.72	8.01	9	17.38	5.89
1/4-length Thickness	239	3.05	1.74	18	2.95	1.07	4	3.05	1.55	65	2.03	1.78	33	5.32	3.40	9	4.45	1.98
1/2-length Width	239	17.41	6.20	18	11.95	3.52	4	11.61	7.03	65	13.20	5.00	33	15.87	6.96	9	17.92	5.34
1/2-length Thickness	239	3.75	2.06	18	3.03	0.98	4	3.27	1.75	65	2.70	1.69	33	5.85	2.68	9	5.40	2.39
3/4-length Width	239	15.21	5.98	18	11.54	3.27	4	11.70	6.72	65	10.91	3.60	33	14.53	6.71	9	17.09	6.42
3/4-length Thickness	239	3.99	2.22	18	3.03	1.08	4	3.49	1.79	65	2.57	1.26	33	6.01	2.71	9	5.70	2.97
Platform Length	198	9.46	5.97	14	7.77	2.80	3	6.76	3.26	60	6.32	2.88	31	9.90	7.62	10	11.81	7.17
Platform Width	274	2.92	2.12	18	2.30	1.46	3	2.74	1.43	64	1.98	1.33	33	4.56	3.14	10	4.98	3.60

Table 6.9: Metric Data on Unretouched Flakes from the Young 1 Assemblage by Flake Type.

Metric Variables	plain flake			blade			core trimming flake			primary flake		
	N	*x*	*sd*	*N*	*x*	*sd*	*N*	*x*	*sd*	*N*	*x*	*sd*
Max. Length	79	21.11	10.54	5	19.25	13.01	1	22.00	-	2	25.70	14.62
Max. Width	79	15.30	8.03	5	10.87	4.87	1	13.32	-	2	24.70	0.42
Max. Thickness	79	4.09	2.98	5	2.53	1.47	1	4.30	-	2	10.51	2.56
L/W Ratio	79	1.49	0.59	5	1.68	0.33	1	1.65	-	2	1.04	0.57
1/4-length Width	9	14.64	3.09	0	-	-	0	-	-	0	-	-
1/4-length Thickness	9	3.98	1.55	0	-	-	0	-	-	0	-	-
1/2-length Width	9	17.57	5.26	0	-	-	0	-	-	0	-	-
1/2-length Thickness	9	4.68	2.27	0	-	-	0	-	-	0	-	-
3/4-length Width	9	15.92	4.90	0	-	-	0	-	-	0	-	-
3/4-length Thickness	10	4.89	3.00	0	-	-	0	-	-	0	-	-
Platform Length	14	9.19	3.36	0	-	-	0	-	-	1	11.81	-
Platform Width	75	2.92	2.70	4	1.71	1.22	1	2.70	-	2	6.45	1.48

Table 6.10: T-test Results of Metric Data between Plain Flake and Blade from the Grand Banks Unretouched Flake Assemblage.

Group 1: Plain Flake — Group 2: Blade

Statistics t-test	Mean Group 1	Mean Group 2	S. D. Group 1	S. D. Group 2	t-value	df	p-value	Valid N Group 1	Valid N Group 2
*Max. Length	22.12	28.20	7.57	5.59	-5.83705	333	1.2611E-08	276	59
*Max. Width	17.15	11.48	6.54	2.47	6.55553	333	2.1127E-10	276	59
*Max. Thickness	4.21	3.66	2.00	1.17	2.03504	333	0.04263825	276	59
*L/W Ratio	1.42	2.48	0.74	0.31	-10.80799	333	1.5372E-23	276	59
*1/4-length Width	14.49	9.38	6.07	2.07	6.29874	260	1.2735E-09	204	58
1/4-length Thickness	2.96	2.59	1.67	0.95	1.60204	259	0.11036652	203	58
*1/2-length Width	15.65	10.40	5.77	2.43	6.73799	259	1.0355E-10	203	58
*1/2-length Thickness	3.60	2.87	1.83	0.93	2.92096	259	0.00379719	203	58
*3/4-length Width	13.70	9.76	4.43	2.30	6.53495	259	3.3583E-10	203	58
*3/4-length Thickness	3.70	2.98	1.66	1.05	3.10456	259	0.00211716	203	58
*Platform Length	7.89	6.03	3.83	1.89	3.54875	269	0.00045662	214	57
*Platform Width	2.63	2.20	1.40	0.96	2.23801	327	0.02589295	271	58

* indicates significant difference between the two groups

Table 6.11: T-test Results of Metric Data between Plain Flake and Blade from the Lone Pine Unretouched Flake Assemblage.

Group 1: Plain Flake — Group 2: Blade

Statistics t-test	Mean Group 1	Mean Group 2	S. D. Group 1	S. D. Group 2	t-value	df	p-value	Valid N Group 1	Valid N Group 2
*Max. Length	22.80	29.81	8.33	10.94	-3.48019392	305	0.000574	288	19
*Max. Width	19.19	13.93	7.59	4.45	2.90684961	303	0.003920	287	18
Max. Thickness	4.42	3.76	2.54	1.29	1.09527131	303	0.274268	287	18
*L/W Ratio	1.28	2.42	0.45	0.78	-9.84838853	302	0.000000	286	18
*/4-length Width	16.15	11.56	6.86	4.53	2.7917158	255	0.005640	239	18
1/4-length Thickness	3.05	2.95	1.74	1.07	0.25743651	255	0.797049	239	18
*1/2-length Width	17.41	11.95	6.20	3.52	3.68865686	255	0.000276	239	18
1/2-length Thickness	3.75	3.03	2.06	0.98	1.46129735	255	0.145165	239	18
*3/4-length Width	15.21	11.54	5.98	3.27	2.56971491	255	0.010747	239	18
3/4-length Thickness	3.99	3.03	2.22	1.08	1.81512922	255	0.070679	239	18
Platform Length	9.46	7.77	5.97	2.80	1.04795767	210	0.295863	198	14
Platform Width	2.92	2.30	2.12	1.46	1.21568105	290	0.225095	274	18

* indicates significant difference between the two groups

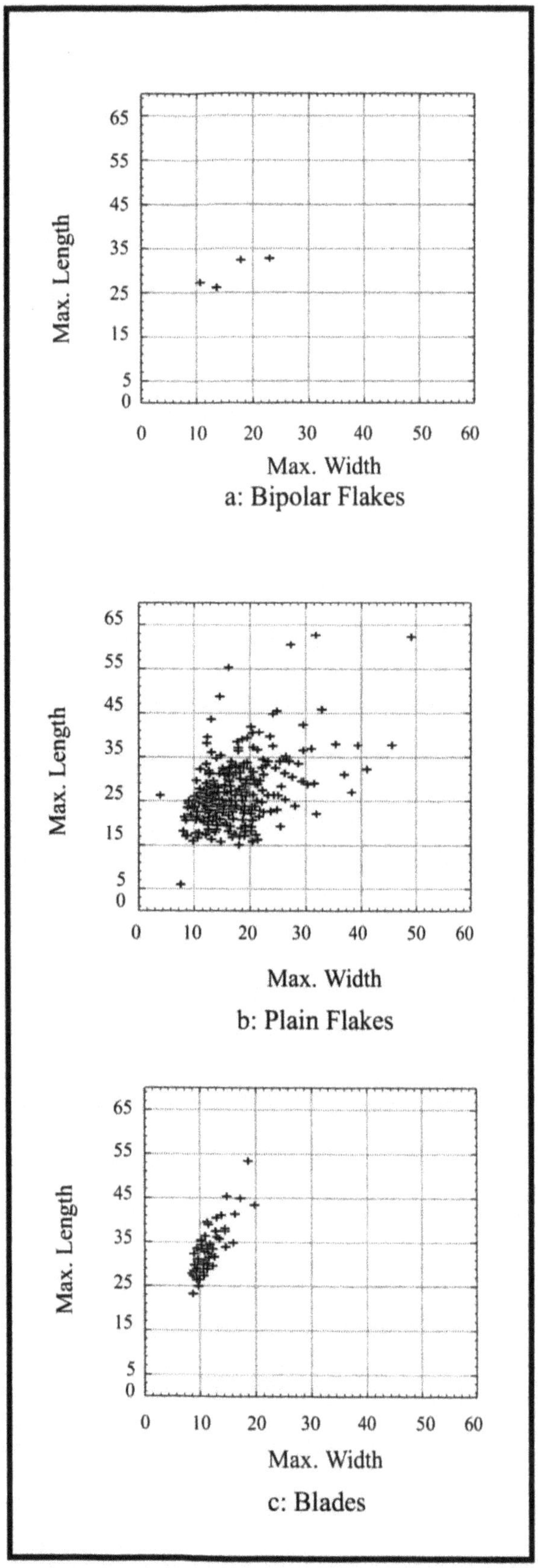

Fig. 6.4: Scatterplot Comparisons of Selected Unretouched Flake Types from Grand Banks.

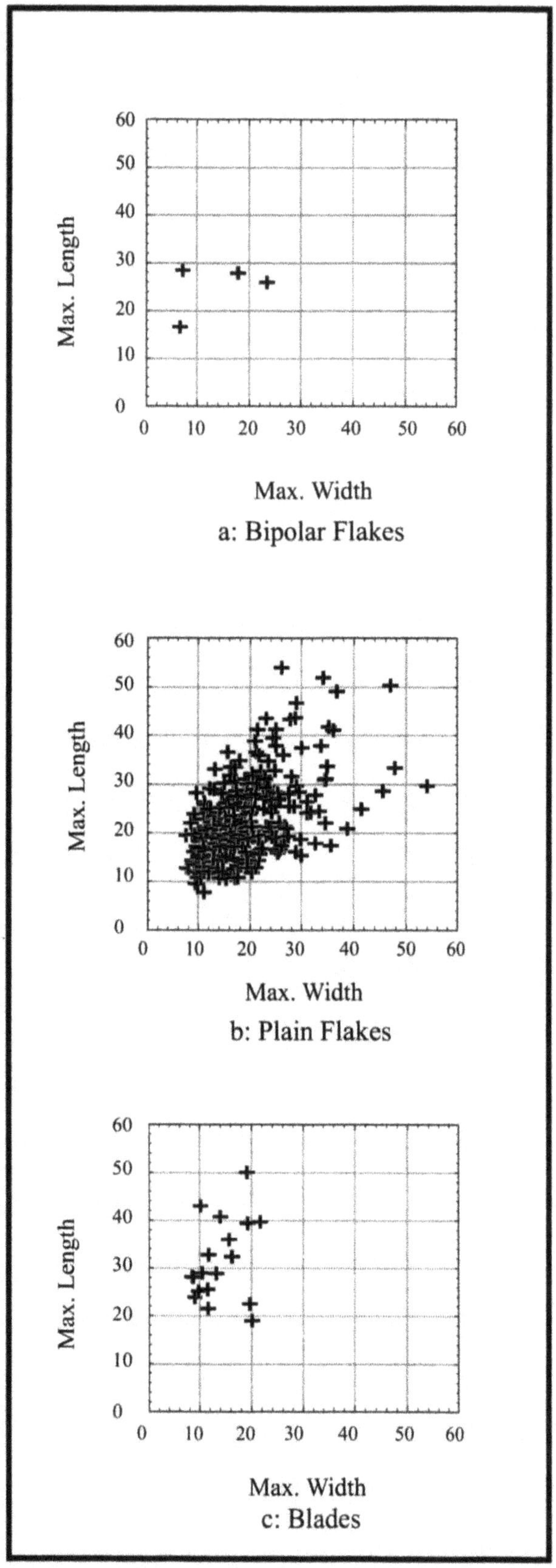

Fig. 6.5: Scatterplot Comparisons of Selected Unretouched Flake Types from Lone Pine.

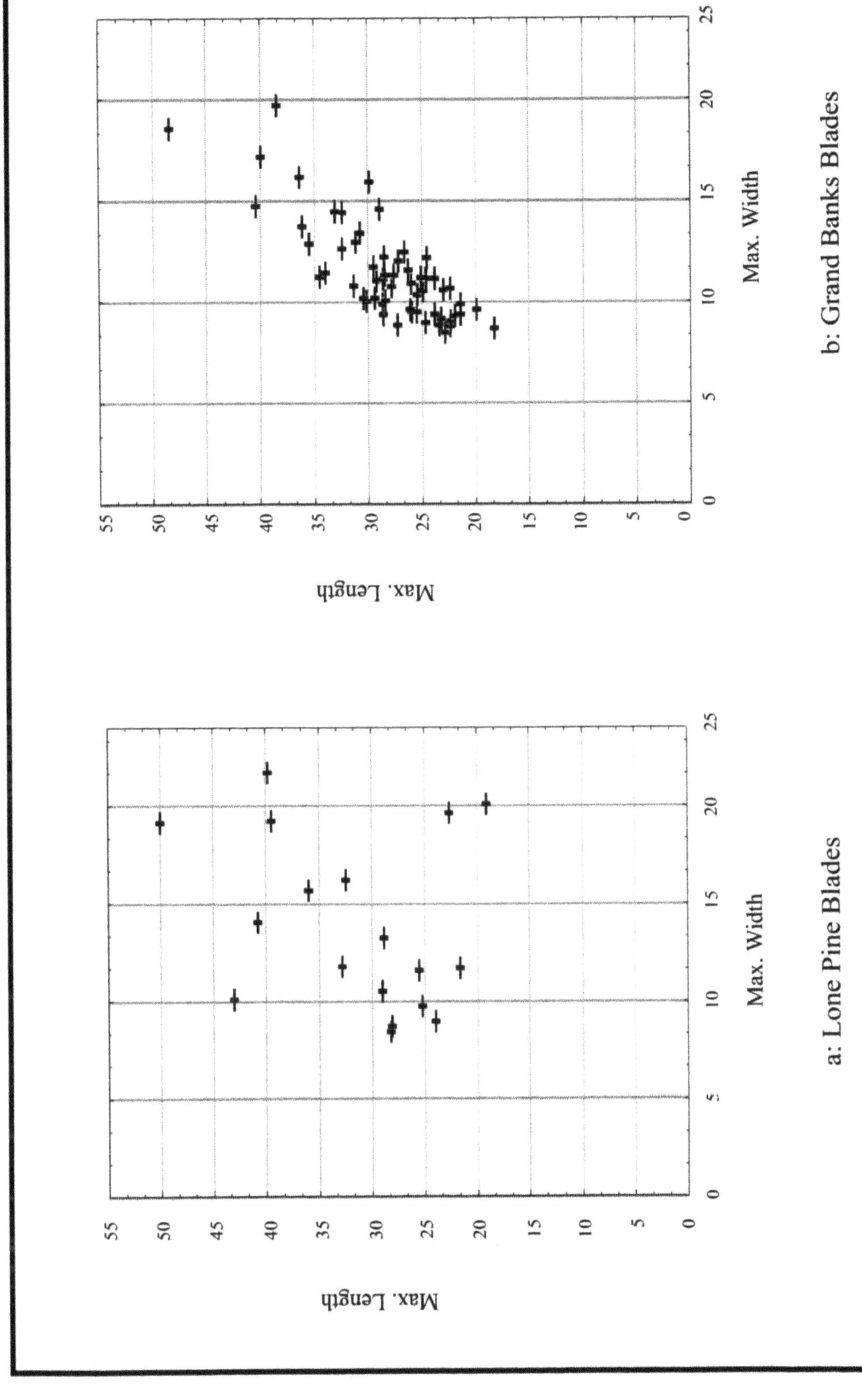

Fig. 6.6: Scatterplot Comparisons of Blades Sizes between Grand Banks and Lone Pine

as long as they are wide. When length or width alone is considered, plain flakes and blades may be similar. Therefore, to enhance the dimensional comparison of plain flakes and blades, the width and thickness at quarterly lengths is also considered.

In the Grand Banks assemblage, blades statistically differ from plain flakes and bipolar flakes in most, if not all, dimensions (Table 6.10). The t-test results show that blades differ significantly from plain flakes in all mean measurements except 1/4-length thickness. The differences in width and thickness at quarterly lengths between blades and flakes are significant, because they indicate that blades must have been produced using a specially designated procedure different from that used for amorphous flake production. These differences can be also illustrated using scatterplots (Fig. 6.4): the majority of blades are 25-30 mm long and 10-15 mm wide with strictly limited ranges, while the majority of flakes are 20-25 mm long and 15-20 mm wide with very extended ranges. The restricted ranges of blade dimensions suggest that the blades or blade-like flakes may not be simply accidental products at Grand Banks.

Bipolar flakes resemble plain flakes in their dimensions, but the length and width ranges of bipolar flakes are more restricted than those of plain flakes (Fig. 6.4). However, this is likely biased by the small sample of bipolar flakes. Nevertheless, there are significant differences in the means of maximum thickness and platform sizes between bipolar and plain flakes. This may reflect the fact that bipolar techniques may produce relatively thicker flakes.

At Lone Pine, the dimensions of blades are similar to those of bipolar flakes but differ from amorphous plain flakes. The t-test suggests that blades have statistically different lengths and widths from plain flakes (Table 6.11, Fig. 6.5). But unlike at Grand Banks, the Lone Pine blades and flakes have similar striking platform dimensions. Furthermore, compared to the Grand Banks blades, Lone Pine blades have a relatively greater range in length and width (Fig. 6.6). This may indicate that more variety in blade-like flake production existed at Lone Pine than at Grand Banks, or it may simply suggest that blades were indeed accidental products.

Technological Variability of the Core Reductions

Three major stages are, in general, recognized in the process of core reduction (prior to tool production): (1) acquisition or selection of raw materials; (2) primary processing, and (3) blank production. The technological aspects of these core reduction stages are discussed in this section, in conjunction with available data, in order to determine variability in these processes.

Table 6.12 shows the types of raw material utilized in the unretouched flake assemblages at the three sites. It suggests that nearly 90%of the unretouched flakes are of local Onondaga chert, although other locally available raw materials (Haldimand and Ancaster chert) were present too. Very low quantities of unknown or non-local chert materials are also found in unretouched flake assemblages. This pattern is identical to that of raw material utilization seen in the core and formed type assemblages discussed earlier. When the raw material type is cross-tabulated with unretouched flake type (Table 6.13), no distinctive distribution of raw material utilization for different core reduction strategies is apparent. All samples of biface thinning flakes and bipolar flakes are restricted to the Onondaga chert.

Although the frequencies of primary flakes and core thinning flakes is very low, the presence of these two flake types is indicative of primary processing. A small percentage of primary flakes and core trimming flakes made on Haldimand and Ancaster chert has been found at the Grand Banks site, indicating that a few such raw materials, in addition to Onondaga chert, were exploited and brought to the site for tool production. Most of these flakes have 30% to 50% cortical coverage on the dorsal surface (Table 6.14). Primary flakes with full cortical coverage comprise a very small percent of the sampled assemblage: 12.07% of primary flakes. This strongly suggests that primary processing took place off-site. Combined with data from core sizes, this suggests that the inhabitants likely trimmed the cortex from nodules or reduced cores to a manageable size and weight at quarries before the cores were brought

Table 6.12: Distribution of Raw Material in the Unretouched Flake Assemblages.

Chert Types	GRAND BANKS		LONE PINE		YOUNG 1	
	N	%	N	%	N	%
Onondaga	543	87.16	630	86.78	126	94.74
Haldimand	44	7.06	32	4.41	7	5.26
Ancaster	18	2.89	29	3.99	0	0.00
Others	18	2.89	35	4.82	0	0.00
Total	623	100	726	100	133	100.00

Table 6.13: Unretouched Flake Type by Raw Material in the Grand Banks Assemblage.

Debitage Type	Raw Material				
	Onondaga	Haldimand	Ancaster	Others	
Primary Flake	52	4	2	-	58
Core Trimming Flake	29	2	2	1	34
Bipolar Flake	4	-	-	-	4
Bifacial Thinning Flake	34	-	-	-	34
Plain Flake	320	27	13	11	371
Blade	83	9	1	2	95
Bladelets	21	2	-	3	26
Combined	543	44	18	17	622

Table 6.14: Cortical Distribution on Primary Flake and Core Trimming Flake.

	Cortical Coverage	GRAND BANKS		LONE PINE		YOUNG 1	
		N	%	N	%	N	%
Primary Flake	100% coverage	7	12.07	2	10.00	1	33.33
	50%-99% coverage	11	18.97	3	15.00	0	0.00
	30-50% coverage	40	68.97	15	75.00	2	66.67
Core Trimming Flake	30-50% coverage	0	0.00	1	2.17	0	0.00
	1%-30% coverage	5	14.71	6	13.04	0	0.00
	no coverage	29	85.29	39	84.78	1	100.00

back to Grand Banks and Lone Pine for blank production. This is supported by the evidence from core trimming flakes, indicative of core preparation in blank production. Almost all core trimming flakes with cortex, at both sites, have less than 30% cortical coverage on the dorsal surface. Core trimming flakes with no cortex are predominant.

Data from the analysis of cores and formed types suggest that the predominant blank production strategy at Grand Banks was to produce amorphous plain flakes and blade-like flakes. Here, I have examined the dorsal scar counts of plain flakes greater than 25 mm in length. Table 6.15 indicates that, at both Grand Banks and Lone Pine, over 50% of plain flakes have a dorsal scar count of over 3. At Young 1, about 70% of plain flakes have a low dorsal scar count (< 3), but this could be affected by the small sample (N=7). The majority of blades, not surprisingly, are characterized by a low dorsal scar count (Table 6.16). However, representative percentages of blades have a moderate number of dorsal scars (between 3 and 5 scars): 29.3% at Grand Banks and 36.4% at Lone Pine. These patterns may indicate that

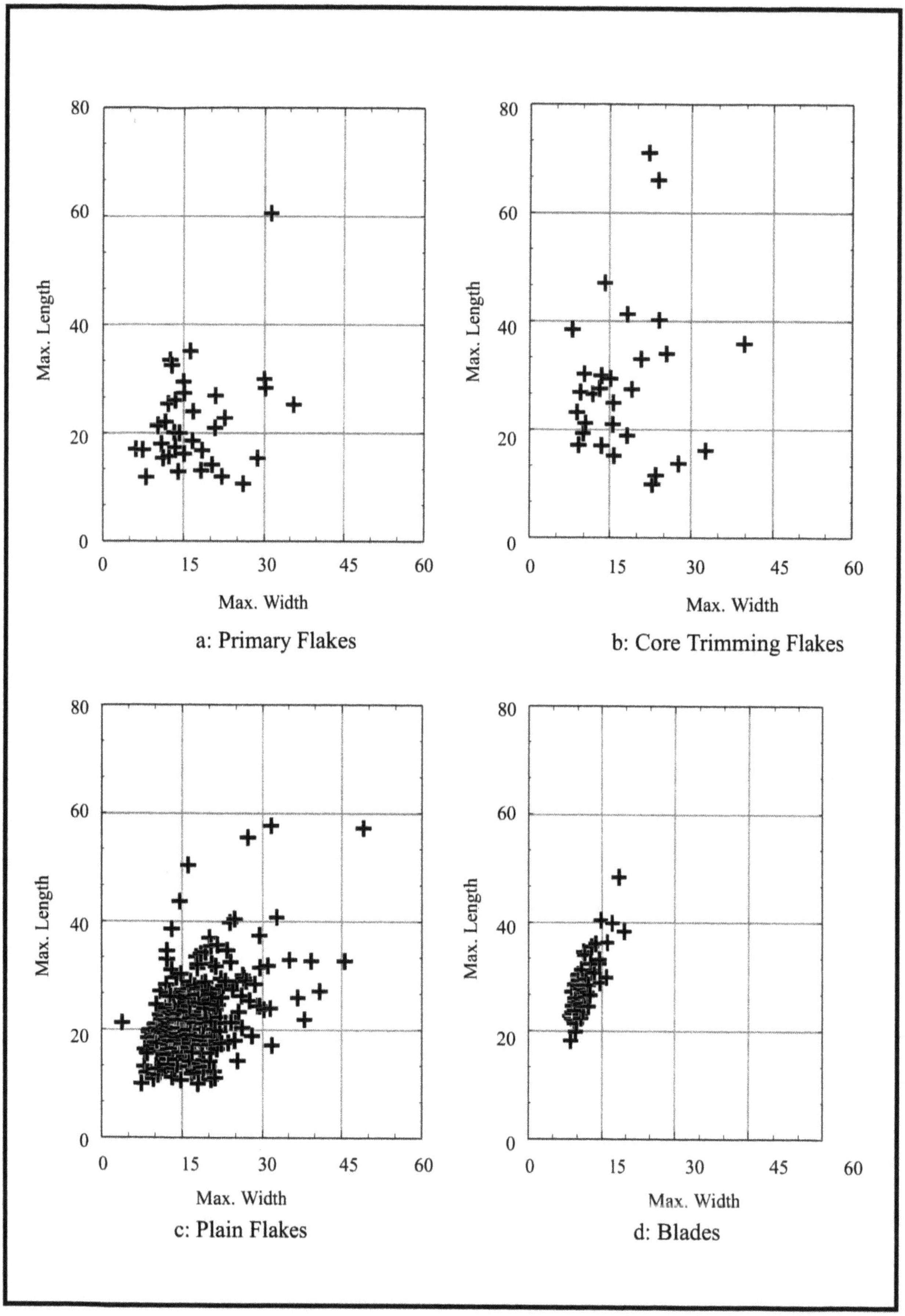

Fig. 6.7: Scatterplot Comparisons of Selected Unretouched Flake Types from Grand Banks.

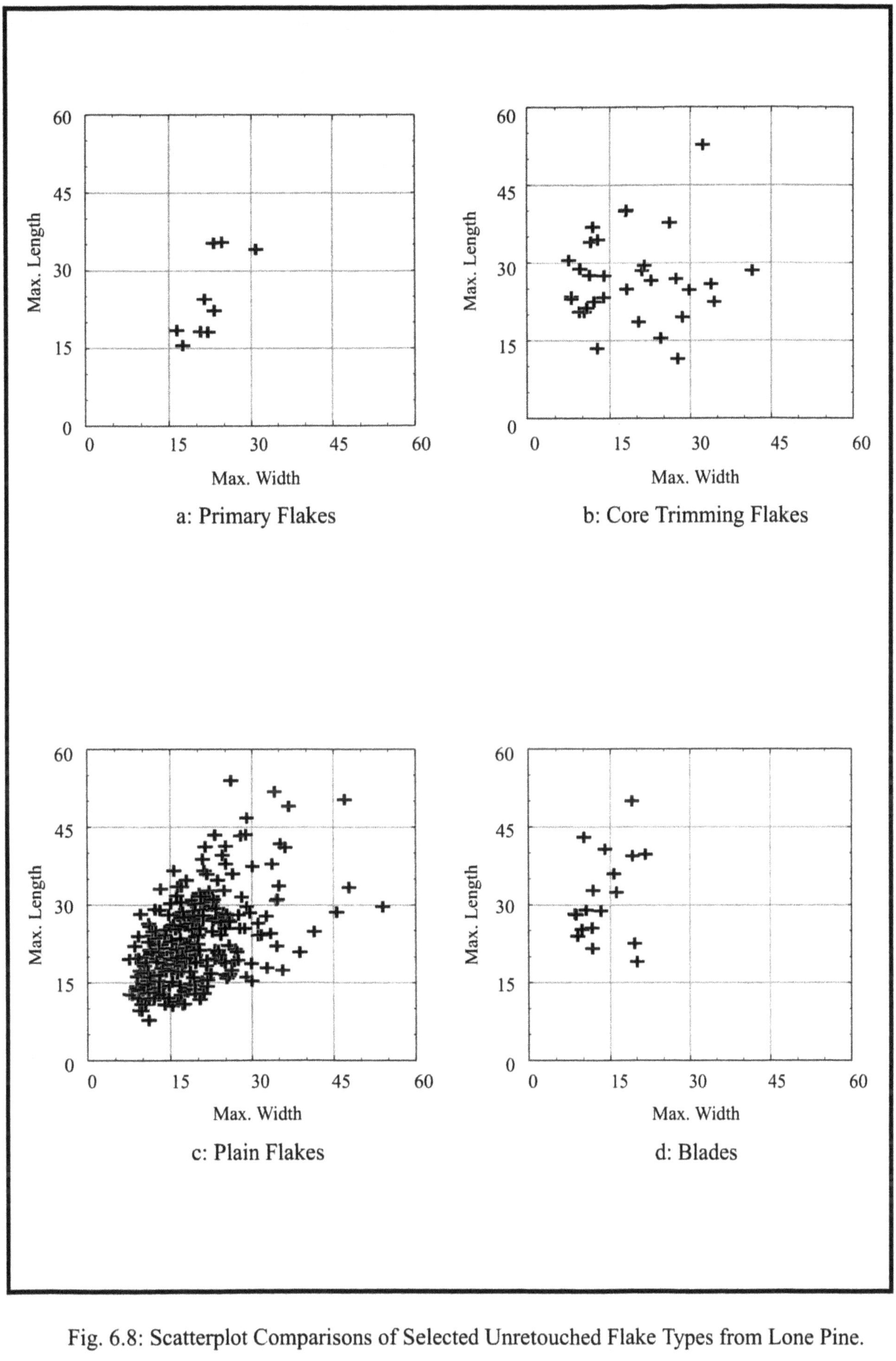

Fig. 6.8: Scatterplot Comparisons of Selected Unretouched Flake Types from Lone Pine.

most blades and amorphous flakes were produced during the later stages of core reduction.

When the lengths and widths of the early stage unretouched flakes (primary flakes and core trimming flakes) and later stage unretouched flakes (plain flakes and blades) were compared, it was found that although the mean length of primary flakes is shorter than that of blades, the dimensional ranges of both primary flakes and core trimming flakes are wider than those of both flakes and blades (Tables 6.7-6.9, Fig 6.7). At Grand Banks we see a reduction in the size of unretouched flakes through the reduction process. This points to a consecutive core reduction process: primary flakes and core trimming flakes are larger in size than plain flakes and blades. The same pattern does not exist at Lone Pine (Fig. 6.8). The small size of primary flakes at Lone Pine might be explained by the use of small-sized core nodules for blank production. However, the two diagrams also exhibit dimensional evidence to suggest the possibility that core trimming flakes were the by-products of blade-blank production. The ranges of lengths and widths of core trimming flakes exceed those of blades, suggesting presumably that larger core trimming flakes were removed prior to blades.

Observations of platform attributes can suggest the types of loading techniques used in core reduction (hard- or soft-hammer percussion, punch, and pressure). The platform type and bulb shape are cross-tabulated. I first compared these attributes within the unretouched flakes produced during the later stages of reduction: bifacial thinning flakes, plain flakes, and blades. These three flake types are assumed to indicate three different core reduction sequences. The pattern of the paired attributes is very interesting (Table 6.17). Within the category of biface thinning flakes, flakes with a lipped bulb and a multiple-faceted platform predominate. Single platform flakes with a lipped bulb predominate for the blades assemblage. This is also the case for plain flakes, followed by single platform flakes with unlipped bulbs. Plain flakes are also associated with other types of platform facet patterns and bulb shapes.

In short, the evidence implies that biface thinning flakes were produced mostly through soft-hammer percussion or pressure flaking, while blades were removed predominantly by soft-hammer percussion. Amorphous plain flakes were probably produced through application of both soft-hammer and hard-hammer percussion, although the former was likely applied more frequently. This suggests that soft-hammer percussion is predominant in the later stages of blank production which contrasts with the pattern suggested for the

Table 6.15: Dorsal Scar Count Distribution on Plain Flakes (> 25 cm in length).

Dorsal Scar Count	GRAND BANKS		LONE PINE		YOUNG 1	
	N	%	N	%	N	%
indeterminate	3	3.41	6	7.14	0	0
few (0-2)	26	29.55	29	34.52	5	71.43
moderate (3-5)	39	44.32	34	40.48	1	14.29
high (>5)	20	22.73	15	17.86	1	14.29
Total	88	100.00	84	100.00	7	100.00

Table 6.16: Dorsal Scar Count Distribution on Blade.

Dorsal Scar Count	GRAND BANKS		LONE PINE		YOUNG 1	
	N	%	N	%	N	%
indeterminate	0	0.00	0	0.00	-	-
few (0-2)	28	68.29	7	63.64	-	-
moderate (3-5)	12	29.27	4	36.36	-	-
high (>5)	1	2.44	0	0.00	-	-
Total	41	100.00	11	100.00	0	0.00

primary reduction process.

I combined primary flakes and core trimming flakes and observed the distribution of cross-tabulating platform and bulb shapes (Table 6.18). The relatively high frequency of crushed platforms and unlipped bulbs suggests that hard-hammer percussion was employed during the early stages of core reduction. This clearly indicates that technological variability existed during the core reduction process.

Table 6.17: Cross-Tabulation of Platform Facet and Shape of Bulb of Major Unretouched Flake Types from the Grand Banks Assemblage.

a: Plain Flake

Platform Facet	Shape of Bulb				
	indeterminate	lipped	pointed	unlipped	
indeterminate	-	1	1	10	12
single	-	75	12	20	107
dihedral	-	5	-	4	9
multiple	1	22	3	14	40
crushed	1	2	2	25	30
cortex	2	6	5	5	18
Combined	4	111	23	78	216

b: biface Thinning Flake

Platform Facet	Shape of Bulb				
	indeterminate	lipped	pointed	unlipped	
indeterminate	-	-	-	-	0
single	-	7	-	2	9
dihedral	-	2	-	-	2
multiple	-	10	-	1	11
crushed	-	-	1	2	3
cortex	-	1	-	-	1
Combined	0	20	1	5	26

c: Blades

Platform Facet	Shape of Bulb				
	indeterminate	lipped	pointed	unlipped	
indeterminate	1	-	-	2	3
single	-	22	2	5	29
dihedral	-	3	-	1	4
multiple	-	4	1	2	7
crushed	-	1	-	7	8
cortex	-	5	2	-	7
Combined	1	35	5	17	58

Table 6.18: Cross-Tabulation of Platform Facet and Shape of Bulb of both Primary Flake and Core Trimming Flake Combined from the Grand Banks Assemblage.

Platform Facet	Shape of Bulb				
	indeterminate	lipped	pointed	unlipped	
indeterminate	8	-	-	3	11
single	1	18	3	21	43
dihedral	-	1	-	9	10
multiple	-	3	-	5	8
crushed	-	-	-	14	14
cortex	2	-	1	7	10
Combined	11	22	4	59	96

Use-Wear Analysis

Grand Banks

A total of 622 unretouched flakes from Grand Banks were examined for use-wear. The results of this analysis indicate a relatively low utilization of unretouched flakes. Ninety-eight flakes were determined to have use-wear, accounting for 15.8% (with +/- factors) of the total analyzed sample. This figure is lower than a previous estimate (22.2%) in a preliminary report (Shen 1995:152) when a smaller sample (N=178) was used. This low utilization of unretouched flakes at Grand Banks may upset the previous conclusion that the Princess Point people relied heavily on "expedient" flake tools. The low rate of utilized flakes, of course, could also be affected by the relatively large number (N=161) of flake fragments included in the analyzed sample. Normally, fragmentary flakes are less likely to display use-wear due to their breakage. It should also be kept in mind that not all of the utilized flake tools retain used traces that allow us to ascertain use-tasks, especially if the pieces were used lightly. Nevertheless, this analysis suggests a lower frequency of use of unretouched flakes than previously thought.

In general, "utilized flakes" in the following discussion refers only to those unretouched flakes with evidence of use-wear, not categorized "type" artifacts. When utilized flakes are examined by flake type (Table 6.19), I find that blanks from both amorphous flake production and blade production, as well as byproducts of biface production, were utilized to about the same extent. In all cases, utilized flakes comprise slightly less than 20% of the sample. Flakes from the early stages of blank production were also utilized, but to a lesser degree, as would be expected (8.6% of primary flakes and 5.9% of core trimming flakes). Of 25 bladelets examined, only one was determined to have been utilized. Given the small sample, it is not surprising that four analyzed bipolar flakes were found to have been unused.

Of the 98 utilized pieces, 12 unretouched flakes (12.2%) bear more than one employed unit of non-prehensile wear. Six pieces have 2 EdUs, four have 3 EdUs, and two have 5 EdUs. All together, a total of 120 employed units involved in use-tasks was found in the examined assemblage.

The Grand Banks unretouched flake assemblage exhibits a preponderance of scraping/shaving tool motions (42.5%), followed by cutting/sawing tool motions (26.7%) (Table 6.20). This pattern resembles that of utilization of formed type artifacts. However, differences in the utilization of the two groups do exist. A number of the unretouched flakes were used in slicing/carving and planing/whittling, which were uncommon applications for the formed type artifacts. In total, transverse tool motions are slightly more frequent than longitudinal motions. Penetration or projection wear and boring and graving wear are found in low frequencies. Rare tool motions such as wedging and picking, are primarily associated with plain flakes. Use-wear found on blades indicates the use of these artifacts for cutting/sawing, scraping/shaving, and planing/whittling, exclusively. Most primary flakes and core trimming flakes were used in scraping tool motions.

Animal substances were likely the dominant

Table 6.19: Utilization Rates of Unretouched Flake.

Unretouched Flakes	GRAND BANKS			LONE PINE			Young 1		
	total N	used	use rate (%)	total N	used	use rate (%)	total N	used	use rate (%)
primary Flake	58	5	8.62	20	4	20.00	3	0	0.00
core Trimming. Flake	34	2	5.88	46	10	21.74	1	0	0.00
bipolar Flake	4	0	0.00	5	1	20.00	x		
bifacial Thinning Flake	36	6	16.67	27	5	18.52	2	0	0.00
plain Flake	370	67	18.11	569	147	25.83	117	20	17.09
blade	95	17	17.89	42	17	40.48	7	2	28.57
bladelet	25	1	4.00	14	1	7.14	2	1	50.00
Combined	622	98	15.76	723	185	25.59	132	23	17.42

x: indicates that such types do not exist in the assemblage

Table 6.20: Tool Motions in the Grand Banks Unretouched Flake Assemblage.

Tool Motion	primary Flake	core trimming flake	bifacial thinning flake	plain flake	blade	bladelet	Combined	Percentage of Tool Motion
	Unretouched Flake Type							
indeterminate	1	1	1	9	2	-	14	11.67
cut/saw	1	-	-	22	8	1	32	26.67
slice/carve	-	-	1	6	-	-	7	5.83
scrape/shave	2	1	4	36	8	-	51	42.50
plane/whittle	1	-	-	4	3	-	8	6.67
projection	-	-	1	2	-	-	3	2.50
bore/drill	-	-	-	1	-	-	1	0.83
grave	-	-	-	2	-	-	2	1.67
chop	-	-	-	-	-	-	0	0.00
hoe/dig	-	-	-	-	-	-	0	0.00
pound/grind	-	-	-	-	-	-	0	0.00
wedge	-	-	-	1	-	-	1	0.83
pick	-	-	-	1	-	-	1	0.83
Total Employed Units	5	2	7	84	21	1	120	100.00

Table 6.21: Contact Material in the Grand Banks Unretouched Flake Assemblage.

Contact Material	primary flake	core trimming flake	bifacial thinning flake	plain flake	blade	bladelet	Combined	Percentage of Worked Material
	Unretouched-Flake Types							
indeterminate	1	2	2	11	2	-	18	15.00
soft animal (SA)	-	-	-	25	5	-	30	25.00
soft vegetal (SV)	-	-	-	5	3	1	9	7.50
medium-soft vegetal (1M)	3	-	1	13	3	-	20	16.67
medium animal (MA)	-	-	3	9	4	-	16	13.33
medium-hard vegetal (2M)	1	-	-	4	-	-	5	4.17
hard animal (1H)	-	-	1	13	4	-	18	15.00
very hard animal (2H)	-	-	-	4	-	-	4	3.33
hard inorganic (3H)	-	-	-	-	-	-	0	0.00
Total Employed Units	5	2	7	84	21	1	120	100.00

materials worked by the unretouched flakes found at Grand Banks. These range from soft animal skin, meat, hide (SA, 25%), to moderately hard fresh bones or dried skin and hide (MA, 13.3%), to hard bones or antlers (1H and 2H, 18.3%) (Table 6.21). Wood substances, including materials of both medium-soft and medium-hard resistance, account for 20.9% of the contact material. Amorphous flakes were used on each resistance category in approximately the same frequency. Blades however, were more likely to have been used on soft to moderate animal substances rather than on wood. Primary flakes were used predominantly on wood.

When tool motions are compared with contact materials (Table 6.22), we see a pattern that differs from the one seen for utilization of formed types. Unretouched flakes were primarily used for cutting/sawing and scraping/shaving of soft and moderate animal substances. The secondary emphasis on wood substances involves mainly cutting/sawing and scraping/shaving. Cutting/sawing motions were also carried out on bone/antler material. Other tool motions such as penetrating, boring, and graving are only associated with materials of hard resistance such as dried wood and bone/antler.

The utilized flakes were then examined by flake size in order to determine preferred blank sizes. Over half of the utilized flakes are of moderate size, between 25 mm and 50 mm in length or width (Table 6.23). The remaining utilized pieces are divided nearly equally between small flakes (less than 25 mm) and large flakes (over 50 mm). The t-test shows there is a significance in the correlation of utilized flakes and their sizes. Complete utilized flakes (N=69) were tested against the complete unused flakes to determine

Table 6.22: Tool Motion by Contact Material in the Grand Banks Unretouched Flake Assemblage.

Tool Motion	Contact Material									
	?	SA	SV	1M	MA	2M	1H	2H	3H	Comb.
indeterminate	14	-	-	-	-	-	-	-	-	14
cut/saw	-	10	6	7	6	-	3	-	-	32
slice/carve	-	4	-	-	2	-	1	-	-	7
scrape/shave	1	15	3	9	7	4	9	3	-	51
plane/whittle	-	-	-	4	1	-	2	1	-	8
projection	1	1	-	-	-	-	1	-	-	3
bore/drill	-	-	-	-	-	-	1	-	-	1
grave	-	-	-	-	-	1	1	-	-	2
chop	-	-	-	-	-	-	-	-	-	0
hoe/dig	-	-	-	-	-	-	-	-	-	0
pound/grind	-	-	-	-	-	-	-	-	-	0
wedge	1	-	-	-	-	-	-	-	-	1
pick	1	-	-	-	-	-	-	-	-	1
Total Employed Units	18	30	9	20	16	5	18	4	0	120

Table 6.23: Flake Size Grade of the Used Flakes in the Unretouched Flake Assemblages.

Size Grade	GRAND BANKS		LONE PINE		YOUNG 1	
	N	%	N	%	N	%
Small (10-25 cm in length)	24	24.74	50	38.76	(no sufficient data)	
Medium (25-50 cm in length)	53	54.64	68	52.71	-	
Large (>50 cm in length)	20	20.62	11	8.53	-	
Combined	97	100.00	129	100.00	-	

the statistical difference of the means. Results (Table 6.24) show that there are significant differences between the two populations in maximum length, width, and thickness, but not in platform length and width. This implies that technological variability was not a consideration in selecting items for utilization. However, size attributes were a consideration. The Grand Banks inhabitants carefully selected appropriately sized blanks for use. The illustrations (Fig. 6.9) demonstrate that the blanks selected for utilization were longer and wider than the average unretouched flakes. The limited ranges in maximum length and width, as compared with the unused unretouched flakes, also suggest the deliberate selection of blanks with certain dimensions. Utilized

Table 6.24: T-Test Results of Flake Sizes Between Unused and Used Flakes from the Grand Banks Unretouched Flake Assemblage.

Group 1: Unused Flake — Group 2: Used Flake

Statistics t-test	Mean Group 1	Mean Group 2	S. D. Group 1	S. D. Group 2	t-value	df	p-value	Valid N Group 1	Valid N Group 2
*Max. Length	22.39	26.82	7.94	7.54	-4.29123	443	2.18E-05	376	69
*Max. Width	15.54	18.38	6.20	8.19	-3.31601	443	0.000988	376	69
*Max. Thickness	4.30	5.12	2.25	2.79	-2.64728	443	0.008404	376	69
Platform Length	7.31	7.52	3.68	3.61	-0.37366	369	0.708872	317	54
Platform Width	2.71	2.76	1.58	1.92	-0.22562	433	0.821604	369	66

* indicates significant difference between the two groups

Table 6.25: The Mann-Whitney U Test Results of Fracture Attributes between Unused and Used Flakes from Grand Banks.

Group 1: Unused Flake — Group 2: Used Flake

Statistics the Mann-Whitney	Rank Sum Group 1	Rank Sum Group 2	U	Z	p-level	Z adjusted	p-level	Valid N Group 1	Valid N Group 2
*Dorsal Scar Count	306939	91339	51684	-3.85713	0.000115	-4.10913	0.000040	714	178
*Dorsal Scar Pattern	312642	85636	57387	-2.0027	0.045218	-2.08574	0.037010	714	178

* indicates significant difference between the two groups

Table 6.26: Dorsal Scar Count Presented in the Grand Banks Unretouched Flakes.

Dorsal Scar Count	Unused Flake N	Unused Flake %	Used Flake N	Used Flake %	Total N	Total %
indeterminate	41	10.88	2	2.90	43	9.64
few	150	39.79	21	30.43	171	38.34
moderate	141	37.40	31	44.93	172	38.57
high	45	11.94	15	21.74	60	13.45
Combined	377	100.00	69	100.00	446	100.00

Table 6.27: Dorsal Scar Pattern Presented in the Grand Banks Unretouched Flakes.

Dorsal Scar Pattern	Unused Flake N	Unused Flake %	Used Flake N	Used Flake %	Total N	Total %
indeterminate	68	18.04	5	7.25	73	16.37
parallel	144	38.20	30	43.48	174	39.01
parallel-opposed	45	11.94	10	14.49	55	12.33
parallel-truncated	47	12.47	10	14.49	57	12.78
parallel-converging	34	9.02	12	17.39	46	10.31
non-parallel	39	10.34	2	2.90	41	9.19
Combined	377	100.00	69	100.00	446	100.00

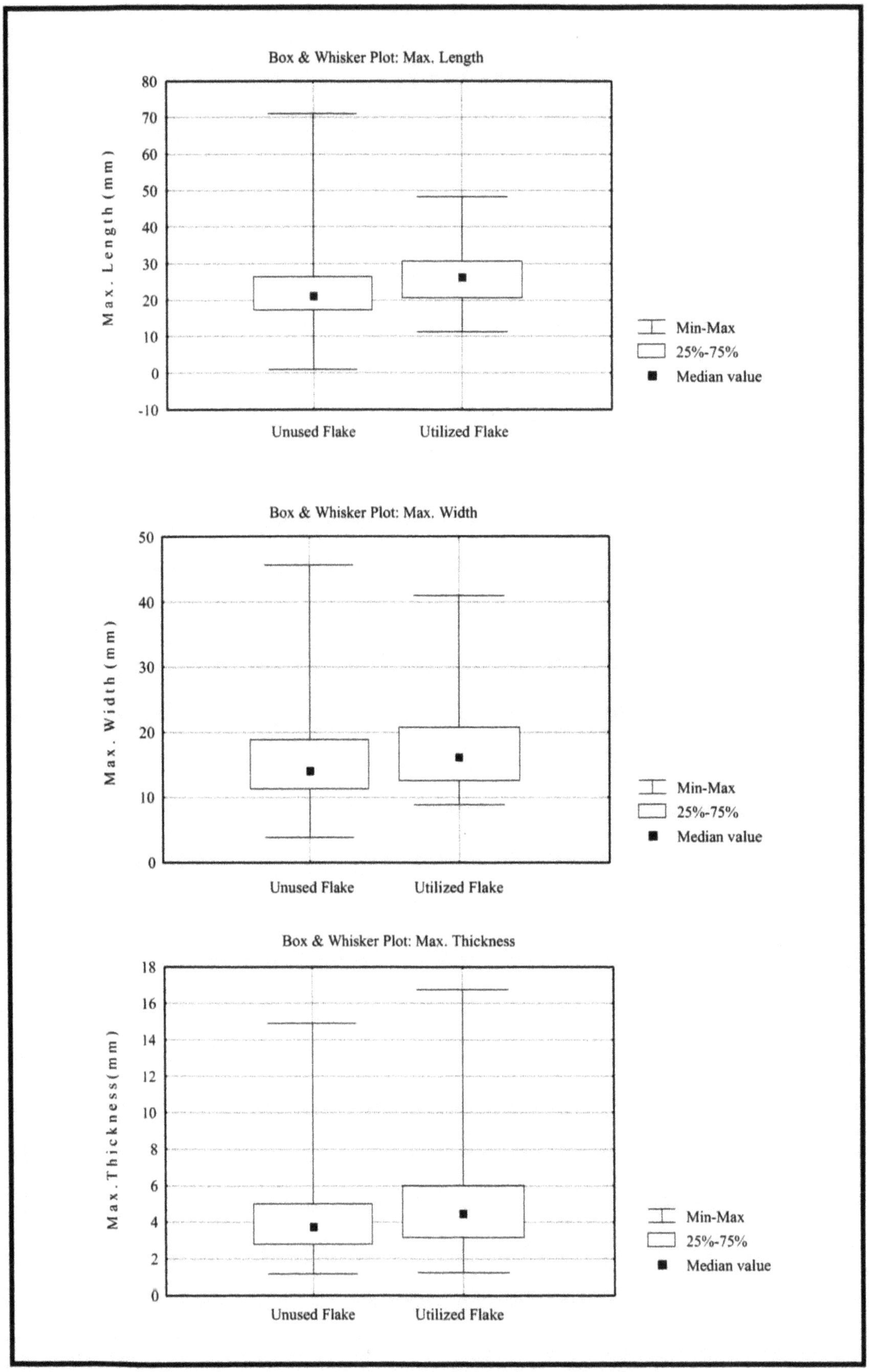

Fig. 6.9: Box & Whisker Comparisons of Blank Sizes between Unused Flakes and Utilized Flakes from Grand Banks.

blanks also have a greater thickness than unused flakes, and the range in thickness of utilized flakes exceeds that of discarded flakes.

Preferred selection of certain blanks for utilization can also be explicated by examination of the dorsal scar counts and scar patterns of both used and unused flakes. The Mann-Whitney U test suggests that there are significant differences between the two kinds of flakes (Table 6.25). Utilized flakes predominantly exhibit higher scar counts (over 3) on dorsal surfaces than the unused ones (Table 6.26), suggesting clearly that the blanks selected were mainly removed during the later stages of core reduction. Examination of the dorsal scar pattern indicates that there is a slightly higher rate of unused flakes (10.3%) showing a non-parallel dorsal scar pattern than exists in the sample of utilized flakes (2.9%) (Table 6.27). The predominant blanks from both used and unused flakes, however, have parallel-opposed, parallel-truncated, and parallel-converging dorsal scar patterning.

Lone Pine and Young 1

A total of 723 unretouched flakes from Lone Pine and 133 pieces from Young 1 were subjected to microscopic analysis, and the utilization rates at both sites are slightly higher than that at Grand Banks (Table 6.19). While the sample of blades from Lone Pine is smaller (N=42) that that from Grand Banks (N=95), there is evidence of greater utilization of blades at Lone Pine (40.5%) compared to Grand Banks (17.9%). Primary flakes and core trimming flakes show relatively higher rates of utilization compared with the sample

Table 6.28: Tool Motion in the Unretouched Flake Assemblages.

Tool Motion	LONE PINE		YOUNG 1	
	N	%	N	%
indeterminate	26	12.3	6	25.0
cut/saw	63	29.7	10	41.7
slice/carve	14	6.6	0	0.0
scrape/shave	63	29.7	7	29.2
plane/whittle	12	5.7	0	0.0
projection	21	9.9	0	0.0
bore/drill	7	3.3	1	4.2
grave	4	1.9	0	0.0
chop	0	0.0	0	0.0
hoe/dig	0	0.0	0	0.0
pound/grind	1	0.5	0	0.0
wedge	1	0.5	0	0.0
pick	0	0.0	0	0.0
Total Employed Units	212	100.0	24	100.0

Table 6.29: Contact Material in the Unretouched-Flake Assemblages.

Contact Material	LONE PINE		YOUNG 1	
	N	%	N	%
indeterminate	41	19.34	8	33.33
soft animal (SA)	34	16.04	7	29.17
soft vegetal (SV)	8	3.77	0	0.00
medium-soft vegetal (1M)	50	23.58	6	25.00
medium animal (MA)	25	11.79	0	0.00
medium-hard vegetal (2M)	31	14.62	1	4.17
hard animal (1H)	17	8.02	2	8.33
very hard animal (2H)	5	2.36	0	0.00
hard inorganic (3H)	1	0.47	0	0.00
Total Employed Units	212	100.00	24	100.00

Table 6.30: Tool Motion by Contact Material in the Lone Pine Unretouched Flake Assemblage.

Tool Motion	Contact Material									
	?	SA	SV	1M	MA	2M	1H	2H	3H	Comb.
indeterminate	25	-	-	1	-	-	-	-	-	26
cut/saw	5	10	6	14	7	12	8	1	-	63
slice/carve	-	10	-	1	3	-	-	-	-	14
scrape/shave	1	11	2	23	4	14	6	2	-	63
plane/whittle	-	1	-	8	-	2	1	-	-	12
projection	9	-	-	-	11	1	-	-	-	21
bore/drill	1	1	-	3	-	1	1	-	-	7
grave	-	1	-	-	-	-	1	2	-	4
chop	-	-	-	-	-	-	-	-	-	0
hoe/dig	-	-	-	-	-	-	-	-	-	0
pound/grind	-	-	-	-	-	-	-	-	1	1
wedge	-	-	-	-	-	1	-	-	-	1
pick	-	-	-	-	-	-	-	-	-	0
Total Employed Units	41	34	8	50	25	31	17	5	1	212

Table 6.31: Tool Motion by Contact Material in the Young 1 Unretouched Flake Assemblage.

Tool Motion	Contact Material									
	?	SA	SV	1M	MA	2M	1H	2H	3H	Comb.
indeterminate	6	-	-	-	-	-	-	-	-	6
cut/saw	1	4	-	3	-	-	2	-	-	10
slice/carve	-	-	-	-	-	-	-	-	-	0
scrape/shave	1	3	-	2	-	1	-	-	-	7
plane/whittle	-	-	-	-	-	-	-	-	-	0
projection	-	-	-	-	-	-	-	-	-	0
bore/drill	-	-	-	1	-	-	-	-	-	1
grave	-	-	-	-	-	-	-	-	-	0
chop	-	-	-	-	-	-	-	-	-	0
hoe/dig	-	-	-	-	-	-	-	-	-	0
pound/grind	-	-	-	-	-	-	-	-	-	0
wedge	-	-	-	-	-	-	-	-	-	0
pick	-	-	-	-	-	-	-	-	-	0
Total Employed Units	8	7	0	6	0	1	2	0	0	24

from Grand Banks. In accordance with the data from Grand Banks, we find extremely low rates of utilization of bladelets and bipolar flakes from Lone Pine. Utilized flakes from the Young 1 sample are limited to 20 plain flakes, 2 blades and 1 bladelet.

Similar to the pattern of flake utilization at Grand Banks, the used flakes from the unretouched flake assemblages from Lone Pine and Young 1 indicate an emphasis on cutting/sawing and scraping/shaving (Table 6.28). The utilized flakes from Lone Pine were also used equally in other tool motions (projecting, slicing/carving, boring, graving). Pounding/grinding, wedging, and picking tool motions are also present at Lone Pine, but not in abundance.

The unretouched flakes from Lone Pine were used primarily in woodworking (38.2% for both 1M and 2M), followed by soft to moderately resistant animal materials (27.4% for SA and MA) (Table 6.29). The Young 1 flakes exhibit utilization on restricted materials, with emphasis on soft animal substances and medium-soft vegetal substances.

Cross-tabulating the tool motions with contact materials from Lone Pine (Tables 6.30), we find that cutting/sawing and scraping/shaving are pri-

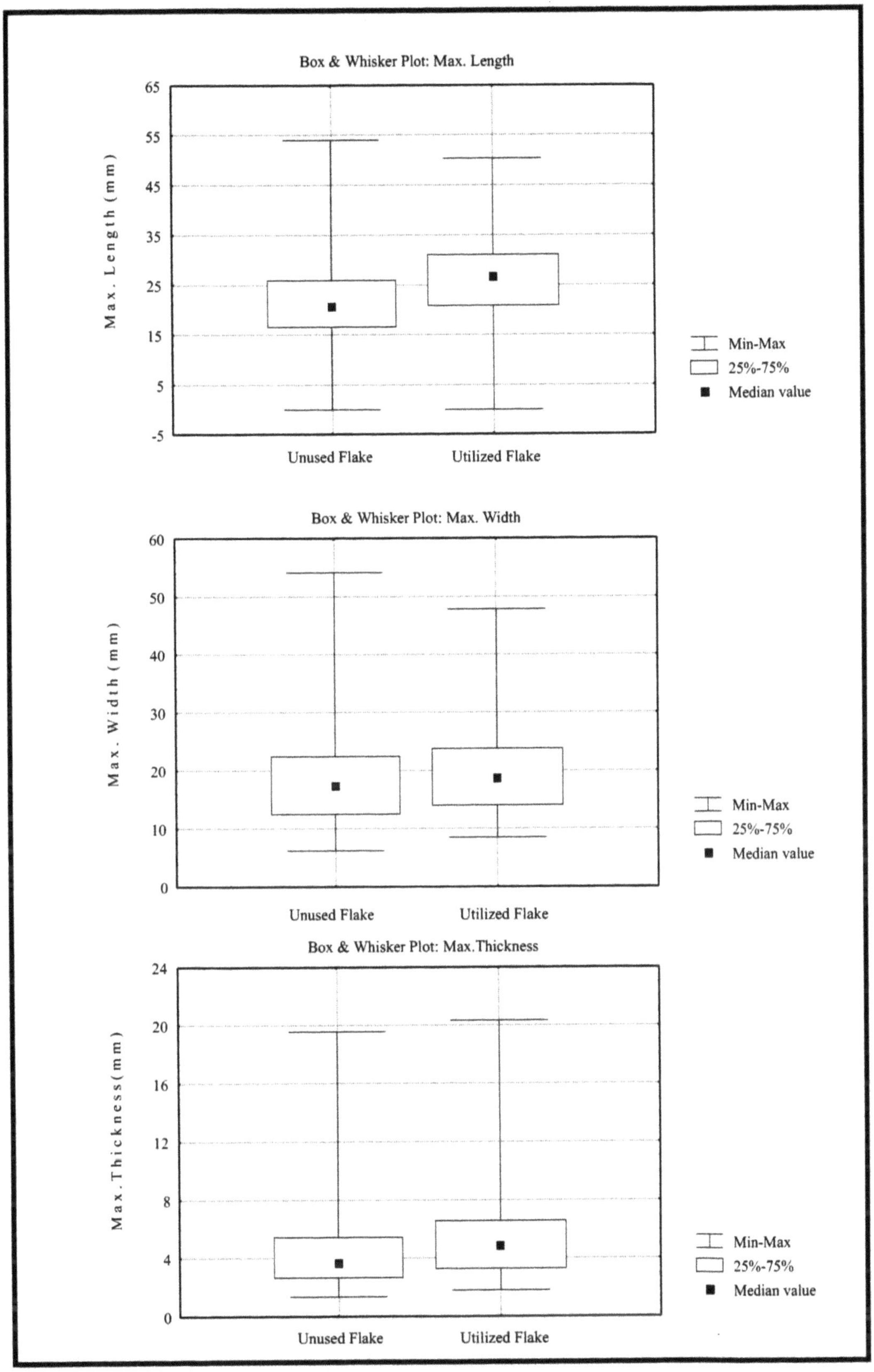

Fig. 6.10: Box & Whisker Comparisons of Blank Sizes between Unused Flakes and Utilized Flakes from Lone Pine.

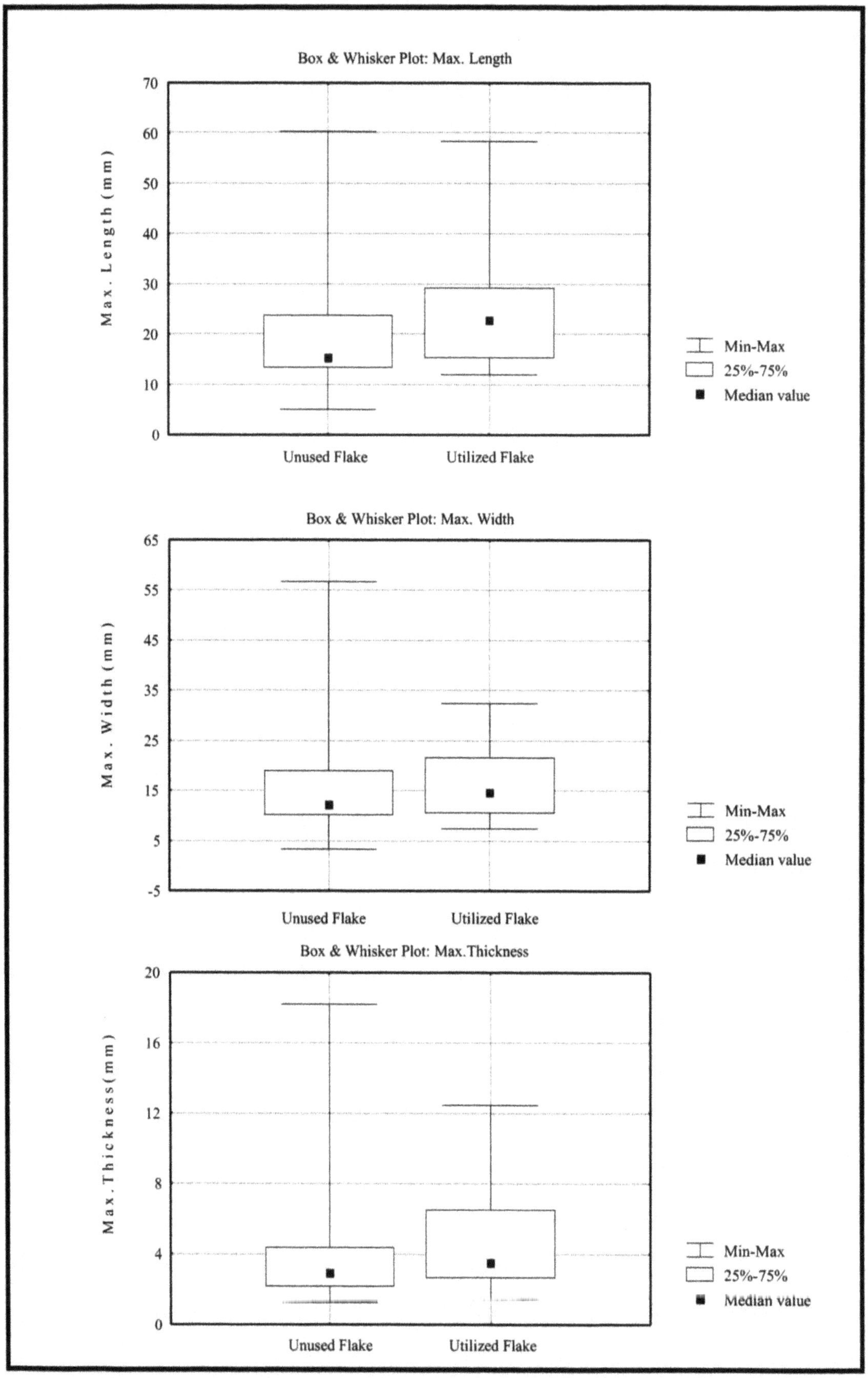

Fig. 6.11: Box & Whisker Comparisons of Blank Sizes between Unused Flakes and Utilized Flakes from Young 1.

marily associated with both soft and hard woods, a pattern which differs from that of the Grand Banks utilized flakes. There is also abundant evidence of use-tasks involving butchering on-site, given the relatively higher rate of cutting, slicing, and carving of soft to moderately hard animal substances. Impact wear on animal substances found on flakes from Lone Pine may indicate use as projectile points. Such flakes normally show hafting wear, indicating the stated tool use. This implies that during this later period, hunting or butchering was also one of the activities carried out at the site. While boring is associated with woodworking, graving is concentrated with bone-working or working with other hard materials. Pounding on stone or other inorganic materials is present but not in abundance. The same is true for the wedging of hard wood.

Much of the wear on the Young 1 unretouched flakes is difficult to determine (Table 6.31). The recognizable wear indicates flakes that were mostly used for cutting/sawing of animal meats or skins, scraping/shaving of skins or hides, sawing and scraping/shaving of medium-soft woods, and cutting/sawing of hard bones or antlers.

As far as the size of utilized flakes is concerned, at both sites, more small flakes (less than 25 mm) were chosen for use (Table 6.25) than at Grand Banks. There appear to be indications, here as well, that utilized flakes were also selected according to certain ideal dimensions. Evidence for this can be found in the fact that at both sites the used flakes have significantly different lengths compared with the unused ones (*t-value*=-4.57882, *df*=359, *p-value*=0.000006 for Lone Pine; *t-value*=-2.08313, *df*=85, *p-value*=0.040244 for Young 1). The diagrams showing the dimensional comparisons between the two kinds of unretouched flakes from the both sites are presented in Figures 6.10 and 6.11.

Prehensile Wear and the Use of Individual Tools

Over half of the prehensile wear identified from the Grand Banks unretouched flakes assemblages are attributable to hand grasping (Table 6.32). Only about a quarter of the prehensile units are identified as hafting wear, in contrast to the formal type assemblage. At Lone Pine a slightly greater number of utilized flakes were used with a hafting device: haft wear accounts for about one-third of the Lone Pine assemblage. This higher frequency of hafting wear may be related to the higher frequency of flake tools used for projection at Lone Pine. It is interesting to note that a certain amount of recognizable prehensile wear is well developed on flake tools with evidence of multiple-utilization.

As I have mentioned in Chapter 4, one core tool from Grand Banks was used repeatedly for a single task. Amongst the unretouched flakes, the same pattern is often observed. Most flakes with multiple use-tasks show evidence of utilization in the same tool motions on similar materials. For instance, all five employed units on Grand Banks flake #3131 were identified as scraping motions on hard bone. The Lone Pine flake #2163 has four employed units, all showing use in planing of fresh wood. It is interesting to note that most of the tools with multiple EdUs show that the same actions were primarily associated with woodworking. One may attribute this to the longer duration of work required to carry out a single task on a hard substance, such as wood or bone, than on a softer substance like meat, skin, or soft

Table 6.32: Prehensile Wear in the Unretouched Flake Assemblages.

Prehensile Wear Type	GRAND BANKS		LONE PINE		LONE PINE	
	N	%	N	%	N	%
general prehensile wear	4	11.43	27	29.67	0	0.00
hand-held wear	23	65.71	36	39.56	3	100.00
haft wear	8	22.86	28	30.77	0	0.00
Total Employed Units	35	100.00	91	100.00	3	100.00

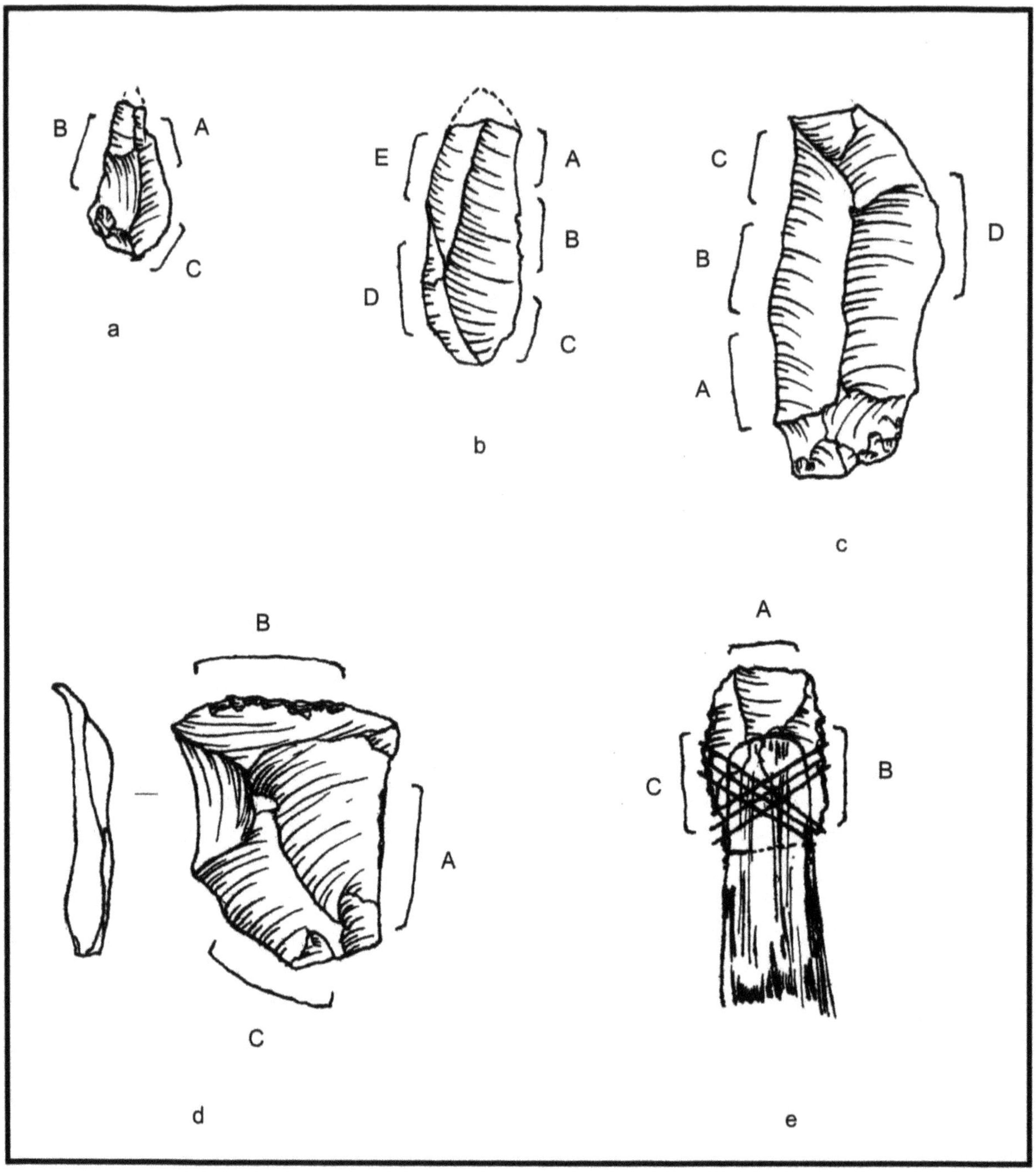

Fig. 6.12: Illustration of Utilized Unmodified-Flake; a: AfGx-3:645.1; b: AfGx-3:5353.1; c: AfGx-3:6886.1; d. AfGx-113:1252.1; e:AfGx-113.862.2.

plant material.

A small flake from Grand Banks (#645.2) possesses two cutting edges, A and B, which were assessed as having been employed for cutting fresh wood (Fig. 6.12a). Exhibiting little polish, the two edges feature a row of denticulated scarring with a rolled-over pattern. Due to the surfacial topography, most medium-sized feather scars are distributed on the dorsal side, although a few can still be observed on the ventral side. Moderate rounding is clearly indicated on the two employed units, even though they might be used lightly. A possible employed unit is detected at location C, where a few scar fractures may be attributed to hand grasping.

An unretouched blade from Grand Banks (#5353.1) was intensively employed in various use-tasks, possessing five employed units along its entire lateral margin except for the broken distal end (Fig. 6.12b). Four of the five EdUs were employed as scraping edges (A, C, D, E), while the EdU at location B is assessed as having been employed for cutting medium-soft wood (Photo 28-30). The wood-cutting edge exhibits features identical to those that we observed on the piece above (Grand Banks #645.2). Wear at locations C and E show similar patterns: heavily rounded edges and bright polishes on the contacted surface (ventral side in this case). The two employed units were possibly employed in scraping animal skins or hides. Use traces at location A are characterized by light rounding, incipient polish on the ventral side, and stepped and feather scarring. I have determined that this employed unit was used to scrape medium resistance animal substances such as dried hide or soft bone. A row of crushed scarring on edge D represents contact with some sort of hard material. Edge rounding and polish may indicate scraping of hard bone. Judging from the functional emphasis on scraping of animal substances, this piece was possibly used by a single user who might have carried out part of a butchering activity and used the tool mostly to fulfill the scraping requirements. This blade was also probably re-used and employed in different activities such as cutting/sawing wood by either the same user or a different user. The extensive use of such a tool is evidenced by the broken tip, which, as I observed, was caused by scraping activity either on A or E.

Another example of an individual tool presumably used by a single individual, is a large flake tool from Grand Banks (#6886.1). Transverse tool motions on fresh wood (1M) were detected on three EdUs on the left margin (Fig. 6.12c). While locations A and B were employed for planing wood (Photo 31-32), location C was used to scrape wood. Very bright polish can be seen on the ventral side, concentrated on the margin but moving towards the center of the piece. Unifacially distributed scarring on the ventral side exhibits rolled-over patterns. Differences in scar fractures among the three EdUs enabled me to distinguish planing activities from scraping activities. Planing scars at locations A and B are relatively larger and shallower than those caused by scraping at location C. Also, planing scars may possess more stepped terminations compared to the ill-defined, feather terminations found on scraping edges. A projecting part of the right margin was employed for holding the piece, indicating prehensile wear found at location D.

The next two examples illustrate the importance of prehensile wear in assessing the function of flake tools. A large flake from Lone Pine (#1252.1) possesses macro-fractures on its distal end. The distal end of this piece has a convex shape in a plain view (Fig. 6.12d). Due to this curvature, conventionally this piece would be catalogued as a "utilized flake," and assessed as a scraping tool with a working edge at location B. However, microscopic analysis shows that the distal end (B) was intentionally blunted by micro-retouch, likely to dull the edge for hand-holding (Photo 35). The actual working edge is at location A, the entire lateral edge of which was employed for cutting or slicing meat with bone contact (Photos 33-34). Prehensile wear is also found at location C, which was possibly pressed by the mid-finger while carrying out this butchering activity. The other flake from Lone Pine (#862.2) could have been mistakenly considered as a cutting tool, judging from the edge damage on the lateral margins (Fig. 6.12e). However, microscopic examination indicates evidence of fracture scarring on both lateral edges B and C attributable to hafting (Photo 36). This postulation is confirmed by the use traces on the distal end A, which was used for wood scraping.

Discussion and Summary

The data available do not make it possible to make precise cultural inferences regarding all aspects of our interest in the reconstruction of core reduction strategies. Through the examination of quantitative and qualitative parameters of unretouched flakes, however, I have been able to reach some conclusions regarding core reduction strategies and the utilization of flake tools.

Technologically, the reconstruction of core reduction strategies at the Princess Point sites, through the unretouched flakes analysis, is consistent with the implications derived from core analysis (Chapter 4). The Princess Point reduction strategy, represented by the "transformed" reduction sequence, contrasts markedly with the biface reduction strategy which was extensively used by earlier inhabitants (Paleo-Indian or Archaic). Biface production, which produced both flakes and, at the end of the sequence, a biface tool from the core "residue," is weakly witnessed on site. Bipolar production is not common, but the presence of bipolar cores and bipolar flakes indicates that there was at least a minimal use of this reduction strategy on-site.

It is determined that blades or some flakes at Grand Banks were likely produced as tool blanks during the early phase of the "transformed" reduction sequence. Although the occurrence of blades is much lower than amorphous flakes and biface thinning flakes, they were primarily removed through the careful preparation of core platforms. The restricted dimensions of blades, which differ from those of plain flakes, suggests that blades should not be considered accidental products at Grand Banks, although blade production did not exist.

Tool manufacture involves primary processing and blank production. Technological variability exists in blank production (production of blades, amorphous flakes, and bipolar flakes). I also observed that there is variability in the core reduction strategy as it has been reconstructed for Grand Banks and Lone Pine. Further, as far as utilized flakes are concerned, it appears that there are definite preferences in the sizes of flakes chosen for utilization.

The utilization of unretouched flakes at Grand Banks represents a pattern that differs slightly from those observed from the formed type and core assemblages. Utilized pieces in the latter assemblages were employed largely in woodworking, while unretouched flakes were employed largely in butchering and bone-working activities. In general, the utilized flakes were used in a much wider range of activities than the formed types and cores both in terms of tool motions and in terms of contact material.

Based on the above summary of core reduction strategy, two implications are apparent. Firstly, the amorphous reduction mode at Grand Banks does not represent an "expedient" technology. Secondly, use of transformed core reduction at Grand Banks may reasonably be considered as a raw material economizing strategy.

There has been a great deal made out of the relationship between core technology and settlement patterns on the North American continent (Johnson and Morrow 1987). Parry and Kelly (1987) clearly demonstrate the correlation between increased sedentism and a decline in the use of prepared cores at most archaeological sites in North America. Amorphous cores, as defined by Johnson (1987:2), are those that "show minimum platform preparation and little regard for setting up subsequent flake removals." These authors argue that the relationship between sedentism and amorphous core technology reflects the importance of labor costs versus material costs in lithic production systems. They suggest that, for most sedentary people, "curated" or portable stone tool-kits were no longer required, and that, instead, villagers produced simple flake tools only when needed (Parry and Kelly 1987). The residents of sedentary settlements can obtain large quantities of raw materials through exchange and direct access. "In these cases, it will not be necessary to spend the extra time and labor needed to produce prepared cores and standardized bifaces, even if the prepared core and biface technologies are more efficient in their use of raw materials" (Yerkes 1989:184). According to these authors, the amorphous core technology is thus regarded as time and labor efficient, or "expedient" technology.

However, the first problem that this model encountered is the availability of raw materials. If the "expedient" core technology was considered as a wasteful strategy in terms of raw material conservation, the use of such technology must rely on the ready availability of raw material. Parry and Kelly (1987:301) themselves also realize that highly mobile hunter-gatherers also made and used "expedient" flake tools in cases in which abundant raw material resources were available. In the regions where suitable lithic materials are difficult to obtain, both mobile and sedentary groups would adopt economized lithic production systems, that is, the prepared core or biface technologies, as the above authors argue. Therefore, the production of "expedient" flake tools is not exclusively linked to the lithic production of a sedentary population.

Secondly, stone tools produced by standardized core technology could also be used as "expedient" tools. Blade production, of course, may be considered a highly standardized core reduction technique (Nelson 1991:68). Within the area including Ohio, Illinois, and the lower Missouri River, Hopewell blades are present in a similar form, produced on conical cores by indirect percussion (Greber et al. 1981; Hofman 1987; Reid 1976; Sanger 1970; White 1963; Morrow 1988).

Yerkes (1983, 1994) and Odell (1985, 1994a) discovered interesting use patterns by applying high-power and low-power use-wear analysis, respectively, to examine these blade implements. Yerkes found, in examining three Ohio Hopewell assemblages (Murphy, LIC-79, and Marietta), that the blades were used for a variety of tasks, and on materials ranging from plants to stone. Further, he proposed that given evidence of weakly-developed microwear traces on most such tools, these blades were used as *expedient tools* (Yerkes 1994:123). Odell (1994a) compared three assemblages from the Smiling Dan settlement and two Napoleon Hollow mortuary sites of southern Illinois, and concluded that different uses of blades were apparent: at the habitation site, blades were used as generalized tools, whereas at mortuary sites they were likely only used for meat cutting or scraping, presumably in ritual ceremonies. These observations suggest that blades, which were produced using standardized core technologies, could also be used as expedient tools.

This study of the Grand Banks lithic industry demonstrates that the application of flake-producing core reduction might not be expedient. Although the production of amorphous flakes was prevalent at the Grand Banks site, the making and use of flake tools first required a careful procedure of blank production and selection.

The Grand Banks lithic assemblages present relatively higher proportions of amorphous flakes and cores, however. One explanation of this is that the amorphous cores and flakes found at the Princess Point sites represent discarded or end products. Since only a certain amount of small core nodules or reduced cores were brought back to site, people needed to maximize flake production on-site. The amorphous core reduction mode at Grand Banks represents the late phase of the "transformed" reduction sequence, necessary to obtain whatever they could produce from the last striking. Thus, the amorphous core technology at Grand Banks likely transformed the prepared core "residues" into amorphous cores. This resulted in predominantly amorphous cores and flakes, left at the site as end products. From this point of view, Patterson (1987:51) correctly observe that amorphous core reduction may be related to situations in which the population did not want specialized stone tools (e.g., bifacial tools or blade tools) and occurred at places where there were limitations regarding raw material sizes, shapes, or quality.

In their study of the lithic industries of the Keatley Creek site in the Interior Plateau of British Columbia, Hayden and his colleagues demonstrate that the amorphous core technology represents an economizing strategy and is "the most efficient use of raw material in terms of procurement, reduction, and the employment of minimal amounts of raw material in any given task" (Hayden et al. 1996:38). Among the variables that they used to indicate such economizing behavior, They described mentioned two significant factors: (1) the remarkably small size of many of the flake tools and cores and (2) the high rate of breakage.

The same economizing behavior, employing the amorphous core technology, is recognized at prehistoric sites in the Lower Illinois Valley (Odell 1996a:202-209, 1989b). Odell (1989b) suggests that the breakage pattern of flakes could be the best indicator of prehistoric tool economizing behavior. The reason for this, as Odell points out, is that "if tasks such as cutting and scraping are necessary but the stone for them is lacking, then one obvious strategy is to fashion new tools out of whatever is currently available (i.e., cores, old tools, debris, etc.) by splintering them further" (Odell 1996a:202). If this is correct, the data from Grand Banks point to a similar pattern of economizing raw material.

Metric data presented earlier suggest that most of the various reduction flakes from Grand Banks are small. The average lengths of complete flakes range from 20 to 30 mm. Flakes, even primary flakes and core trimming flakes, in excess of 50 mm in length are very rare. The two common flake tool types, scrapers and retouched pieces, have an average length of about 27 cm. This may imply either that most flakes are removed from small cores or that large flakes could have been reshaped into small flake tools if they were broken during initial use.

The breakage rate of various reduction flakes at Grand Banks is presented in Table 6.1 and Figure 6.13. It makes sense that the relatively high breakage rates are found only in plain flake (60.6%) and blade (31.1%) categories since these two types of flakes are found to be the most likely tool blanks. In contrast, very low rates of breakage are found in primary flakes (13%) and core trimming flakes (0%) because the two types would have been discarded during the initial processing prior to blank production. It also makes sense that only 16.4% of bifacial reduction flakes are fragmentary since only a small proportion of this type of flake would have been selected as tool blanks. Only nine bipolar flakes are identified and all are complete. The identification of classic bipolar flakes is based on the overall characteristics of bipolar reduction and thus only complete ones can be recognized with confidence. It should be noted, of course, that there are some other agencies (human trampling, the flintknapping process, etc.) that could have caused the breakage of the flakes as well. Bearing this in mind, the breakage pattern points to the possibly intensive use of flake tools at Grand Banks in order to economize on raw materials. That is, one way to maximize the utilization of raw materials on-site is to repeatedly use still-useable flake tools. The more often flake tools were repeatedly used, the higher the chance that they could have been broken.

Furthermore, the small size and the breakage of cores at Grand Banks are also indications of economization of raw material. Almost all of the amorphous cores at the site were used to produce flakes until the exhaustion of cores. The exhausted cores, as indicated by the small size of core facets and by their battered platforms, suggests that the tool makers at Grand Banks could not afford to throw away their cores before the cores became entirely useless. After these cores had become unusable for producing free-held flakes, they could have been smashed into bits to retrieve every last morsel of usable tool material. The relatively high frequency (55.3%) and small sizes of many core fragments in the Grand Banks core

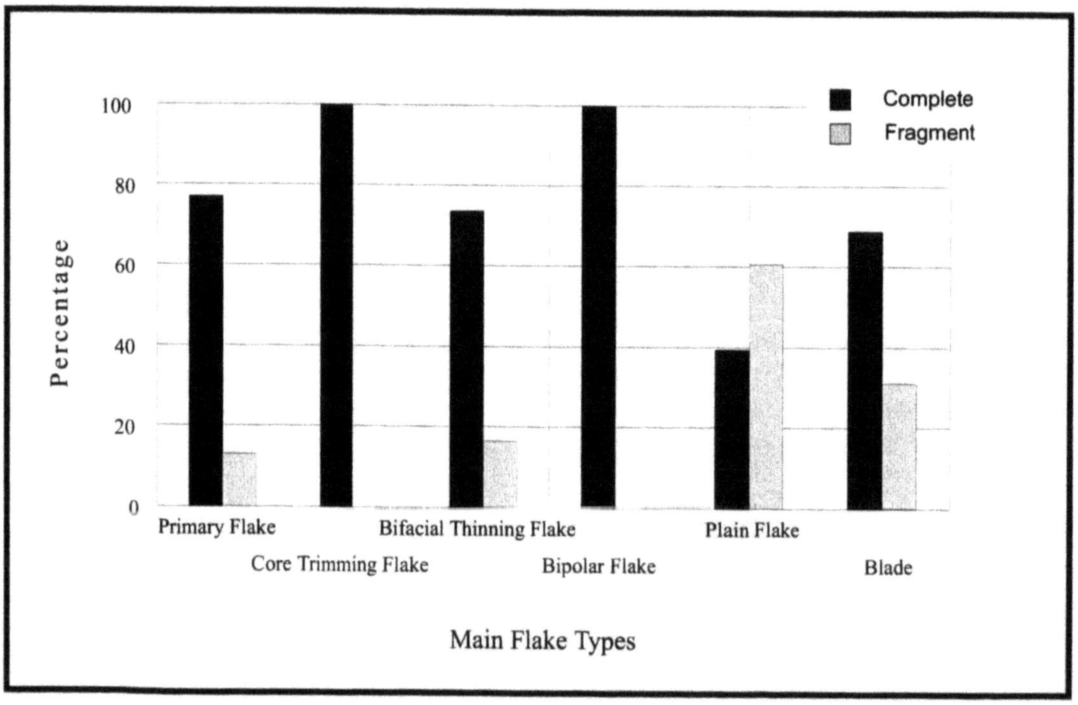

Fig. 6.13: Breakage Rate of Grand Banks Flakes.

assemblage points to such an economizing scenario. The use-wear analysis demonstrates that such small core fragments were often directly used as tools (Chapter 4).

Traditionally, it is assumed that bifacial or blade production represents an economizing reduction strategy since bifaces could have served as multipurpose tools and might, therefore, have been regarded as "curated" or durable tools (Kelly 1988; Shott 1986, 1989). Moreover, because blades can provide elongated employable units, they have a high ratio of usable edge to total material (Nelson 1991; Clark 1987; Hofman 1987; MacDonald 1968; Sheets 1978; Morrow 1987). Why did people at Grand Banks choose amorphous flake production to economize on raw materials, instead of bifacial or blade production? There are probably two explanations for this. First, for some tool makers blade or bifacial production could have been wasteful (Hayden et al. 1996). Hayden et al. (1996) propose that most prepared cores, such as Levallois cores and especially blade cores, are actually wasteful of raw material because of the high risk of failure at all stages, the initial need to shape cores, and the need for specific sizes, shapes, and high quality raw materials. Analysis of cores (Chapter 4) demonstrates that most prepared cores at Grand Banks are not found in their exhausted conditions. By contrast, the amorphous core technology would have provided maximum flexibility in terms of the production of different sizes and shapes of blanks for most flake tools, including scrapers, retouched pieces, notches, and drills (Hayden et al. 1996:37).

A second explanation is that during the Princess Point period, the amorphous core technology at Grand Banks may have been employed as a form of generalized tool production. With this strategy of stone tool production, various simplified and unstandardized flake tools were produced and used in a wide range of use-tasks as a response to a series of social changes during the transition to food production. I will return to this later.

Whichever was the case, the amorphous core reduction at Grand Banks most likely represents an economizing behavior. I do not intend to refute the recognition that amorphous core technology can be a wasteful raw material strategy. Discussion of economizing behaviors should be done case-by-case. Naturally, amorphous core reduction could have been employed in resource-rich areas as an "expedient" technology in a general sense (Parry and Kelly 1987; Johnson 1989). What I suggest here, however, is that other inferences, and the significance of this core technology, should not be overlooked. Amorphous core technology should never be subjectively considered as exclusively "wasteful" and/or "expedient." Fortunately, the notion that amorphous core technology may actually represent an economizing behavior has been gradually accepted by some of the above mentioned authors. It would be significant in any further study to test this hypothesis by considering the overall lithic production (which should be studied along with refitting techniques), rather than core morphology alone. This undoubtedly would provide insights into our understanding of core technology, settlement organization, and economizing behavior.

CHAPTER 7

ANALYSIS OF FLAKING DEBRIS

There are two objectives in the aspect of lithic analysis in this chapter: (1) to present spatial distributions of flaking debris, from which (2) a spatial pattern of lithic production loci at Grand Banks can be derived and interpreted.

As stated in Chapter 3, flaking debris is defined as the waste of flaked stone production. Waste products are identified on the basis of two criteria: (a) their extremely small size, which makes utilization or further modification impossible, and (b) their angular or irregular shapes resulted from flintknapping blowoffs. In order to verify that flaking debris were in fact discarded as unusable items, I selected a small sample of flaking debris and examined these pieces under the microscope for use-wear. Out of 56 pieces examined, only two show ambiguous traces of edge fractures and light polish. These are not significant enough for me to assess them according to any use-wear categories but indeterminate ones. It should be noted that fragments of unretouched flakes may fall into the defined dimensions of flaking debris (<10 mm in length or width). However, during our sorting process we carefully excluded unretouched flake fragments from flaking debris unless no clear distinction between flaking debris and such small flake fragments could be drawn.

Flaking debris makes up the majority of the Grand Banks lithic assemblage (63.6%) (Table 4.1). More than 11,000 pieces of flaking debris were recovered from 35 m^2 excavation areas. Such higher numbers are in great part attributable to the fact that flotation was employed in the recovery of a significant portion of the artifact assemblage. Chips smaller than 2 mm in size were frequently collected from flotation soil samples. It was observed during our cataloguing process that smaller chips (< 5 mm in size) are predominant although no systematic size distributions of the flaking debris have been recorded due to time constraints. By contrast, the mass of flaking debris by weight accounts for 17.7% of total mass of the Grand Banks lithic assemblage, and nearly 25% of both the Lone Pine and Young 1 assemblages (Table 4.2). The average mass of flaking debris at Grand Banks is 0.12g per piece, pointing to the very small size of flaking debris.

Given the nature of flaking debris, no formal analysis is necessary for individual pieces. However, the spatial patterning of flaking debris may be informative. It is believed, on the basis of flintknapping experiments, that flaking debris is refuse resulting from any type of reduction mode, or from any stage of core reduction. However, most chips were likely produced from pressure flaking during biface production or tool modification (e.g., resharpening), whereas chunks are byproducts produced from hard- or soft-hammer percussion modes. These small pieces, mostly found in soil samples, were unlikely to be re-deposited post-depositionally so that their distributions may reflect the spatial patterning of lithic production. Although unretouched flakes could have been re-deposited on site during the process of modification and utilization, their spatial association with flaking debris may also be informative when reconstructing the lithic production process. As our initial investigation of the Grand Banks site formation process indicates the stability of the floodplain site during the Princess Point occupation, the spatial analysis I present here will be based on the assumption that secondary deposition of flaked stone artifacts has been minimal. Therefore, the following presentation focuses on data of flaking debris from two excavation areas of the Grand Banks site, as well as the association of debris with unretouched flakes. A complete list of the flaked stone artifact distribution by excavation square is presented in Table 7.1.

Table 7.1: Distribution of Grand Banks Lithic Artifact by Excavation Squares.

Squares	Formed Types	Cores	Unretouched Flakes								Flaking Debris		
			Primary Flake	Core Trimming Flake	Biface Thinning Flake	Bipolar Flake	Plain Flake	Blade/Bladelet	Burin Spall	**Sub-Total**	Chip	Chunk	**Sub-Total**
Area A													
725-669	**5**	-	1	-	43	-	66	4	-	**114**	297	28	**325**
725-670	**7**	**1**	-	3	40	-	171	5	-	**219**	269	22	**291**
725-671	-	**1**	4	1	22	2	261	6	-	**296**	342	18	**360**
726-669	**6**	**1**	6	1	15	-	121	5	-	**148**	233	17	**250**
726-670	**5**	**3**	-	1	13	-	201	2	1	**218**	212	26	**238**
726-671	**6**	**4**	3	1	11	-	188	2	-	**205**	203	7	**210**
727-669	**5**	**5**	3	2	32	1	283	7	-	**328**	539	61	**600**
727-670	**6**	**4**	7	3	36	1	356	15	-	**418**	586	39	**625**
727-671	**11**	**3**	3	1	23	-	194	21	-	**242**	339	34	**373**
728-669	**9**	**3**	3	2	58	-	282	12	-	**357**	944	87	**1031**
728-670	**9**	**3**	4	3	39	-	222	11	1	**280**	401	40	**441**
728-671	**7**	-	3	1	36	-	220	10	-	**270**	419	55	**474**
729-669	**2**	**1**	3	1	14	1	109	9	-	**137**	215	34	**249**
729-670	**4**	-	-	-	59	-	145	-	-	**204**	398	24	**422**
729-671	**10**	**2**	1	-	100	-	268	2	-	**371**	387	34	**421**
730-669	**1**	**2**	-	-	21	1	95	8	-	**125**	299	30	**329**
730-670	**7**	**3**	4	4	26	-	202	12	-	**248**	189	38	**227**
730-671	**3**	-	3	2	16	-	134	9	-	**164**	85	12	**97**
Area B													
729-680	**2**	**2**	1	1	10	-	40	-	-	**52**	46	1	**47**
729-681	**5**	-	3	1	46	-	108	2	1	**161**	241	29	**270**
729-682	**2**	**1**	2	-	4	-	58	5	-	**69**	48	13	**61**
729-683	**1**	-	-	1	2	-	26	1	-	**30**	32	3	**35**
729-684	**2**	**1**	-	-	0	1	31	4	-	**36**	42	4	**46**
729-685	**3**	**1**	1	2	1	-	38	7	-	**49**	6	5	**11**
729-686	**2**	-	-	1	3	1	13	1	-	**19**	21	0	**21**
730-680	**7**	**2**	2	1	14	-	292	8	-	**317**	28	5	**33**
730-681	**4**	**1**	1	1	10	-	83	1	-	**96**	69	12	**81**
730-682	**2**	-	-	-	4	-	43	3	-	**50**	16	5	**21**
730-683	-	**1**	-	-	2	1	24	1	-	**28**	27	4	**31**
730-684	-	**1**	1	-	16	-	90	-	-	**107**	102	3	**105**
730-685	**4**	-	1	-	29	-	101	1	-	**132**	313	23	**336**
730-686	-	-	1	1	63	-	165	2	-	**232**	325	35	**360**
730-687	**2**	**1**	1	-	32	-	138	1	-	**172**	334	46	**380**
730-688	-	-	-	-	1	-	60	-	-	**61**	107	12	**119**
730-689	**5**	-	1	1	27	-	162	6	-	**197**	197	47	**244**

Spatial Distribution of Flaking Debris

Flaking debris counts are used for the examination of spatial patterning. Because the cultural deposits of Area A and B where flaking debris were recovered are about equal in thickness (10cm), counts are approximately proportional to densities. Due to the different nature of chips and chunks, I first present their spatial distributions by counts, consecutively (Fig. 7.1 & 7.2). They both illustrate a similar pattern with slight differentiation. Area A has a complete coverage of flaking debris, displaying higher density in its center part. It appears that for both types of flaking debris, the highest density is located in one southern square of the area, where most waste products were concentrated. This square (728-669), of course, is surrounded by other squares with relatively high occurrences of both chips and chunks. However, chips are also concentrated on the western side of the area (Fig. 7.1), whereas chunks are concentrated in the southeastern corner of Area A (Fig. 7.2).

In Area B, the spatial pattern is clearly presented. In general, waste products have a lower concentration in Area B than in Area A. They are unevenly distributed in the southern and northern areas, separated by a zone of low density. However, chunks have a relatively higher concentration in the northern part, although one square (730-688) presents a significant decrease of chunk presence between the two higher density squares.

Distribution of the two kinds of debris combined is illustrated in Figure 7.3. From this configuration, six localities where there are concentrations of flaking debris are clearly shown. This spatial distribution may indicate a possible spatial patterning of on-site lithic production since the definition and the recovery method of flaking debris reinforce the notion that these waste products are lithic production micro-refuse whose presence *in situ* would provide a rational pattern of lithic production loci on-site. Therefore, I define six lithic production loci at Grand Banks, as shown on Figure 7.3.

Spatial Pattern of Lithic Production

As compared to Area B, Area A exhibits a higher probability of being an area of major lithic production. The three production loci in Area A (A1, A2, and A3) are not separated by clear boundaries. Locus A2 is the largest and has the highest density of production refuse. I infer that this area was used to a greater degree during core reduction and tool production. This could mean that the area was used for a longer duration or that it was used more intensively.

By contrast, Area B does not appear to have been an area of major lithic production although 3 loci may be defined in Area B. These are small compared to the loci in Area A, and are clearly separated from one another. Based on this, I infer that Area B does not represent an overall production sector of the settlement, and that these production loci represent places which were probably occupied for a short period during lithic production.

This spatial patterning of lithic production inferred from data regarding flaking debris is confirmed by the distribution of unretouched flakes (Fig. 7.4). The unretouched flakes distribution is similar to that of flaking debris in Area B; the three production loci encompass the areas of the highest density of unretouched flakes. Some differences may be noted at Area A. Unretouched flakes are not well presented at Locus A3 and are only concentrated in the northern part of Locus A1. The highest density of unretouched flakes is found at Locus A2 where the highest density of debris is also recognized.

Bearing in mind that unretouched flakes may possibly be removed from their original location, it is interesting to note that their spatial patterns are generally correlated with the distribution of flaking debris. These production loci derived from spatial distribution of flaking debris, therefore, can be rationally accepted.

I will now examine these loci on the basis of the types of unretouched flakes in order to generate probable functions for each production locus. A detailed distribution for each type of unretouched flake is listed in Table 7.1. Given the massive

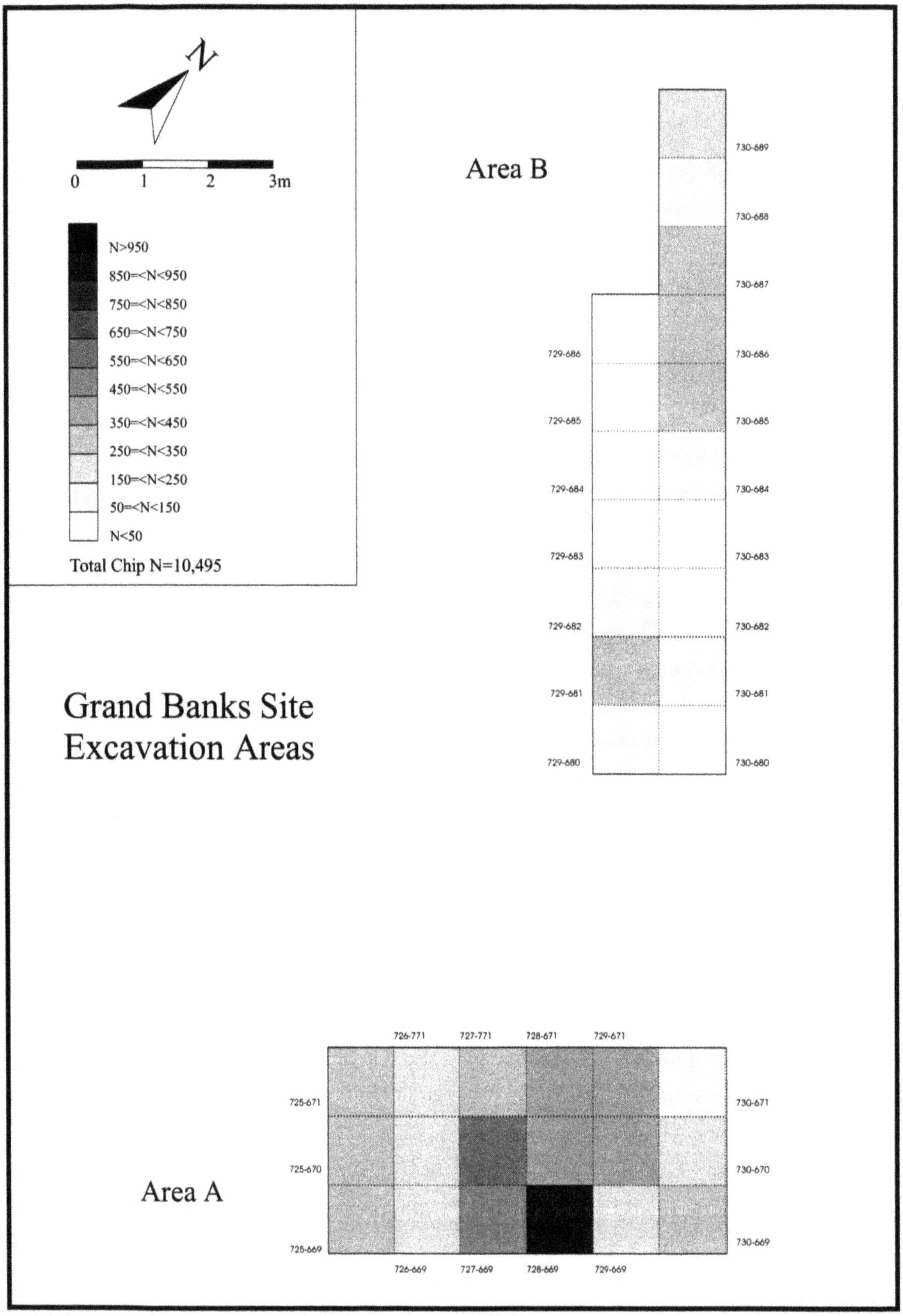

Fig. 7.1: Spatial Distribution of Chips.

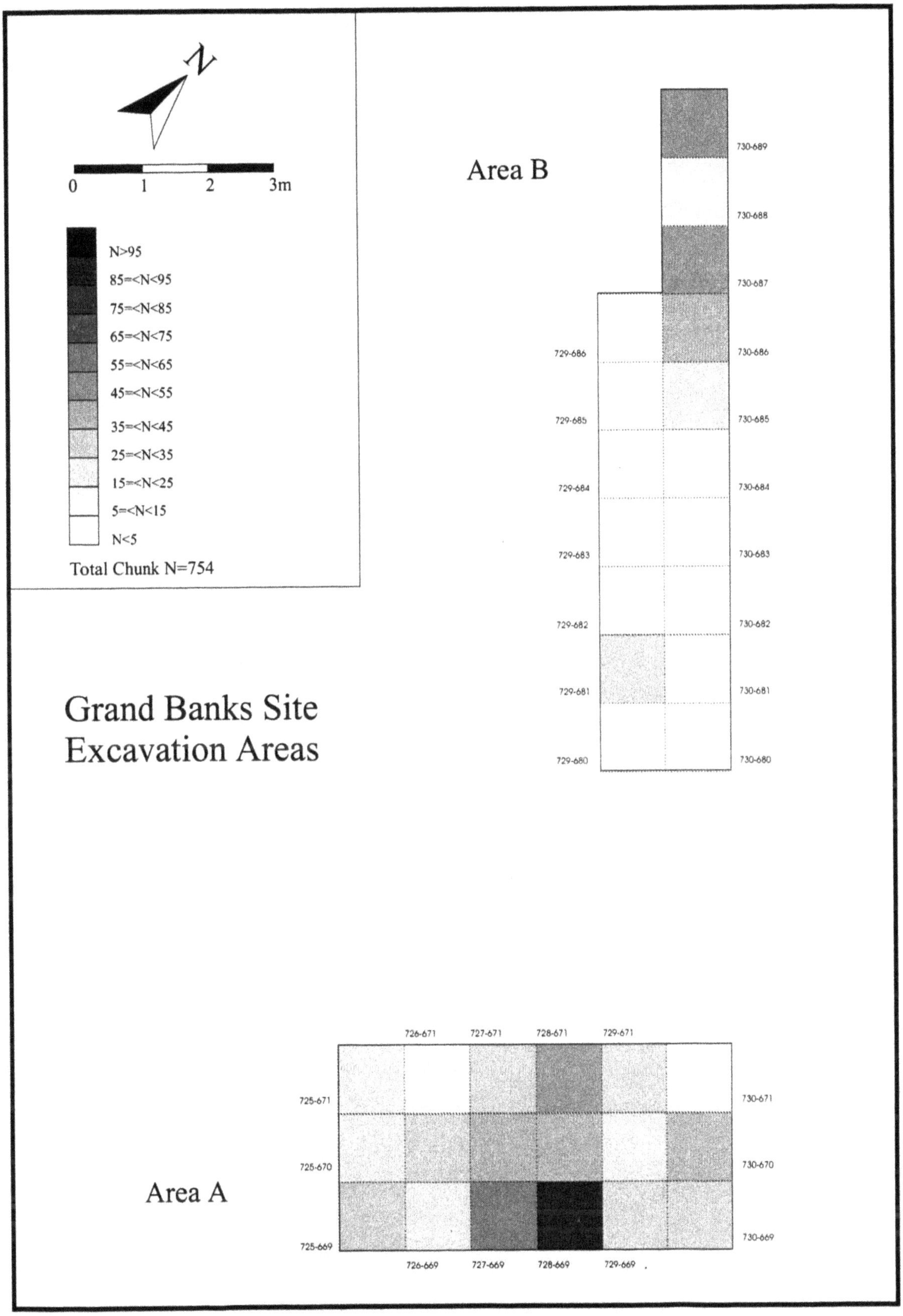

Fig. 7.2: Spatial Distribution of Chunks.

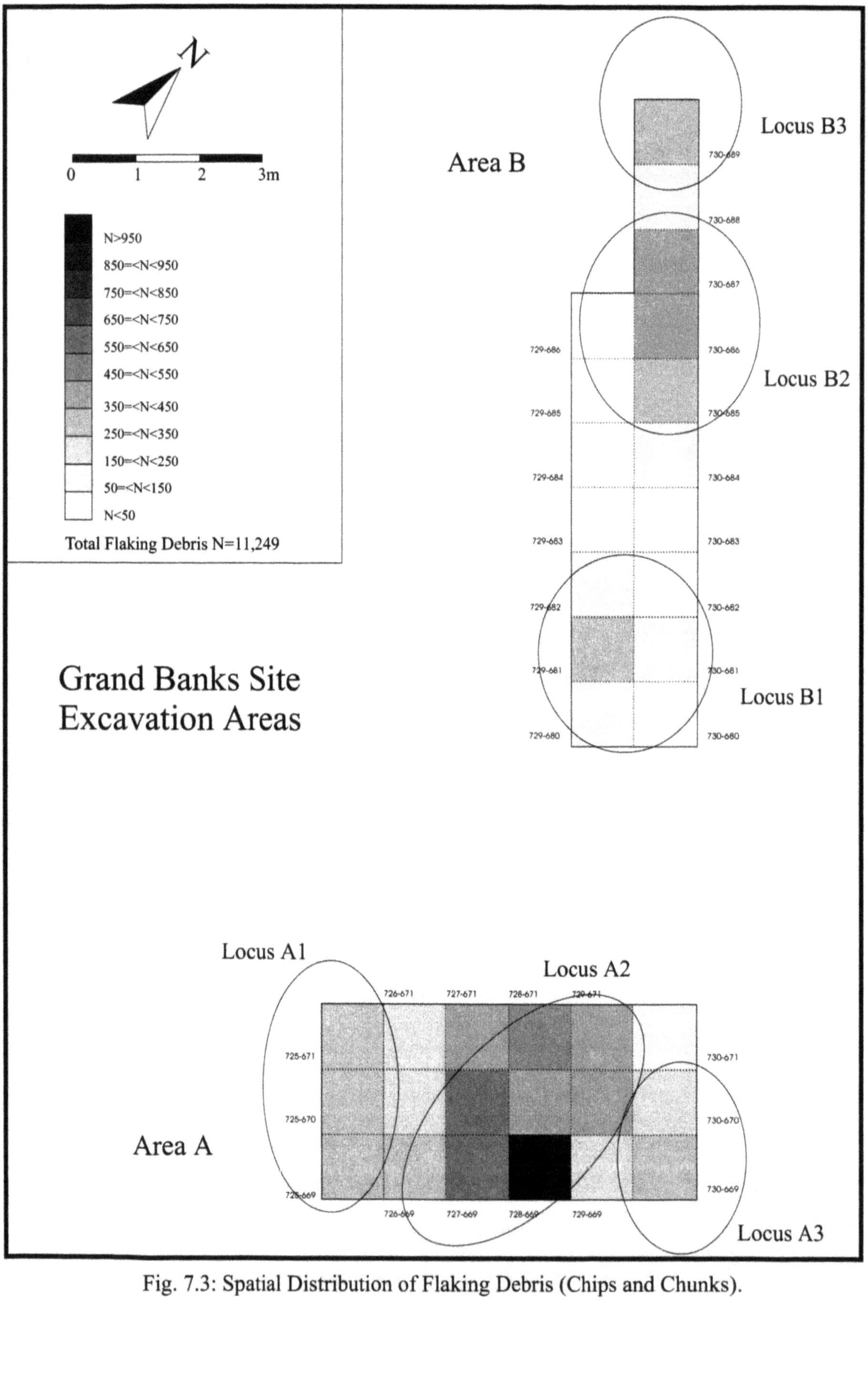

Fig. 7.3: Spatial Distribution of Flaking Debris (Chips and Chunks).

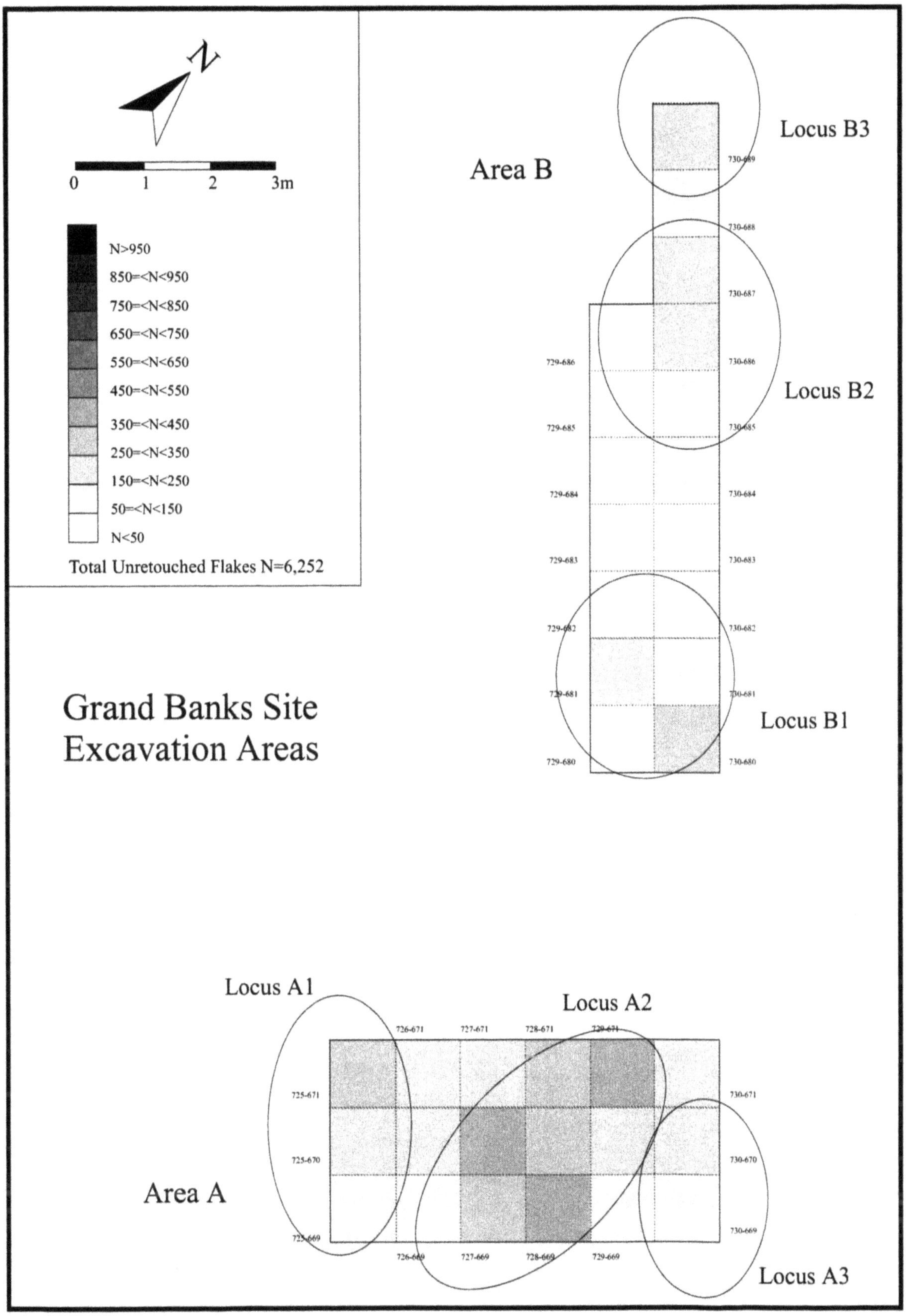

Fig. 7.4: Spatial Distribution of Unretouched Flakes.

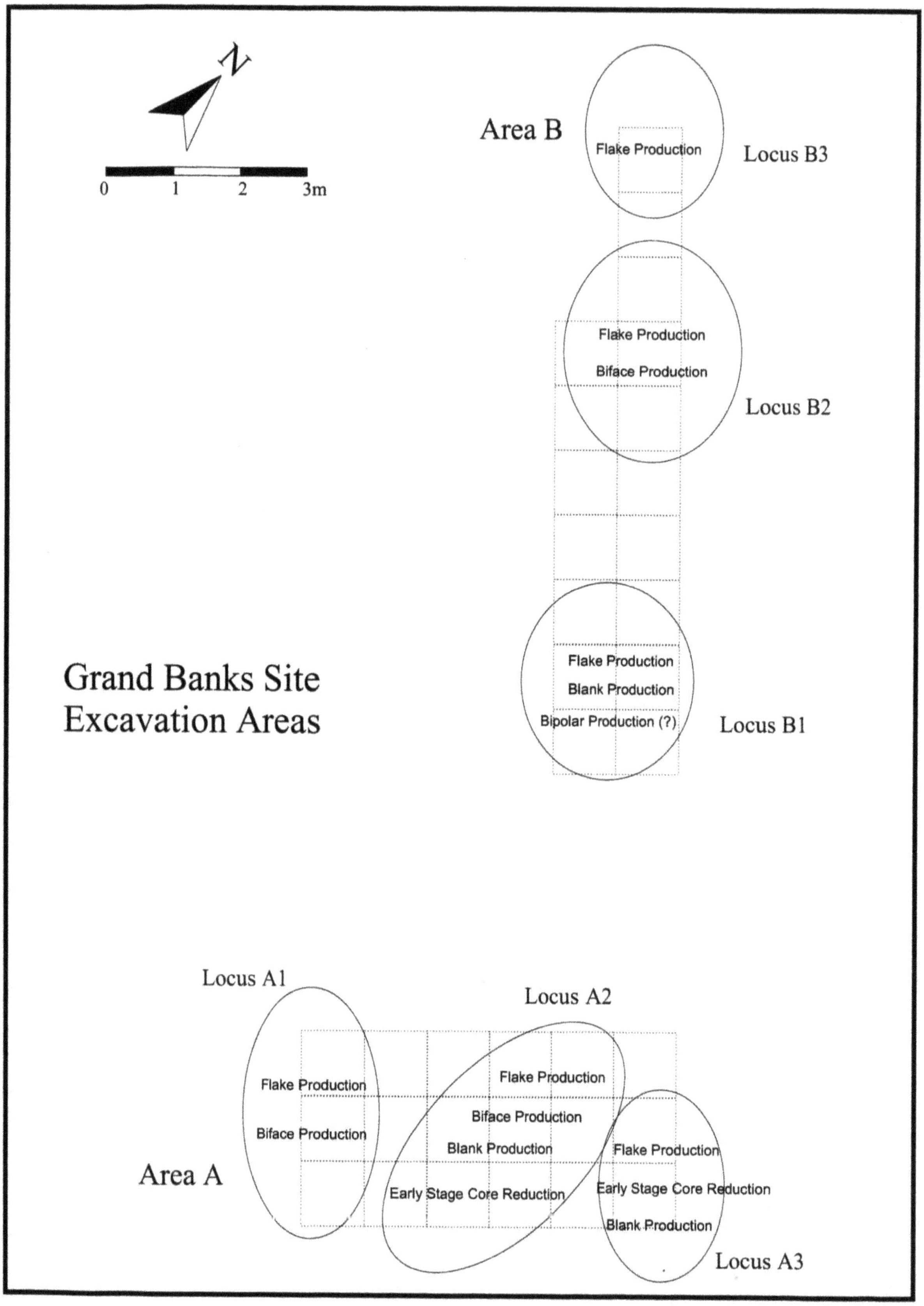

Fig. 7.5: Spatial Pattern of Lithic Production at Grand Banks.

presence of plain flakes, the distributions are in accordance with the general spatial patterns of unretouched flakes shown on Fig. 7.4. On this basis, we can assume all six production loci represent places for amorphous flake production or tool blank production.

Primary flakes and core trimming flakes are mainly concentrated at Locus A2 and at Locus A3. A few of them are present at the margin of area between Loci A1 and A2. These products of core reduction are rarely, if ever, present at Area B loci (Table 7.1). As these two kinds of unretouched flakes are considered to result from the early stages of core reduction, it is reasonable to regard A1 and A2 as production loci where the early stage production took place. Such activities are rarely seen at Area B.

Bifacial thinning flakes are predominant at Loci A1 and A3, and especially at Locus B2 (Table 7.1). At Locus B2, plain flakes have a comparatively lower incidence, with lower presence of cores and the occurrence of some bifacial tools. Locus B2, therefore, can be referred to as a production place where, probably, the later stage of bifacial tool production took place. The same holds true for Locus A1 where a clear decrease of plain flakes is observed and a few cores and some bifacial tools are present.

While the population of blades and bladelets is small, an uneven distribution in their spatial patterning can still be seen. Blades are primarily associated with flaking debris at Loci B1, A3, and A2. Interestingly, the two prepared cores are also found within Locus B1, while the majority of prepared cores are in Locus A2. This may indicate that blades or blade-like blanks were produced, although rarely, at Loci A2, A3, and B1. Such production loci could be interpreted as onsite blank production areas.

Given the unrepresentative data on bipolar flakes, the spatial patterning of bipolar reduction connot be reconstructed. However, four out of five bipolar cores were found at Locus B1, presumably indicating the concentration of bipolar production at the southern part of Area B.

According to the above interpretations, the spatial pattern of lithic production is reconstructed as shown on Fig. 7.5. To sum up, Locus A2 is regarded as the main production area, where almost all the processes involved in core reduction and tool production were carried out. Flake production, as a predominant core reduction strategy at Grand Banks, took place at all loci. Some early stage reduction processes may have been carried out in the southern part of Area A. Biface production took place mainly at Area A, but some also in the central part of Area B. In the central southern to eastern part of Area A and the southern part of Area B, some blade-like flakes were removed through prepared core reduction. There is a possibility that the southern part of Area A might also have been used for bipolar production.

It may be argued that this spatial pattern of lithic production is intuitive. This model of spatial patterning of production, however, is rigorously based on flaking debris distribution. However, because of the limited excavation area, the production clusters inferred above may well be affected by edge effects. Area B, for instance, could represent the margins of another main production locus. In addition, it may be argued that the three production loci in Area A should not be separated because the artifact densities in the bounding squares are relatively higher as compared to that of Area B. It would not be surprising that the six inferred clusters would change shape considerably, or that some of them could disappear, if future exacavations continue exposing surrounding areas. With these problems in mind, this spatial pattern of lithic production is preliminary. Nevertheless, the results, based on data available to date, suggest that there is indeed potential patterning of lithic production at the Grand Banks site. At least production activity in Area A is different from that in Area B. Although this spatial pattern may be altered through integration of new data, this picture provides a preliminary understanding of the lithic production and a guidance for our future fieldwork.

CHAPTER 8

DISCUSSIONS AND COMPARISONS

The previous four chapters have presented analytical results from each class of flaked stone artifacts. The lithic assemblages have been characterized from a number of perspectives in an effort to describe the Grand Banks lithic industry. In this chapter, I will first consider the variability of tool use patterning at the Grand Banks site through an intrasite comparison. A second objective of this chapter is to provide an intersite comparison in order to explore the transformation of lithic production during the Princess Point period.

In this study, tool use is regarded as a component of the lithic production system (Chapter 1). In the previous chapters, I have made general intersite and the intrasite comparisons of aspects of lithic typo-technology. Thus, the intrasite comparison at Grand Banks presented here will focus on stone tool use. The same emphasis will also be given to the intersite comparison. The results of use-wear analysis of a lithic assemblage from the HH site will also be summarized in this chapter.

Use-Tasks and Use-Related Activity

Clarification of the terminology used in the interpretations of stone tool use is needed. Most use-wear analysts advocate the use of the terms "activity," "action," "woodworking," or "butchering," etc., to interpret tool use as derived from observations of use traces. As I mentioned earlier, activities such as woodworking or bone/antler-working represent a series of components of human behavior. A tool used for slicing soft animal substances, as determined by microscopic analysis, might also have been employed in bone-working-related activities such as cleaning bone. Similarly, it is impossible to link a tool that has been worked on wood to a specific activity; for example, making a wooden spear as tool-making or making wood-craft artifacts as household manufacture. In short, there are serious limitations applying use-wear results to specify or determine dynamic activities in the past.

However, this certainly does not mean that interpretations of tool use cannot go beyond presentations of "tool motion" and "contact material." At present, careful microscopic analysis of use traces on stone tools can ascertain the physical action used (tool motion) on the worked object (contact material). The tool use-task has been defined to assist an understanding of such use-related activities. A *use-task* is an action directly involved between a specific tool and worked object. *Use-related activity* represents a series of similar use-tasks for a possibly similar purpose. With this definition of use-related activity in mind, the defined terms below have specific meanings only, rather than serving as broad definitions of activity. The category of "butchering/meat-preparation" is inferred from use-tasks related to actions executed on soft to medium animal substances. Unfortunately, hide-working is difficult to distinguish from this category and is therefore included within "butchering/meat-preparation" use-related activity. "Woodworking" is limited to any actions on wood materials, and "bone/antler-working" to any actions on bone or antler materials as determined by microscopic examination. "Plant-working" refers to use-tasks involving plant substances.

Assuming an understanding of activity dynamics, this terminology is used in the following discussion to help in the comparison of general tool use patterns among the lithic assemblages. These use-related activities, as combined use-tasks, are used to draw a picture of how a cluster of flaked stone tools have been employed in a particular lithic production system. However, it should be noted again, at present the interpretation of use-tasks through the microscopic examination of flaked stone tools is still subject to error. Therefore, the discussion of flaked stone tool use pattern below should be considered as an exploratory exercise, rather then as a conclusive result.

Tool Use at Grand Banks: an Intrasite Comparison

In this section, I combine all the utilized pieces from formed types, cores, and unretouched flakes, and refer to them by the general term "tools." One hundred and forty-nine worn tools out of nearly 1,000 samples yield a total of 230 employed units. Interestingly, two areas possess about equal numbers of employed units (116 EdUs in Area A and 114 EdUs in Area B). However, given a much lower density of artifacts in Area B (Chapter 7), tool use at this part of the site should be considered to be relatively intensive.

A series of drawings are presented here to illustrate the variation between the two areas. In general, tool motions present a similar pattern between the two areas although slight differences in tool use are observed (Fig. 8.1). Tools were commonly used for scraping/shaving and cutting/sawing at both areas. Scraping/shaving tool motions are more frequently present in Area A than in Area B, whereas cutting/sawing motions display about equal frequencies in both areas (Fig. 8.1). Area B has a greater emphasis on slicing/carving than Area A, perhaps implying that Area B possesses more tools used for butchering/meat-preparation use-related activities. Planing/whittling tool motions were used more often in Area B than in Area A. However, projecting/penetrating and graving tool motions took place more often in Area A than in Area B. Finally hoeing/digging and picking tool motions are found only in Area A, while chopping and wedging are seen only in Area B.

The next step was to compare tool motions in forms of more general categories. Figure 8.2 clearly shows that the predominant tool motions at both areas are transverse motions, secondarily followed by longitudinal motions. Tipping tools, used for drilling, projection, and graving, are also present at both areas in moderate frequencies. The dissimilarity of tool use between the two areas is also reflected in this graphic. The majority of transverse motion tools are found in Area

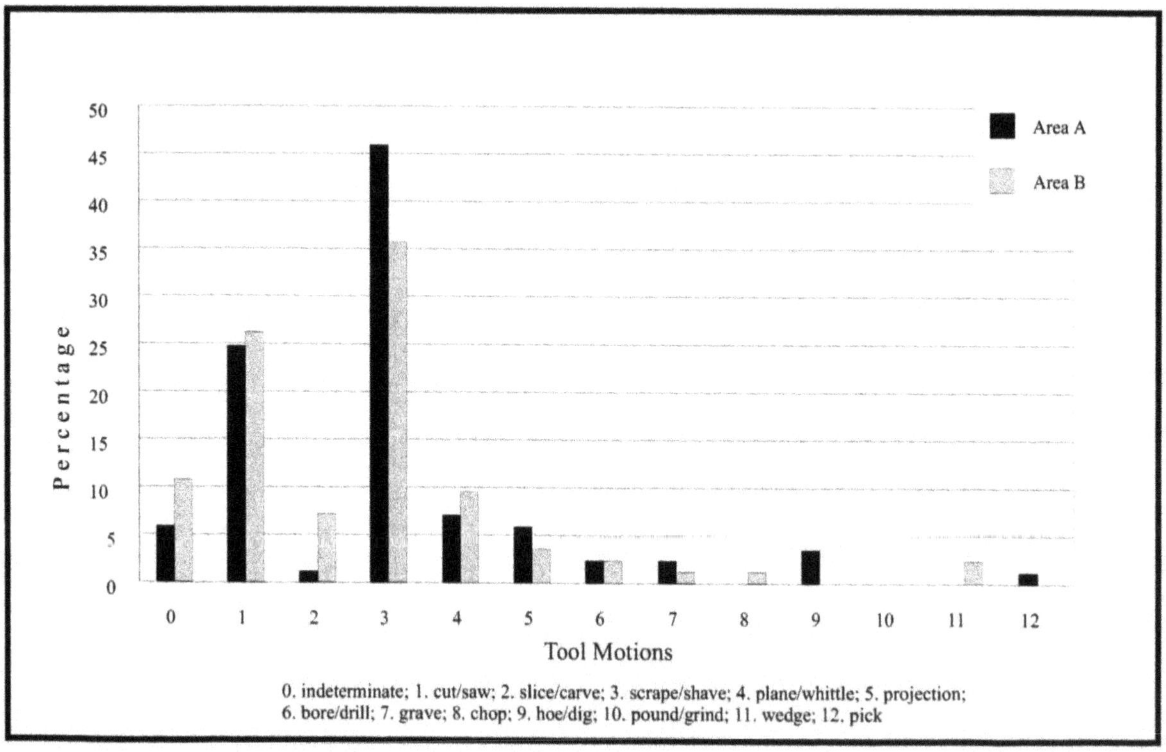

Fig. 8.1: Detailed Tool Motion Comparisons of Tool Use between Area A and Area B at Grand Banks.

A, whereas more longitudinal and tipping motion tools are found in Area B than in Area A. Although other combined tool motions (chopping, hoeing/digging, wedging, picking) are essentially equally present at both areas, individually each particular tool motion is strongly present in each of the two areas (refer to Fig. 8.1).

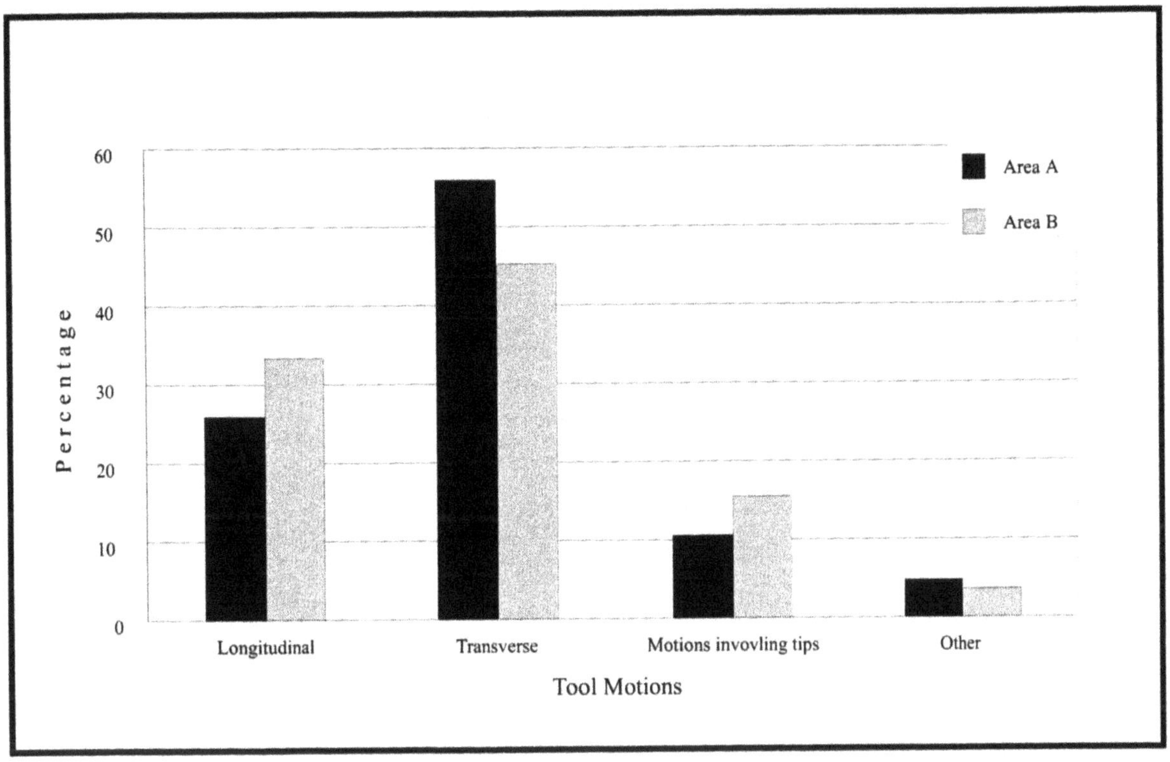

Fig. 8.2: General Tool Motion Comparisons of Tool Use between Area A and Area B at Grand Banks.

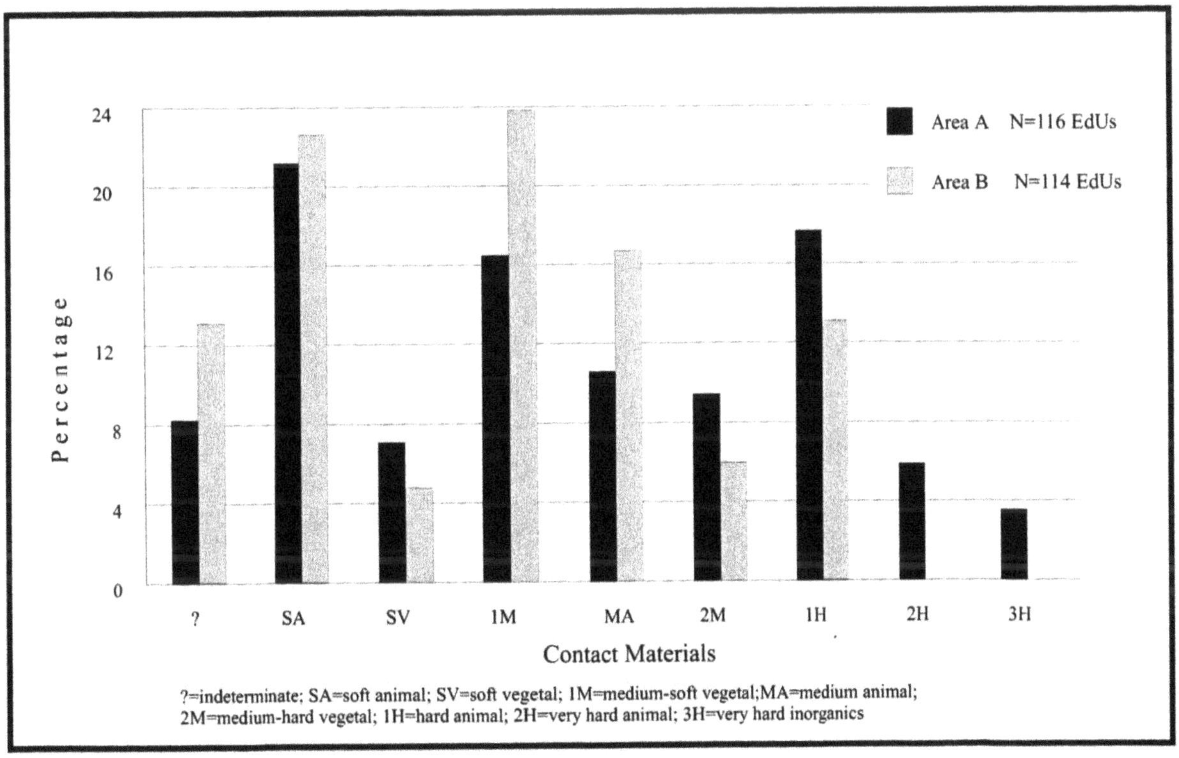

Fig. 8.3: Contact Material Comparisons of Tool Use between Area A and Area B at Grand Banks.

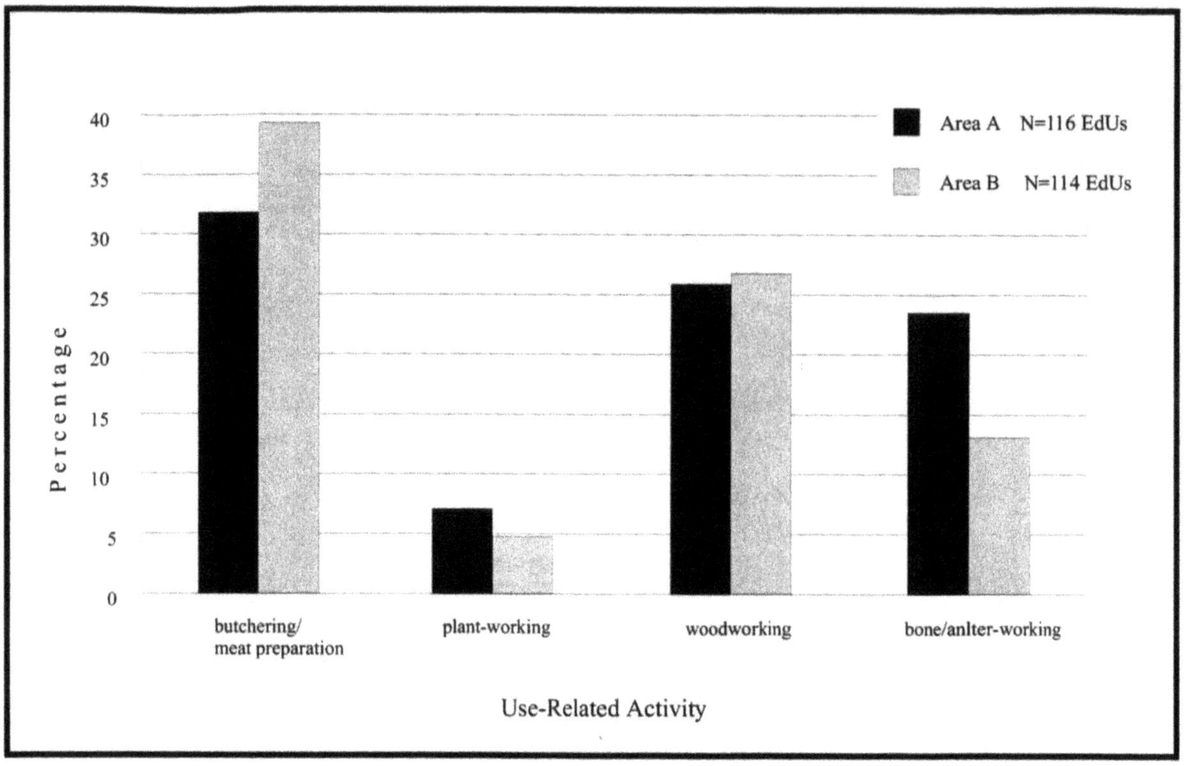

Fig. 8.4: Use-Related Activity Comparisons of Tool Use between Area A and Area B at Grand Banks.

Inter-area variations of contact materials also existed, as is shown in Figure 8.3. Both areas have high proportions of soft animal substances (meat, skin, hide, etc.). However, hard bones/antlers are limited to Area A, despite some occurrences of fresh bone/antler-working in Area B. Moderately resistant animal substances (MA) (i.e., hard skin/hide or meat processing with bone contact), occur with greater frequency in Area B. As far as wood material is concerned, a concentration on fresh wood-working (1M) is observed in Area B. In Area A woodworking focuses roughly on both fresh wood (1M) and hard wood (2M). Plant contact materials are present in both areas, though in low percentages. Only at Area A were very hard materials and inorganic contact materials found.

A general pattern of tool use at Grand Banks could be reconstructed in terms of inferred use-related activities. The most common use-tasks at Grand Banks were the scraping of soft animal substances. Secondary use-task emphases were scraping/shaving of fresh wood or of fresh bone. Cutting and sawing were primarily associated with soft animal and fresh wood substances. Therefore, it can be inferred that both butchering/meat-preparation and woodworking are the major use-related activities at Grand Banks, regardless of areas. Area A, however, has roughly equal proportions of activities: butchering /meat-preparation, woodworking, and bone/ antler-working (Fig. 8.4). Area B, on the other hand, seems to represent, to some degree, a specialized place for butchering/meat-preparation, although a significant amount of woodworking was also undertaken. Bone/antler-working is also present in Area B, but much less than in Area A. Plant-working was undertaken equally at both areas, whereas use-tasks involving inorganic materials such as soil or stones took place only at Area A.

Tool Use at HH, a Late Middle Woodland Site

In Chapter 3, I have briefly discussed the reasoning for including the HH site in this study. A collection of 204 pieces were examined microscopically from the HH lithic assemblage, a late Middle Woodland site from the study region. The samples were chosen only from the artifacts catalogued as "utilized flakes." According to the typology put forth in this study, the analyzed "utilized flakes" are further identified as retouched pieces and unretouched flakes. Of the 204 pieces examined, 16 retouched pieces and 67 unretouched flakes were determined to have been used (Table 8.1). Tools at HH comprise just less than half of the analyzed sample (40.68%). However, this figure could be skewed, because the higher proportion of used pieces may be due to the sample being drawn from the conventional "utilized flakes" category. The pieces in the "utilized flakes" category show a greater tendency towards being used as tools.

The common tool motions at HH are scraping/shaving, cutting/sawing, and slicing/carving, in that order (Table 8.2). Impact wear, which represents traces of projecting forces, is also well preserved at this site. In combination, longitudinal tool motions dominate the tool use at HH, comprising nearly haft of all tool motions (38.9%). Soft animal substances (SA) are the most substantial of contact materials (28.89%) at the site, followed by the less hard bone/antler substances (1H) (18.9%). By contrast, combined wood materials (1M and 2M) are comparatively poorly preserved, accounting for only 17.8% of total contact materials. According to wear types, the most common use-tasks were the cutting or slicing/carving presumably of fish. However, butchering small game animals or meat-preparation was also an important activity at this late Middle Woodland site. Scraping and shaving of wood are more common than cutting and sawing of wood. Drilling, pounding and wedging use-tasks are mainly restricted to hard wood (2M), but drilling bone/antler (1H and 2H) also occurred.

The results suggest that the tool use at this late Middle Woodland site represents a possible specialization in butchering/meat-preparation use-related activities. In comparison, woodworking occurred at a lower frequency. Plant-working was extremely rare; only three episodes of cutting grass or other soft plants occurred at the site. According to prehensile wear, 5 out of 26 prehensile wear units show hafting wear types.

Table 8.1: "Utilized Flakes" from the HH Lithic Sample.

Refined Category	Unused	Used	Used Rate	Total
	N	N	%	N
retouched Pieces	38	16	30.2	54
unretouched Flakes	83	67	44.3	150
Combined	121	83	40.7	204

Table 8.2: Tool Motion by Contact Material in the HH Assemblage.

Tool Motion	Contact Material									
	?	SA	SV	1M	MA	2M	1H	2H	3H	Comb.
indeterminate	8	-	-	-	-	-	-	-	-	8
cut/saw	-	9	3	1	-	1	8	-	-	22
slice/carve	-	7	-	-	6	-	-	-	-	13
scrape/shave	3	10	-	4	1	4	6	2	-	30
plane/whittle	-	-	-	3	-	-	1	-	-	4
projection	2	-	-	-	4	-	-	-	-	6
bore/drill	-	-	-	-	-	1	-	-	-	1
grave	1	-	-	-	-	-	1	-	-	2
chop	-	-	-	-	-	-	-	-	-	0
hoe/dig	-	-	-	-	-	-	-	-	-	0
pound/grind	-	-	-	-	-	1	-	-	-	2
wedge	1	-	-	-	-	1	-	-	-	1
pick	-	-	-	-	-	-	1	-	-	1
Total Employed Units	15	26	3	8	11	8	17	2	0	90

Hafted tools are thus poorly represented. However, those tools with identifiable hafted wear were primarily utilized for projecting. Given the relatively small sample size, this conclusion regarding the HH flaked stone tool use is preliminary.

An Intersite Comparison

An intersite comparison was then undertaken to determine a general trend in tool use patterning from a late Middle Woodland site (HH), through two transitional sites (Grand Banks and Young 1), to an early Late Woodland site (Lone Pine). The results are presented here through a series of bar-chart graphics in which four aspects of stone tool use pattern are examined: tool type, tool use, use-related activities, and intensity of tool use.

The intersite comparison is based on an assumption that all sites in the discussion represent similar ecological adaptations. However, data from Young 1 should be regarded as exceptional. It may be skewed either by its unrepresentative samples or by its possible site function as a short-term campsite (Chapter 2, also see Stothers 1974a). Although Young 1 data are also included in the graphics, interpretation of Young 1 data should be made with caution.

The HH, Grand Banks, and Lone Pine sites are considered to be long-term occupations (either year-round or seasonal with regular returns). In chapter 2, I note that Lone Pine and Grand Banks represent riverine habitation sites, with a possible year-round settlement. However, interpretations of precise site function are not available because of a lack of integrated botanical and faunal analysis (Smith and Crawford 1997). Based upon limited faunal and botanical evidence, HH was interpreted to be "a single occupation macroband encampment" with late summer and fall occupation (Woodley 1996:138):

> Of the recovered floral and faunal material, most of the identified species should have been available year round. Exceptions to this are the turtle and bullfrog, the passenger pigeon bone, and the berry seeds, all of which indicate a mid to late summer occupation. Nut collection and nut processing would definitely have been a fall activity. This presents a mixed picture of the inferred season of occupation but with good evidence for a late summer and fall occupation (Woodley 1996:137).

However, variability of site function may exist among the discussed sites for a number of reasons. First of all, the time span from A.D. 300 to A.D. 1,000 was a critical period that witnessed the emergence of food production in the Northeast. Such a major subsistence shift, of course, may affect the site function in subsistence/settlement adaptations from the early to later periods. Second, the location of the sites is also a factor that may have influenced the differentiation in site function. Grand Banks is a flood plain occupation, while Lone Pine is an upland forest occupation. Although the two are close enough to have shared the same surrounding resources, the different physical settings could entail slightly different strategies in food procurement that in turn influenced tool production. Unfortunately, because of limited excavations at Lone Pine, detailed information on spatial patterning of lithic production is not yet available. The HH site is located at the western end of Lake Ontario, 30 km from Lone Pine and Grand Banks (see Fig. 1.1). This difference in locations directly affects the use of raw material sources. Apparently, the HH site is far from Onondaga chert sources (ranging from 30 km to 50 km), but closer to Ancaster chert sources (less then 10 km) (see Eley and von Bitter 1989). As a result, Ancaster chert is emphasized at HH (Woodley 1996). To what degree the procurement of different raw material might affect lithic production at HH and Grand Banks is unclear because of the lack of a regional study. However, the HH site is, so far, the only Middle Woodland site representing a probable long-term occupation, and it also immediately precedes Princess Point in the study region. It was for this reason that this study includes the HH site in the intersite comparison.

With this in mind, the interpretations of temporal trends in lithic production discussed below reflect a preliminary outline for changes in lithic technology in the study region. Much more work is needed, however, to clarify these tentative trends in lithic production.

Tool Types

A comparison of tool types indicates that there was a shift from the manufacture and use of bifacial tools to flake tools during the Princess Point period (Fig. 8.5). Considering the modification of tool blanks into formed types as a process of tool manufacture, three general types of tools can be determined according to a manufacturing trajectory: bifacial tools, flake tools, and other tools. Bifacial tools are those made mainly on core nodules through retouching or thinning the surfaces into formed types such as projectile points, bifaces, and unifaces. Flake tools are those primarily made on flakes (or blades) through retouching or sharpening of the edges. Formed types of flake tools include scrapers, drills, truncations, retouched pieces, and notches, etc. "Other" tools are those not falling into the above two categories, or those that are made using special techniques. Burins and microliths are "other" types of tools according to this definition.

According to the artifact types of the HH lithic assemblage presented in the original report (Woodley 1996:62, Table 11), I combined the projectile points and bifaces into the bifacial tool class, and lumped scrapers, drills, and the 54 retouched pieces that I had identified from the "utilized flakes" category into the flake tool class. The other tool class from the HH site includes the originally defined "wedges" and "knives." These tool classes were then compared with the ones from Grand Banks and Lone Pine. The Young 1 tool types were excluded from this comparison because sample size is too small (N=4). The results are presented in Figure 8.5.

It is clearly shown in this illustration that the earlier assemblages are dominated by bifacial tools. A dramatic decrease in the presence of bifacial tools took place between Grand Banks and Lone Pine. However, flake tools show a consistent increase from the late Middle Woodland, through the transitional period, to the early Late Woodland. Flake tools dominate in the Lone Pine assemblage, as bifacial tools do in the HH assemblage. The "other" tool category has also decreased throughout time.

It has been noted in the analysis of formed types and unretouched flakes, however, that the evidence of bifacial production at Lone Pine is greater than at Grand Banks. A higher percentage of biface thinning flakes and higher frequencies of technological attributes indicative of bifacial core reduction have been noted in the Lone Pine

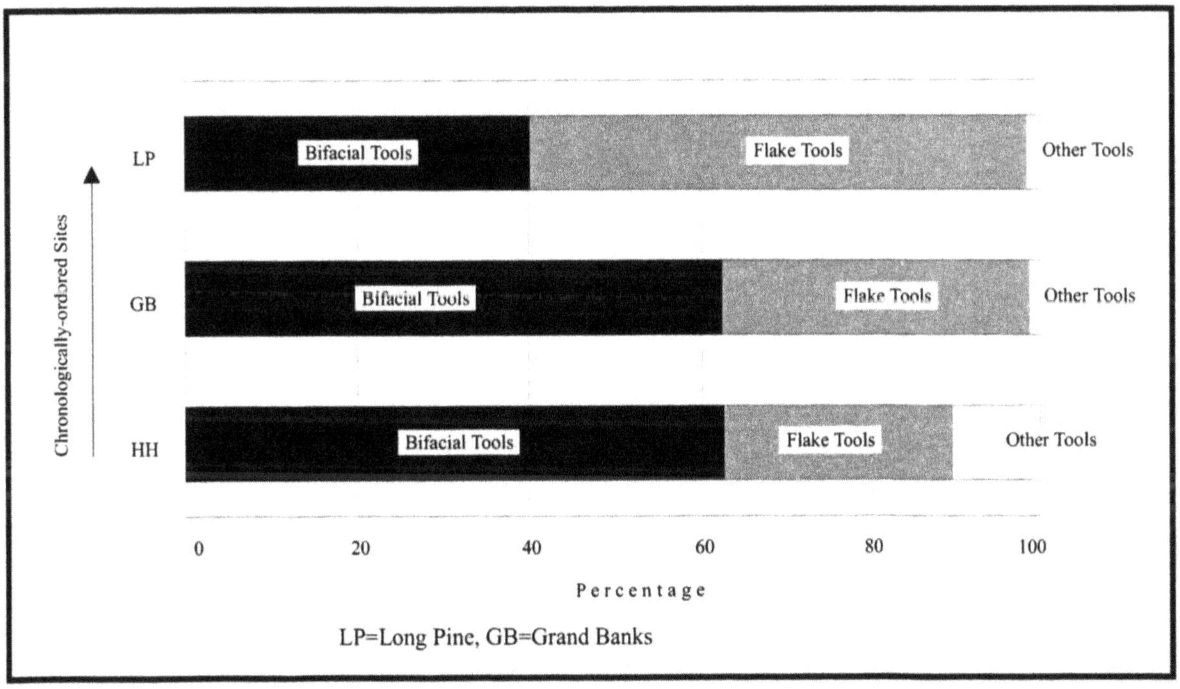

Fig. 8.5:Tool Type Comparison by Sites.

assemblage. Two possible explanations exist for this. The first possibility is that the Lone Pine bifacial tool assemblages have not yet been fully recovered in the limited test excavations. The other possibility is that, at Grand Banks, substantial biface production loci, where biface core reduction byproducts would be found, have yet to be exposed. The testing of these hypotheses will require extensive data from further investigations. At present, however, the second hypothesis is more likely, given that the Grand Banks lithic production occurs within discrete production loci on site. Nevertheless, it is significant to note the tendency towards an increase in flake tools during the transitional period.

Tool Use

A trend in tool motions is represented among the four lithic assemblages (Fig. 8.6): except for the Young 1 tools, which present less variation of tool motions, the later tool assemblages (Lone Pine) present a greater variety of tool motions than the earliest assemblage (HH). In general, cutting/sawing and scraping/shaving were common tool motions in all lithic assemblages. However, apparently slicing/carving, which likely involved butchering/meat-preparation activities, underwent a dramatic decrease from early to later periods. On the other hand, an increase in planing/whittling tool motions, most of which were presumably associated with woodworking, can be observed throughout time. For these reasons, the graphic of combined tool motions does not display significant changes in either longitudinal or transverse tool motions. However, tool motions for which tip areas are used, such as projection, drilling, and graving, occurred at a much higher frequency at the early Late Woodland site (Lone Pine).

Another aspect of tool use is contact materials. The comparative graphic (Fig. 8.7) of the 8 resistance-grades exhibits a clear trend toward the increase of both fresh wood (1M) and hard wood (2M) use-wear over time. Corresponding to this change, both soft animal substances (SA) and hard animal substances (1H) demonstrate a clear decrease. The HH assemblage has the lowest frequency of plant contact material (SV) (i.e., grass and crops etc.). The plant contact material has the highest occurrence in the Grand Banks tool assemblage, and is moderate in the Lone Pine assemblage. Only at the later assemblages (Grand Banks and Lone Pine) are inorganic contact materials such as soil and stone detected from the analyzed samples.

Use-Related Activities

At this point I will surmise on site use-related activities from the tool use. Such activities are inferred from the examination of use-tasks, as I have presented above. Cutting/sawing, slicing/carving, and penetrating of soft animal substances (SA) as well as meat processing with bone contact (MA), are indicative of butchering/meat-preparation activities. In addition, the scraping of animal skin or hide would also be considered as a part of butchering/meat-preparation activities. Projecting into animal skins with bone contact would normally represent hunting activities. However, since only a few incidences of this occurred, I have included these types of activities within butchering/meat-preparation. The various tool motions employed on wood substances (1M and 2M) will be considered here as woodworking, as will tool motions on bone/antler be considered bone/antler-working. Thus a general trend in use-related activities at the four sites can be determined (Fig. 8.8).

A clear-cut increase in woodworking is observed. Correspondingly with this, a noticeable decrease in butchering/meat-preparation activities was undertaken from the Middle Woodland period, through the transitional period, to the Late Woodland period. A decrease of bone/antler-working is also apparent. Plant-working occurred substantially more at late sites (Grand Banks and Lone Pine) than at the earliest one (HH), although Lone Pine did have less plant-working than Grand Banks. Tools used for other purposes, as represented in the "inorganic-related" category, may be related to ceramic manufacture, house construction, or pit-digging. Tools employed for soil digging or that come in contact with stone or other hard inorganic materials are observed at Grand Banks and Lone Pine, but not at HH and Young 1.

Intensity of Tool Use

Intensity of tool use is measured in this study through the number of employed units per tool

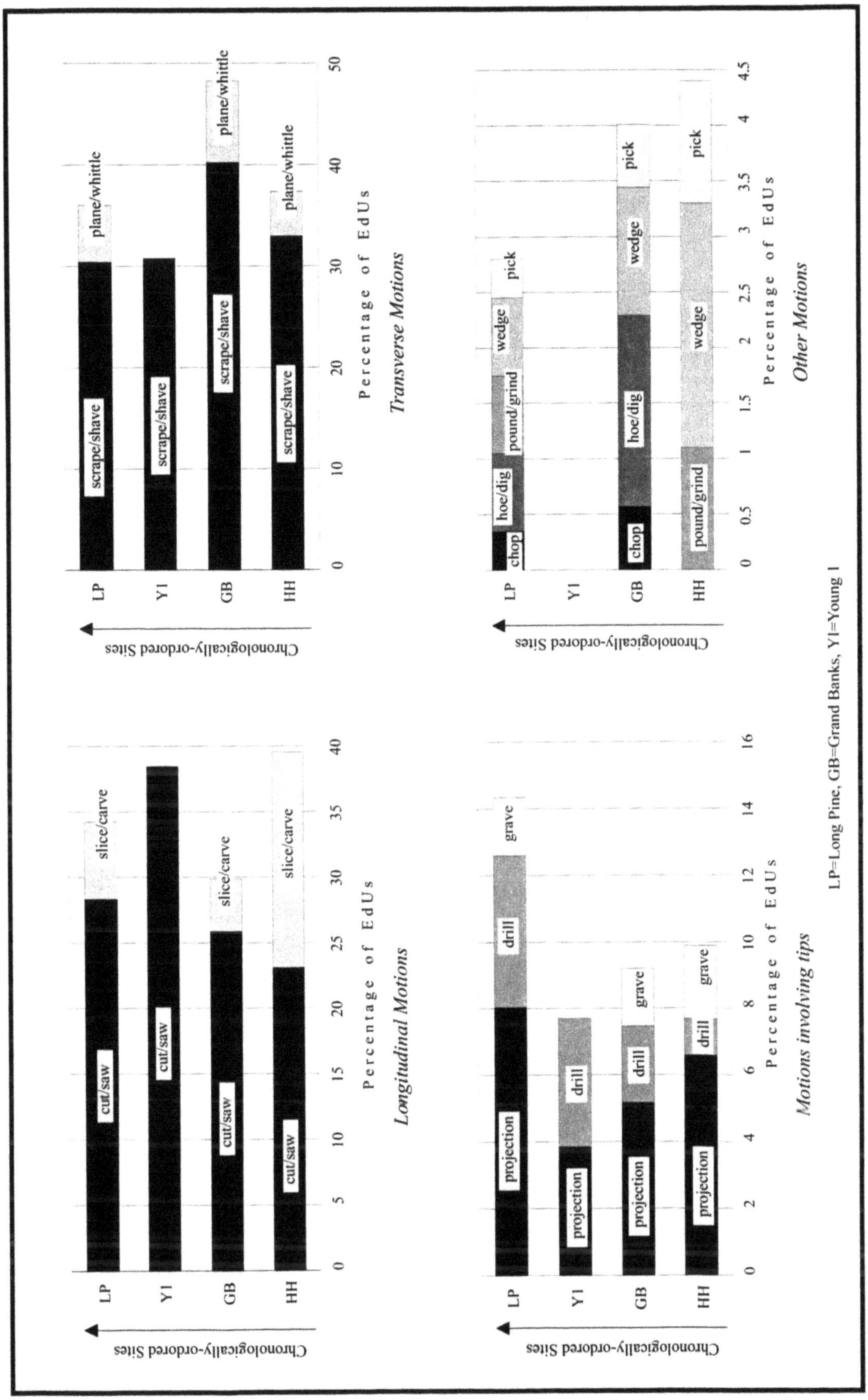

Fig. 8.6: Tool Motion Comparisons of Tool Use by Sites.

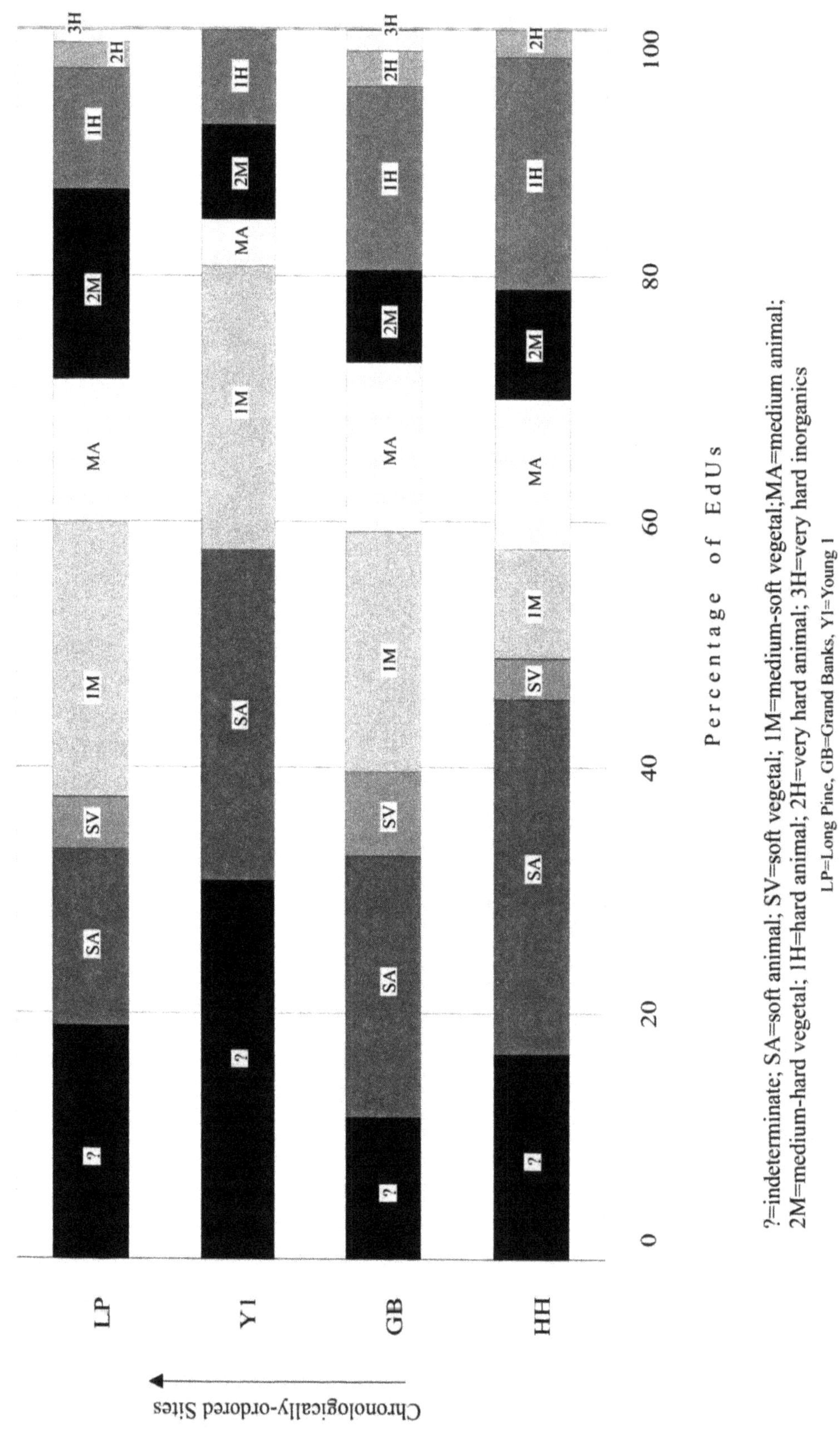

Fig. 8.7: Contact Material Comparison of Tool Use by Sites.

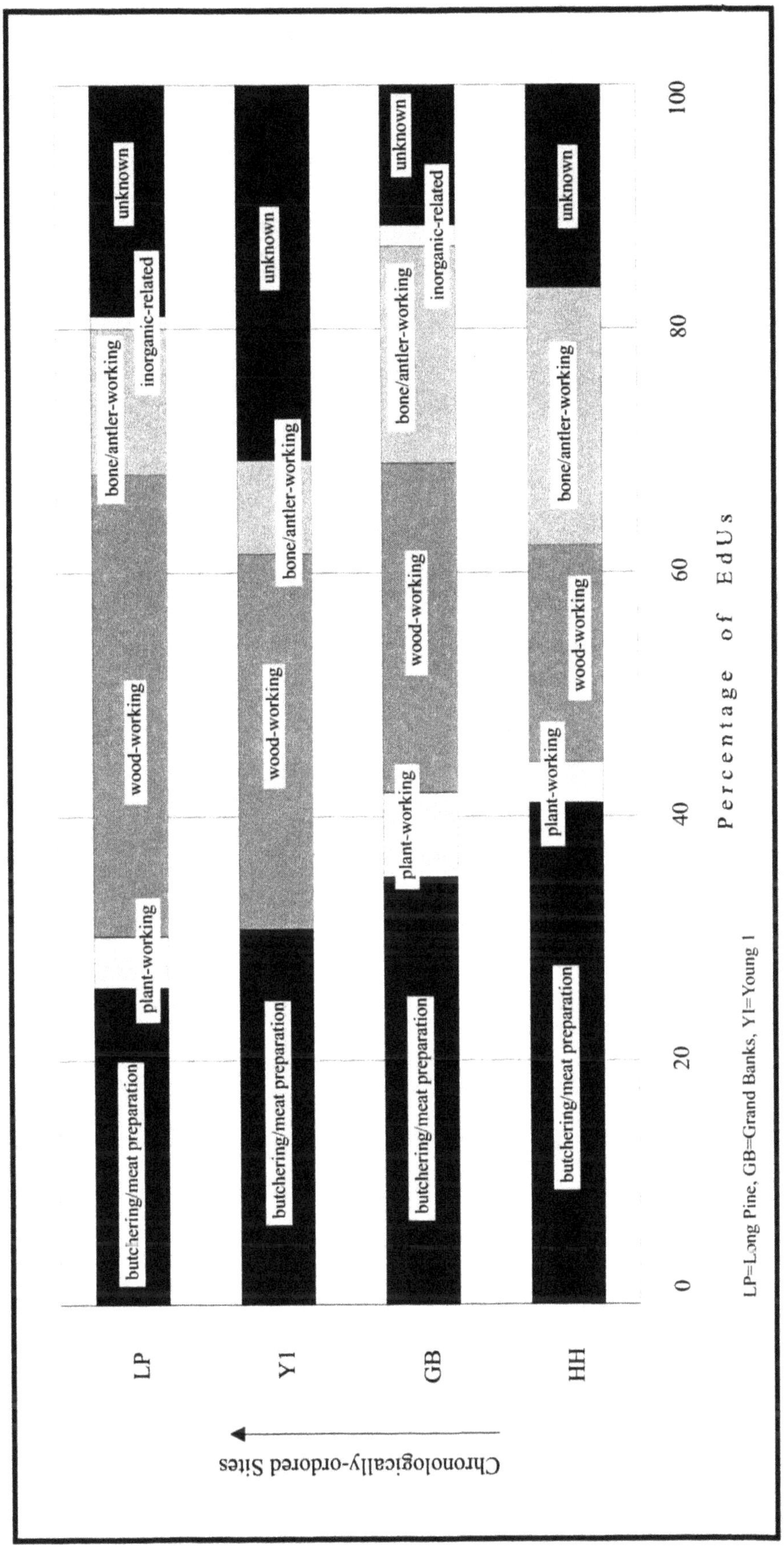

Fig. 8.8: Use-Related Activity Comparison by Sites.

and the quantity of hafted tools. According to Odell's model (Odell 1996:194-201, 1994), sedentary people would not only use hafted tools more often than the mobile groups, but would also utilize them more intensively by employing more employed units. The logic underlying this model is that with increased risk of failure for local resources, tool users would have been more likely to optimize the tools. Whether the Lower Grand Banks Valley case would fit into this model is unknown yet because of limited data. Whatever the case, general trends in the use of hafted tools and multiple-use tools are observed.

Previously, I have introduced some individual tools with multiple employed units (Chapters 4

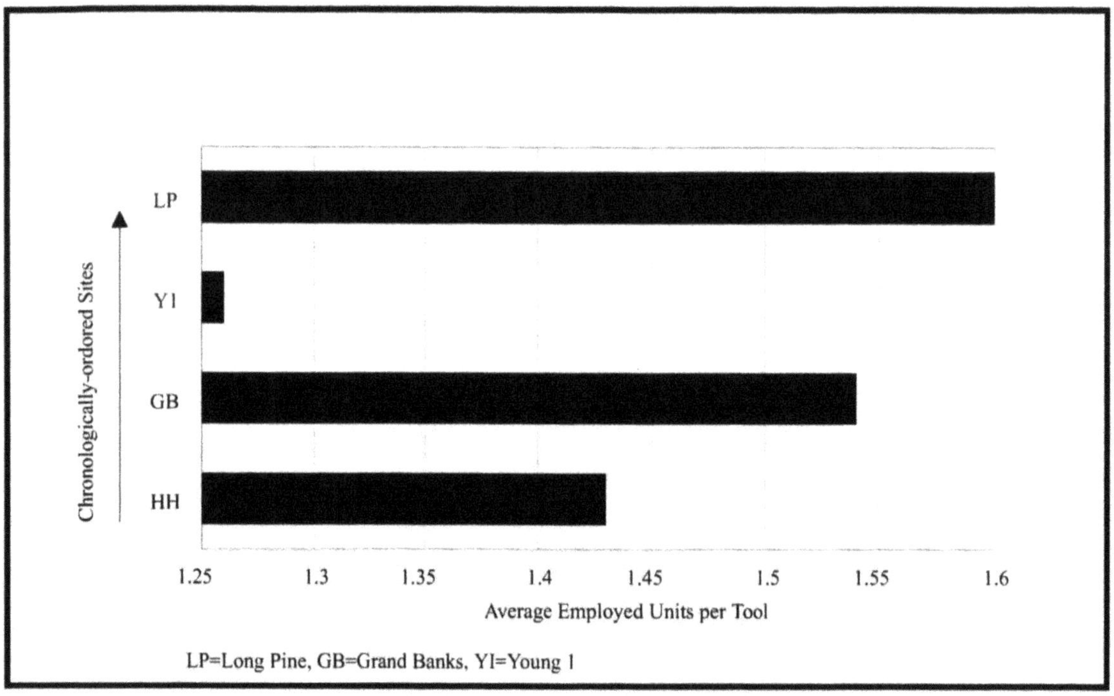

Fig. 8.9: Comparison of Average Employed Units per Tool by Sites.

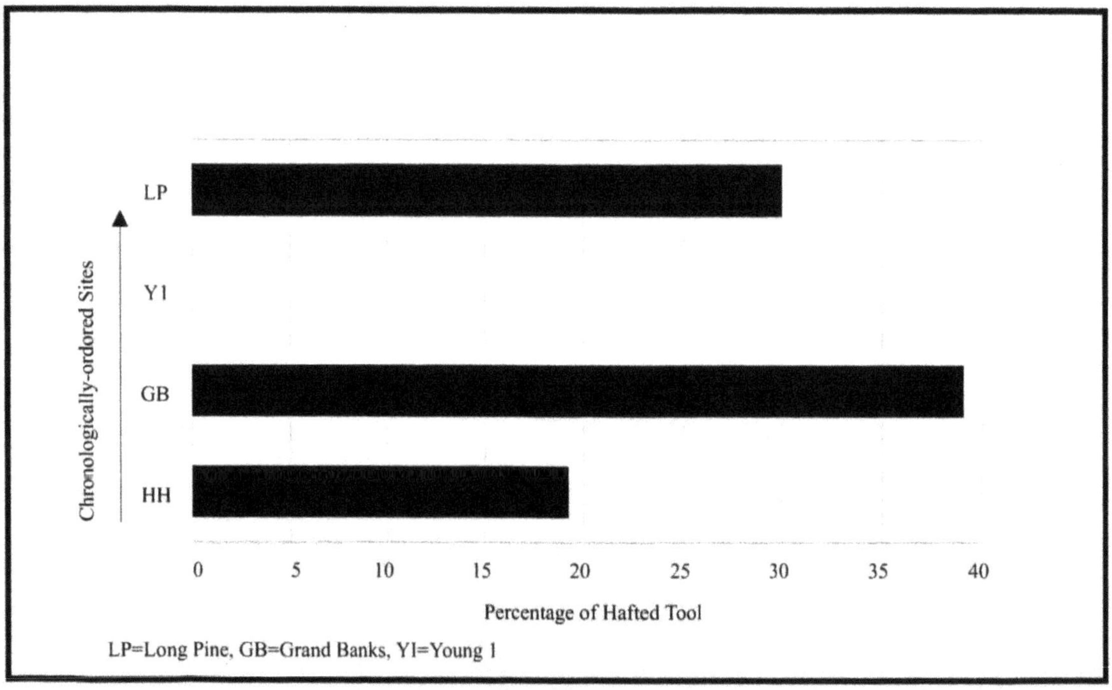

Fig. 8.10: Comparison of Hafted Tool by Sites.

and 6), that may suggest some degree of intensive tool use. Here I have compared the average employed units per tool among the four assemblages (Fig. 8.9). It should be noted that no great diversity in employed units per tool is seen among the analyzed assemblages. However, from early to later periods, there is a consistent increase in average of employed units per tool from 1.43 at HH, through 1.54 at Grand Banks, to 1.60 at Lone Pine. The lowest employed units per tool at Young 1 is, again, considered as an exception.

In addition, there is also a dramatic increase in the number of hafted tools over time (Fig. 8.10). Hafted tools were determined according to utilized pieces with hafting wear. Although the Late Woodland site (Lone Pine) has fewer hafted tools than the Princess Point site (Grand Banks), both assemblages present significantly higher frequencies of hafted tools compared to the HH assemblage. I have observed that most of the hafted implements at the Princess Point site are formed type tools that were employed primarily for scraping or shaving woods (Chapter 6). Some of the hafted tools at Lone Pine are unretouched flakes that were used as projectile points in hunting activities. It is apparent that an increase in intensity of tool use was occurring during this transitional period.

From the above discussion, the temporal trends in lithic production, from late Middle Woodland to early Late Woodland in the study region, can be summarized as follows: (a) decrease in use of bifacial tools and a corresponding increase in use of flake tools; (b) increase in variations of tool use-tasks; (c) decrease in proportion of butchering/meat-preparation activities and bone/antler-working, and correspondingly increase in proportion of woodworking and possible plant-working; and (d) increase in the use of haft tools. As I mentioned earlier, because of a lack of clear understanding of site function, the determined trends should be considered a preliminary outline for changes in lithic production during the transition from the Middle to Late Woodland in Ontario. Confirmation of these trends is much needed as comprehensive data become available.

A further explanation of these changes is also needed. Because only isolated site cases are available for intersite comparison at present, it could be argued that the differentiation of tool use patterns at the investigated sites is attributable to possible different site functions. If this is the case, we can see that at HH, and probably at Young 1 as well, inhabitants employed more bifacial tools in relatively restricted range of use-tasks, concentrating on butchering/meat preparation and bone/antler-working activities. Use-wear analysis points to a possible specialization of tool use at HH in fish processing and presumably small game butchering. In contrast, bifacial tools at Grand Banks and Lone Pine were replaced by flake tools. The occupants of these two sites employed their tools for a great variety of activities on-sites. Although I stressed earlier that an increase in woodworking at these two sites is apparent, both butchering/meat-preparation and bone/antler-working were still essential to the inhabitants. The butchering/meat-preparation activities at Grand Banks and Lone Pine continue to comprise nearly 25%-35% of total tool use-tasks, as compared to about 42% at HH. While about 20% of tool use-tasks at HH involved bone/antler-working, at Grand Banks these constitute use-tasks present about 15% and at Lone Pine about 10% (Fig. 8.8). In addition, other use-tasks that are not seen in HH and Young 1 are present at Grand Banks and Lone Pine. This variability of tool-use patterning implies that the two earlier sites (HH and possibly Young 1) may represent a pattern of specialized on-site tool use, while the two later sites (Grand Banks and Lone Pine) may represent a generalized pattern of on-site tool use.

Whether the variability of tool-use pattern is attributed to different strategies of resource procurement, or to the seasonality of settlements, or to the area's palaeoecology, remains unknown. What seems clear, however, is the change in food procurement strategies during the period from HH, through Grand Banks, to Lone Pine. At this time, intensive corn-based agriculture became established (Smith and Crawford 1997; Crawford et al. 1997; Crawford et al. 1998). The different site functions might be related to such changes in the subsistence strategy, which in turn influenced settlement. Thus the transition to agriculture may, although not in all cases, cause a generalized pattern of stone tool use in the food-producing societies, as seen at Grand Banks and Lone Pine. Taking into account the emergence of food production during this period, it is not surprising to see the increase in plant-working, woodworking, and other activities as well as reduction in meat-focused activities.

CHAPTER 9

INTERPRETATION AND CONCLUSION

This study is based on the plausible assumption that, during the Princess Point period, intensified horticulture was introduced to the lower Great Lake region. The mixed economy (horticulture, hunting, gathering, and fishing) brought changes to the social and economic structures of prehistoric societies in the Lower Grand River Valley.

The results of lithic artifacts recovered from the Grand Banks site reveal a regional manifestation of the Princess Point lithic technology. The Grand Banks lithic industry is now better understood through the reconstruction of core reduction strategy and stone tool use patterning. The fact that there are clear changes in lithic manufacture and use during the transition from the late Middle to early Late Woodland cultures is evident through an intersite comparison (Chapter 8). The general trend in the study region involves by a shift from a specialized to a generalized pattern of lithic production. In the following discussion, I first address certain issues concerning the pattern of lithic production that may be significant in terms of the introduction of food production. This is followed by a discussion of how the transformation of lithic production took place during the Princess Point period, and why. Although interpretations in this preliminary research may be limited by the data available to date, I provide plausible models and suggest new lines for future research.

The Pattern of Lithic Production

The basic traits of the Princess Point lithic industry have been described in some detail in the preceding chapters. In this section, I focus on three issues: the adoption of bow-and-arrow technology, the economizing behavior of flake tool production, and generalized stone tool production. These issues, developed from an understanding of the new lithic data presented earlier, are explored in the context of the shift in subsistence patterns to a mixed economy in the study region.

The Adoption of Bow-and-Arrow Technology

Adoption of bow-and-arrow technology is traditionally regarded as one of the major changes in lithic technology during the Middle to Late Woodland transformation (e.g., Ford 1974; Hall 1980; Muller 1986). Muller (1986, 1987), for example, suggests that this new technology "may have led to the dispersal of Middle Woodland peoples into upland areas and may have lessened their interdependence by, in effect, creating a circumstance of superabundance" (Muller 1987:266). To Muller, the consequence of adoption of bow-and-arrow technology could entail major economic changes, including the introduction of maize production. However, others also hold the view that the changes in the subsistence economy during this period could be responsible for the adoption of bow-and-arrow technology (e.g., Christenson 1986; Shott 1989, 1996). Using an offshoot of Optimal Foraging Theory -- the Diet Breadth Model (Winterhalder 1981), Shott (1996b) suggests that the adoption of the bow-and-arrow hunting practice seems to have been a response to the decrease in diet breadth caused by a greater dependence on starchy seed crops, especially maize. Hunters might have concentrated on a limited number of prey species in order to decrease pursuit time. One way to accomplish this effort was to increase the range and accuracy of weapons.

In most of eastern North America, replacement of large notched or stemmed projectile points with small triangular points occurred around A.D. 200 to 700 (Justice 1987; Ritchie 1961). This shift is commonly linked to the introduction of bow-and-arrow technology to the region (Shott 1993; Blitz 1988; Christenson 1986; Griffin 1983; Hall 1980; Kelly et al. 1984; Morse and Morse 1990; Muller 1986). Ethnographic sources document that historic or proto-Iroquoians employed the bow-and-arrow technology in hunting. Archaeologically, this study demonstrates that most projectile points recovered from Grand Banks were small triangular points, representing a remarkable change in projectile point manufacture in the region (Ellis and Ferris 1990; Ritchie 1961). Therefore, a question arises as to whether or not the Princess Point people employed the bow-and-arrow in their hunting practices.

Answering this question, however, is not an easy task. First of all, it is difficult to distinguish arrow points from dart/spear points by their forms. Recent studies of the morphological form and size of projectile points in relation to their function suggest that there is no simple positive association between large notched or stemmed points and darts or between small triangular points and arrowheads (Thomas 1978; Odell 1988; Patterson 1985; Shott 1993, 1996, 1997). In an attempt to define arrowheads, some analyists apply discriminant function analyses for identification of projection points (Thomas 1978; Patterson 1985; Knight and Keyser 1983; Shott 1997). These studies indicate that morphological forms of triangular projectile points could be identified either arrow or dart/spear points. The traditional view of small triangular points as arrowheads is now open to question.

Second, arrowhead status may not be appropriately judged by the form of a projectile point. A study by Odell (1988) implies that unretouched flakes and blades can function as arrow points as long as their sizes and shapes are appropriate. He demonstrated that there is a clear size bimodality for the archaeological specimens showing projection use-wear. The small-sized group (arrow points) is mainly composed of unretouched flakes and blades as well as small-sized morphological projectile points, while the large-sized group (dart points) is entirely made up of morphological projectile points (Odell 1988:248-349). Examples of asymmetric pointed, retouched or unretouched flakes (or microliths) that were used as arrow points are commonly found in the Old World (Clark et al. 1978; Henry 1995). In the Grand Banks data, there are 2 retouched pieces and 3 unretouched flakes which show use-wear evidence indicating use as projectile points. These pieces fall into Odell's small sized group, and presumably function as arrow points.

Third, debates on the timing of the introduction of bow-and-arrow technology remain. Traditionally the development of the bow-and-arrow technology is assumed to have occurred ca. A.D. 200 to 700 in most areas of North America (Hester 1973; Blitz 1988; Seeman 1992; Shott 1993). However, the fact that small notched or stemmed points could have been employed as arrow points suggests that the bow-and-arrow technology could have been adopted in prehistory earlier than previously thought (Shott 1993, 1996b). Relying on nontypological data, some authors push early bow-and-arrow technology back to 1,000-2,000 B.C. (Aikens 1970; Webster 1980; Odell 1988; Patterson 1993). Shott (1993, 1996b) proposes a continuous model instead of the sudden replacement model and suggests that the hunting toolkit may have changed by expansion and that hunting tools may have been reduced in size over time. Based on his extensive use-wear analysis, Odell (1988) suggests that the first bow-and-arrow technology could have occurred as early as in the Late Archaic/Early Woodland in the Lower Illinois Valley, and that the Middle Woodland seems to have been a transitional period for such technological change, a time during which both bow-and-arrow and spear-and-dart forms of hunting were practiced. He explains:

> The existence of such a period would account for the stylistic differences: spearpoints were manufactured to traditional, standardized forms including familiar point types (Snyders, Gibson, Manker, Steuben); arrowpoints were manufactured primarily on unstandardized flakes, blades, and blocky fragments, which frequently were not even modified for the purpose. According to this model it took a few hundred years, until the late Late Woodland period of the Lower Illinois Valley starting ca A.D. 700, for missile tips to have become standardized to the familiar Klunk, Koster, Madison, and Schild points that are easily rec-

ognizable to the archaeologist as arrowheads (Odell 1988:350).

If this model is correct, I suggest that we should carefully examine archaeological data from the Middle Woodland of southwestern Ontario for evidence of this possible technological change. If unretouched flakes or blades played an important role in the transition from the spear-and-dart to bow-and-arrow forms of hunting, then the Grand Banks data may point to the Princess Point period as the probable end of such a transition in the study region, since retouched and unretouched flakes were infrequently employed as arrowheads at Grand Banks. Further work is needed to test this hypothesis.

Given the facts discussed above, the high frequency of triangular projectile points at Grand Banks may not simply indicate the use of bow-and-arrow technology during the Princess Point period. While it is still plausible that the adoption of the bow-and-arrow may provide a more efficient hunting technology in a mixed economy, questions still remain as to whether or not the subsistence change is responsible for such technological change in southwestern Ontario.

The Economizing Behavior in Flake Tool Production

A unique reduction sequence has been recognized in the Grand Banks lithic industry: the "transformed" core reduction (Chapter 4 and 6). With this strategy, people at Grand Banks applied an amorphous core reduction mode during the later phase of the reduction sequence after produced desired tool blanks with the prepared core reduction mode, in order to maximize the flake production. This raw material economizing behavior is supported with lithic data from Grand Banks. However, due to a lack of systematic study of lithic artifacts from the Middle Woodland in the region, whether this kind of economizing strategy was employed at archaeological sites prior to the Princess Point period is unknown. Nevertheless, the question is prompted: why did people at Grand Banks employ the raw material economizing strategy when raw material resources were local-ready available?

There are not yet satisfactory answers to this question since there are few studies addressing the issue. It is likely that tool users might have economized their resources under specific circumstances or local conditions. In both independent case studies by Hayden (Hayden et al. 1996) and Odell (1996a), it is suggested that the amorphous core technology was adopted as an economizing strategy at sites that were located at rich raw material areas. At Elizabeth blufftop mounds in the Lower Illinois Valley, Odell (1996a) suggests that it would have been a struggle for these Middle-Archaic people to have procured suitable material once they had run out of supplies, because high-quality chert was probably not readily available on the blufftop and it would have required an inconvenient journey down into the nearby creek bottom (Odell 1996a:214-215). At Keatley Creek in British Columbia, there would have been considerable constraints on the amount of raw material that could have been brought to the winter villages due to the need to transport large quantities of food and supplies from the mountains to the winter village. Thus "stone was therefore used in an extremely economical fashion" (Hayden et al. 1996:39).

The constraints on the amount of raw material transported per trip and the number of trips to the sourcing areas likely existed at Grand Banks as well. As I pointed out earlier, results from core analysis suggest that only small core nodules were probably brought back to the site. If there are constraints on the limitation of trips to the sourcing sites along the north shore of Lake Erie, then people at Grand Banks may have maximized the raw material on hand. It is reasonable to assume that tool users at Grand Banks faced such constraints because the intensive field labor of food production (i.e., corn) may restrict the number of trips to the chert sources. People would have saved as much time as possible from traveling to the quarries in order to invest more time and energy on field clearing, planting, and harvesting of corn. One strategy to reduce the number of trips and to transport maximum amount of material in one trip is to trim off excess parts of the core nodules at the quarry and to bring back only the reduced cores ready for further production (Henry 1989b). This is clearly evidenced in the Grand Banks lithic assemblage (Chapter 4 and 6).

This "time-stress" explanation claims that the introduction of field horticulture might have affected the form of core reduction. However, Huron Iroquoian societies developed a clear division of labor, with women working in the fields near or at homes and men traveling a great deal in hunting, fishing, and trading pursuits (Trigger 1987:32-45; Heidenreich 1971:158-218). Given this division of labor, "time-stress" may not have been a significant factor, although I still consider that the amount of chert materials brought back to the villages by men could have been restricted, since large quantities of food and supplies need to be transported. Other explanations for such economizing behavior should also be considered. One possibility would be that Princess Point horticulture or other subsistence activities took people away from the chert quarries, but verifying this model requires a regional catchment study. Seasonal mobility is another possibility; Princess Point people perhaps only occupied the Grand Banks site for a limited time of the year. However, Grand Banks was unlikely to have been only a winter occupation (see Chapter 2). At present, lithic data more likely point to the "time-stress" model, which is proposed based on the assumption that the intensified core-field horticulture took place at the year-round base camp of Grand Banks. Further work is needed to evaluate this model.

The Generalized Stone Tool Production

Lithic production is defined in this study as a form or pattern of lithic material modification process in terms of both manufacture and use. Thus, I define generalized stone tool production as an expression of a cultural system in which people predominantly produce simple and unformalized flake tools for their general needs. This pattern of lithic production is in contrast with the lithic production based primarily on bifacial reduction and emphasizing a multiple-purpose use of bifacial tools (see below). The generalized stone tool production at Princess Point is characterized by the following traits, according to the analytical results presented earlier:

(1) Typologically, the Grand Banks lithic assemblage displays less diversity and complexity of tool types. The intentional modification into certain formed types is minimal.

(2) Although bifacial tools at Grand Banks slightly outnumber edge-modified flake tools in frequency, the use-wear analysis suggests that Grand Banks people preferred edge-modified flake tools over bifacial tools as far as utilization is concerned. Flake tools were utilized to a great extent.

(3) Core reduction strategy at Grand Banks aimed at producing flake tools. This pattern of tool production is significantly different from that employed in Paleo-Indian and Archaic assemblages, which were normally characterized by bifacial tool production through standardized procedures (Deller and Ellis 1992; Storck 1983, 1984, 1997; Ellis et al. 1990).

(4) Flake tools (including modified and unretouched flakes) were produced through careful blank production and selection. The early phase of the "transformed" core reduction was likely used to fulfill this process of blank production, while the late phase of the reduction sequence was carried out to maximize flake production.

(5) The flake tool use-pattern suggests that unretouched flake tools were employed in general use-tasks, while modified flake tools were likely employed for certain kinds of special-purpose use-tasks (i.e., scraping wood). Blades appear to have the specialized use-tasks of cutting, slicing, and carving soft animal substances.

(6) In general, flake tools in the Grand Banks assemblages appear to have been engaged in a wider variety of use-related activities, ranging from butchering/meat-preparation, woodworking, bone-working, to plant-working, as compared to the assemblages from an earlier period (i.e., HH). I will return to this point later.

The presence of the generalized lithic production during the Princess Point period raises the question of how the shift to generalized stone tool production occurred in the study region. In the next section I offer an exploratory account for this change in lithic technology during the transition to food production.

Transformation of Lithic Production During the Princess Point Period

Lithic technology in prehistory can be considered as human behavior involved in the production of stone tools to meet needs, including most subsistence needs (Binford 1977, 1978; Nelson 1991; Torrence 1989; Schiffer 1992; Odell 1993). Changes in lithic technology throughout time did not occur without reasons; the causes of the changes are of primary interest to archaeologists.

Technological Variability and Lithic Production

It has been generally agreed that there is a great diversity of lithic technology among hunter-gatherer groups. Such technological variability is to some degree represented by the different organization of lithic production. That is to say, in a given society the decision-making of tool makers concerning the manufacture and use of stone tools (e.g., tool design, tool use-pattern, tool use-life, tool discard, etc.) depends largely on their physical environment and social constraints. Explanations of technological variability and its socio-economic impact on hunter-gatherer groups have concentrated on settlement mobility (Binford 1977, 1979, 1980; Shott 1986; Odell 1996a; Kelly 1983; Ebert 1979), raw material availability (Bamforth 1986, 1991; Andrefsky 1994), time-stress (Torrence 1983), and subsistence diversification (Vierra 1995).

Because mobility has traditionally been regarded as a strong characteristic of hunter-gatherer societies, their settlement organization has long been presumed to have affected the nature of lithic technology. Binford (1977, 1979, 1980) summarized the two types of settlement organization: residential mobility and logistical mobility, and considered "curational" and "expedient" technology in relation to these two forms of settlement mobility. The "curational" technology is represented by the production of reliable tools and toolkits that "can include advanced manufacture, transport, reshaping, and caching or storage" (Nelson 1991:62). This technology is focused on preparing for the use of tools under unpredictable conditions. Binford (1977:35) claim that this type of technology is commonly associated with logistical mobility; people (collectors) would have occupied regular base camps from which smaller groups would have been sent out to exploit the surrounding resources. In contrast, the "expedient" technology is represented by the production of stone tools and toolkits that are "technologically simpler and formally less patterned because tool manufacture is an immediate response to the specific task at hand" (Bamforth 1986:38; see also Binford 1979; Nelson 1991). Residential mobility, a system by which people (foragers) would have regularly moved from one area to another, is generally associated with "expedient" technology. Given that the nature of this exploitation is restricted to the resources in the immediate area, foragers are expected to have responded with a limited range of tool forms accomplishing a wide range of tasks (Shott 1986, 1989). Shott (1986), for example, suggests that more highly mobile hunter-gatherers would have produced fewer kinds of tools for a wider range of tasks than hunter-gatherers who did not move around a lot. This would have reduced the amount of material that they would have to carry. As a result, generalized tool production should be expected from more highly mobile groups.

Bamforth (1986, 1991), in disagreement with Binford and Shott, argues plausibly that the adoption of "curational" technology should be attributed to a lithic raw material shortage. It is likely that the rate of tool maintenance and recycling, two forms of curation, will increase when localized resources are not available so that the "cost" of replacing worn tools is therefore high (Bamforth 1989:40). By contrast, a simple flake-tool technology can be used by both foraging and collecting societies as long as appropriate local conditions (Bamforth 1991). According to the model of raw material availability, Jeske (1989) claims that as access to raw material resources became more competitive, people would employ curated strategies. These strategies would result in standardization of artifact form, reduction of tool size, and extension of tool use-life. The same observation is also made by Odell (1994b, 1994c), who notices that an increase in hafted tools -- presumably a method of use-extension -- existed among archaeological assemblages in response to a lack of locally available raw material resources. These authors suggest that biface or blade production, as well as employment of hafted tools, developed in response to raw material scarcity. Based on these assumptions, one would expect that adoption of flake tool production would be dependent on the abundance of lithic materials in the locality.

An alternative explanation of technological variability is proposed by Torrence (1983, 1989). Drawing from principles in evolutionary ecology, she suggests that adoption of a favorable technology should be based on the efficient use of time by reducing resource acquisition and processing costs. Torrence introduces the concept of "time-stress" to explain the relationship between the length of time in resource exploitation and the cost for the return, and discusses curation in association with the production of tools in advance of use. She sees these aspects of curation as the results of time-stress and the problem-solving complexities of scheduling different activities around one another.

More recently, Vierra (1995) proposes that there is a strong correlation between technological variability and subsistence diversification. This model, supported by a cross-cultural sample of hunting-gathering societies, predicts increasing technological diversity and complexity with increasing subsistence diversification. It suggests that, as the number of subsistence foods increase, one can expect a corresponding increase in the number of tool types (Vierra 1995:217).

While these explanatory hypotheses of technological variability within hunting-gathering societies are plausible and arguable on the basis of case-by-case situations, the debate continues. To test each model discussed above using the Princess Point data is not possible, and in any case, beyond the scope of this study. My primary concern here is not to clarify the technological variability among mobile hunter-gatherers, but to examine whether there is a general trend in the technological change from mobile hunter-gatherers to relatively sedentary horticulturists over time. Since the settlement organization between early horticulturists and collectors is similar, the former would be expected to adopt "curated" strategies in lithic production according to the "mobility model." However, due to relatively high sedentism and probably localized resources, the "expedient" technology is also argued as a feature of lithic production in early farming societies (Parry and Kelly 1987). Before proceeding further, there is a need for clarification of these two concepts.

Curation, since its introduction by Binford (1977, 1979), has been defined in several ways, but all of these definitions focus on special use or care of certain tools or on special tool designs (Bamforth 1986; Bleed 1986; Nelson 1991). Recent criticism of the term suggests that there have never been clear relationships between curation and mobility or raw material availability (Odell 1996b; Hayden et al. 1996; Nash 1996; Bamforth 1991; Bleed 1986). Nelson clearly confirmed that "curation" and "expediency" are not mutually exclusive systems, but planning options (Nelson 1991:65). Therefore either curation or expediency is "multifaceted, containing elements that may or may not be employed, depending upon situation and need" (Odell 1996c:48). It is suggested that the use of the terms "curation" and "expediency" may not be very helpful in the study of technological variability.

If in some societies, regardless of settlement organization, people have special needs that require a high degree of use and care of specialized tools or toolkits, then such specialized lithic production would likely be represented by a high degree of tool type diversity and complexity, standardized manufacture, or multiple-purposed and portable tool designs. Nelson (1991) mentions that bifacial or blade technology affords the great advantage of reliability for such special needs. Blade or bifacial products, produced through standardized core reduction, are dependable when needed. Kelly (1988) suggests that bifaces are also multiple-purposed and portable tools, preferred by most mobile hunter-gatherers. On the other hand, the need for over-designed products could result in a great diversity and complexity of toolkits (Bleed 1986). In addition, the need for tools with long use lives can be satisfied by the specialized production of highly standardized bifaces (Shott 1989).

On the other hand, generalized lithic production is represented by a preponderance of flake tools rather than bifacial tools and by unformalized instead of standardized production. The results of such generalized stone tool production in an archaeological context should be reflected in the following indices:

- high percentage of flake tools relative to bifacial tools;
- less diversity and complexity of tool types;
- higher percentage of utilized flaketools than utilized bifacial tools;
- greater variety of use-tasks in association with flake tools.

Thus, the generalization of lithic production means that people produce simplified and unstandardized stone tools (flake-tools) to perform a wide range of use-tasks that likely correspond to subsistence diversification.

It may be argued that the use of specialized vs. generalized lithic production may not differ from the use of "curated" vs. "expedient" technology. However, as stated earlier, lithic production is reflected by human decision-making concerning the ways in which stone tools are manufactured and used, while "curation" or "expediency" refers to strategic technology that may influence, or be influenced by, lithic production. Therefore, distinguishing generalized from specialized stone tool production is mainly based on the forms or patterns of *how* stone tools are manufactured and used, not *why*. The choice to standardize lithic products or not is a matter of whether people have different (special or general) needs for tools or toolkits*.

Importantly, both this specialized and generalized lithic production could require "curated" or "expedient" strategies under different conditions. The specialized lithic production can represent the "expedient" strategy when blades, "standardized" products, are used as "expedient" tools for immediate and general needs (e.g., Yerkes 1994; Odell 1994a). Likewise, the generalized lithic production can not always be equated with "expediency," but can represent a planned strategy in manufacturing flake tools in advance of use, preparing for unpredicted circumstances. Archaeologically, distinguishing generalized from specialized lithic production by using the aforementioned indices is a relatively straightforward way to present and compare technological variability among different cultural systems throughout time.

It is generally agreed that from approximately A.D. 500 onwards, a dramatic change in lithic technology from standardized biface technology to unformalized core reduction occurred in the Eastern Woodlands (Johnson and Morrow 1987; Parry and Kelly 1987; Yerkes 1987; Jeske 1990; Odell 1996). There was a decline in the diversity and complexity of stone tool assemblages from archaeological sites in the Northeast (Fitting 1975; Griffin 1983; Mason 1981; Ritchie 1965; Ellis and Ferris 1990). The comparative study presented in the last chapter clearly indicates that in our study region, changes in stone tool production during the Princess Point period resulted in an increase in the use of flake tools and a corresponding decrease of bifacial tool use.

This study demonstrates that the Princess Point period lithic industry in the Lower Grand River Valley represents generalized lithic production. Stone tool production of hunter-gatherers in earlier periods of southern Ontario's prehistory is characterized by highly standardized manufacture, represented by the high frequency of bifacial tools. Various authors often comment on the standardization of reduction sequences as an aspects of Paleo-Indian lithic production in southern Ontario (Storck 1983, 1984, 1997; Deller and Ellis 1992; Ellis and Deller 1990). In addition to coping with resource constraints, "a need for flexibility in the transported tool blanks such that each blank could be potentially made into various *complex* tool type configurations away from quarries may have favored this standardization" (Deller and Ellis 1992:135 italics original; see

* There is an instant difference between specialization of lithic production and lithic craft specialization. The former refers to an organized system of lithic production, while the latter refers to craft skills pertaining to the production of lithic products by craftsmen.

also Ellis 1984). Besides the standardization of tool production, there has been a greater diversity and complexity of stone tool types in early rather than later periods in southern Ontario (e.g., Ellis and Ferris 1990). So far, there is no evidence indicating that generalized lithic production took place in the Middle Woodland or earlier in southern Ontario (see Ellis and Ferris 1990), although Fox (1981) states that there was possibly a decrease in the frequency of bifacial tools after the onset of the Middle Woodland period. Therefore, it is reasonable to assume that a change from specialized to generalized lithic production occurred during the Princess Point period, which, not surprisingly, corresponds to the shift from standardized to unformalized core technology in most areas of eastern North America.

Explanations for such changes of core reduction technology vary a great deal. Again, the prevailing models have focused on two variables: mobility and raw material availability, although they differ in which variable is considered to be the determining factor. The mobility model suggests that the replacement of "prepared core" technology by "amorphous core" technology was caused by increased sedentism (Parry and Kelly 1987; Johnson and Morrow 1987). The raw material availability model claims that the use of "amorphous core" technology should depend on locally available, stockpiled raw materials since this core technology is regarded, sometimes subjectively, as a wasteful strategy of raw material utilization. It seems that both increased sedentism and locally abundant, high quality raw materials are two simple factors that may partially explain the change in the Grand Banks case. However, I argue that the appearance of generalized lithic production in Princess Point was much more complicated than can be explained by mobility or availability of raw material alone, because the Princess Point culture was the first complex society to be engaged in a mixed economy of horticulture and hunting-gathering in the Lower Grand River Valley. Before I consider the factors that may influence the transformation of lithic production, let me first examine the relationship between food production and lithic production.

Relationship between Food Production and Lithic Production

Focusing on the role of subsistence in the changes in lithic technology (Davis 1978; Vierra 1995), I consider the variability in lithic technology to be a direct response to the change in subsistence strategies. It is hypothesized here that, based upon the lithic data from Grand Banks, subsistence diversification may be associated with a pattern of generalized lithic production in the study region. Changes in subsistence strategy appear to affect mobility. Raw material availability may act as an external factor that influence the form of lithic production, but it is not significant enough to cause a dynamic change in lithic technology. In this sense, I agree with Sassaman's contention that lithic technology is socially constructed (Sassaman 1994). Tool production and design embrace economic decisions in social strategies. It is "not only about access to raw materials, portability, and tool efficiency, but also access to social resources, that is, access to people" (Sassaman 1994:100). Introduction of agriculture has brought in technological, economic, and ideological changes within the early farming societies; it to a great degree changed relation of production. Therefore social relations in lithic production should be taken into considerations when changes in lithic technology during transition to agriculture are examined.

Both the Levant of the ancient Near East and the Middle Yangtze River Valley of China are good examples of indigenous agricultural origins (Henry 1989; Bar-Yosef and Belfer-Cohen 1989, 1991, 1992; Bar-Yosef and Meadow 1995; Crawford and Shen 1998). One aspect of changes in lithic technology in these areas is the innovation of ground stone implements that are directly response to agricultural practices. As Wright (1991, 1992, 1994) suggests, intensification of labor in plant-food processing is reflected by the development of ground-stone tools and the increased use of these stationary implements (e.g., grinding slabs and handstones, pounding tools, mortars and pestles). New types of flaked stone tools, such as sickle blades and elongated picks, also appeared and continued to be used during Neolithic times of the ancient Levant. Use-wear studies on sickle blades and picks indicate that they were used for harvesting cereals

(Unger-Hamilton 1991). The sickle blades and elongated picks developed in response to farming activities. In the east Coast of China, where domesticated rice was evident in 7000 BP (You 1994; Crawford 1993), a shift from tool kits that emphasized food gathering to those that emphasized food production was observed (Chang 1986:192-208). Tools for hunting and fishing decreased sharply, while stone implements related to cultivation increased in number and type. Interestingly noted, burial evidence shows that sets of stone tools are found in most male individual burials, indicative of ideals of personal processions and labor allocation related to lithic production (Shanghai 1987).

Lithic studies of the Early Formative of ancient Mesoamerica (Parry 1987, Hole 1986) indicate a trend towards simplification (less diversity and complexity) in lithic techno-typology. There is a consistent decrease in biface production and an increase in amorphous flake tools in Guila Naquitz lithic industry in the Oaxaca Valley (Hole 1986). Flaked stone tool assemblages are dominated by flakes that are, in Parry's own words, "little better than handfuls of gravel" (Parry 1987:1). Parry's analysis (1987) based on household units demonstrate that, although both bifacial and amorphous flake production existed in the Early/Middle Formative villages, bifacial production was restricted to one or a few households. Flake production was organized in each household. The implications of this observation are that the individualized production represented by a household economic unit played an important role in lithic production during the transition to agriculture.

Similar changes in North American lithic technology took place in the U.S. Southwest as well as in the U.S. Southeast, where there are manifestations of "secondary" origins for maize-based agriculture. It is not surprising to see that a shift from prepared to amorphous core technology appeared here nearly 2,000 years after it first occurred in Mesoamerica. This was about A.D. 500 when intensified field agriculture spread through all parts of the Southwest and Southeast, after the introduction of corn from the south (Smith 1989, 1992, 1993, 1995). At this time the sociopolitical structure in Southwestern villages (from Basketmaker III onwards) underwent changes towards cultural complexity (Upham et al. 1989; Minnis and Redman 1990; Wills 1991, 1995). The change in lithic technology can be exemplified from the lithic assemblages of Black Mesa (Parry and Christenson 1986) and the Lower Chaco River (Chapman 1977). My point here is that, although such changes in lithic technology appear to be consistent in North American prehistory, they did not develop without reasons. While sedentism may be a factor associated with this change as Parry and Kelly (1987) correctly stated, the fact is that a primary factor driving this change in lithic technology could have been the shift in primary subsistence strategy. In other words, the economic foundation affected the degree of sedentism.

A recent study by Stafford (1999) suggests that in southern Scandinavia of Eupore, changes in lithic technology during transition to agriculture is represented by (1) an increase in de-standardized flake production and a decrease in blade production, and (2) newly innovated polished axes replacing flaked core axes. Stafford presents multi-factors to explain the switch from blade to flake technology. The factors he identified include reduced resources (e.g., antler), use availability of working edges on flake tools, increased sedentism, and localized control of raw materials. Accessibility to raw materials reflects undoubtedly social relations of lithic production, which certainly affects forms and patterns of lithic production. It is of interest to note that localized control of access to lithic resources was also observed in North America (Nassaney 1992, 1996).

The cross-cultural comparisons presented here postulate a logical relationship between food production and lithic production. The data from these case studies suggest that changes in lithic technology in association with changes in subsistence strategies were made possible by multiple driving forces. It is also possible to generalize that in the Old World lithic technological changes emphasize more on the innovation of new implements whereas in the New World the trend was towards the generalization of stone tool production. I argue that social structures and relations of production in early farming societies developed independently from the two continents, apart from resource availability and sedentism etc., may have something to do with the patterns of lithic production that differ.

Evaluation of the Model

In their reviews of agricultural origins, Gebauer and Price (1992:3) note that there is no single, accepted, general theory for the origins of agriculture. Advocating a multifactorial approach, Crawford and Smith (Crawford et al. 1998; Smith and Crawford 1997) suggest that the emergence of food production in southwestern Ontario should be considered to incorporate sedentism, population increase, and competition over localized resources, among other issues. Unquestionably, economic and social life of early farmers has been largely conditioned by the introduction of corn-based horticulture. The occurrence of generalized stone tool production at Princess Point, as one of a series of consequences arising from the origins of agriculture, could have been caused by a combination of external factors (e.g., sedentism, population increase, availability of raw materials) and internal factors (time-stress, productive organization, land-use), rather than by any single factor. If these elements are enabled by the emergence of food production, patterns of lithic production are thus affected.

Sedentism and the Use of Localized Resources

Introduction of food-producing technology may lead to a greater degree of sedentism and an increased use of local resources (e.g., Muller 1987). Under such circumstances, unformalized production of flake tools is more likely favored than standardized bifacial reduction, as Parry and Kelly note:

> For relatively sedentary peoples who do not move long distances residentially or logistically, stone tools must only insure that some amount of useable stone be available at the locations where it is needed. If raw material is abundant, and readily yields edges of adequate sharpness and durability for the task at hand, then there is no need to manufacture portable lithic tools. In such cases, there is no temporal or spatial difference in the location of raw material and the location of tool use, so portable stone tools are not required to bridge this distance. Instead, the tools merely have to fulfill a specific short-term task, and their function and raw material are the only factors affecting their shapes. Expedient flake tools are quite adequate in those circumstances (Parry and Kelly 1987:300).

This explanation of generalized lithic production based on increased sedentism and availability of raw material is rational, but, as I argued earlier, oversimplified. First, settlement organization of early farming societies is not so very different from that of large, complex, logistical-mobile collectors (e.g., Song and Shen 1999). Seasonal hunting or gathering was still an important event for early farmers as a supplementary economic strategy. Stone tools were also much needed for preparing for such events, in some cases, under unpredictable conditions. In other words, the curational strategy would have been much needed as a planning option. Use of "expediency" to describe the technological organization in most sedentary villages, therefore, appears inappropriate. The study of the Grand Banks lithic industry demonstrates that the flake tool production also implies a curated technology; caring for tool designs and use. For example, flake tools were carefully removed through the prepared core reduction mode. Modified flake tools were most likely hafted and employed primarily for woodworking, while unmodified flake tools were used for various activities.

Second, although there is high-quality local supply of raw materials (Onondaga chert) available, the generalized stone tool production at Grand Banks represents a raw material economizing strategy, rather than an "expedient" technology as argued above. I suspect that the constraints that people faced, which forced the adoption of economizing raw material procurement, developed from the intensified agriculture to which the Princess Point people were committed.

My argument, however, does not refute the notion that, under certain circumstances, increased sedentism and the use of local resources are major external factors acting as the catalyst in the shift from specialized production to generalized production in the late prehistory of the Eastern Woodland. Mobility and raw material availability are undoubtedly important in the organization of lithic technology. However, they would not be the only factors in the transformation of lithic production, especially when a great degree of sedentism and the use of local resources could have been consequence of agricultural origins.

Increased Localized Population

An increase in localized population is commonly associated with the emergence of food production. Increased population likely resulted in a segregation of lithic production. It is reasonable to assume that standardized manufacture of stone tools can be easily and well controlled by a small group if the need for relatively high mobility exists. Once people settled down on a permanent or semi-permanent basis, the socio-economic structure and productive organization would have been re-organized in terms of an increased population. Within a relatively large population, labor allocation and division would be expected, especially when pottery manufacture or other craft production is a part of the daily production. As a consequence, standardized lithic production may not be favored by the whole population of such extended societies. Production of specialized stone tools (bifacial tools) may remain under the control of a few, while most of the populations would favor unformalized production. Ethnographic evidence supports this. Heidenreich (1978:383) mentioned that the Huron-Iroquoians may not have been full-time specialists in "manufacturing or economic pursuits," but "it is obvious that some individuals were more adept at making pottery, pipes, or flint items than others." Evidently, simplified tool manufacture, as represented by amorphous core reduction, placed everyone in these communities in a position ensuring easy access to a means of production able to deal with a wide range of activities.

It is not possible at present to test this hypothesis with the Grand Banks data, since the data regarding population pressure are not available. However, settlement organization at Grand Banks is considered to be a permanent or semi-permanent occupation, thus a significant density of population is expected (Smith and Crawford 1997). Whether these assumptions are truly reflected in the study region is unclear. Much more work is required to assess the role of increased population as one of the factors in the de-standardization of stone tool manufacture.

Alternative "Time Stress"

Time stress, according to Torrence (1983, 1989), can be defined as constraints that limit an individual's ability to exploit subsistence resources and thereby restrain the individual's ability to produce or reproduce other economic and social products. Torrence (1983) argued that among hunter-gatherers, people who depended on subsistence resources that are seasonal and less accessible (e.g., mobile animals) have higher time-stress than people who relied on less-seasonal or non-seasonal and more accessible resources (e.g., immobile plants). Her hypothesis predicts that more sedentary collectors or early farmers would be expected to employ more generalized technologies with less tool diversity and complexity whereas highly mobile foragers would be expected to employ more specialized technologies with greater tool diversity and complexity. Therefore, Torrence's hypothesis may imply that generalization of tool production took place in early farming societies due to a less degree of time-stress. The mixed economy in sedentary farming societies would have provided year-round food supplies.

While Torrence's time stress refers mainly to the "window" of opportunity of resource procurement, I have to point out that there is an other kind of time constraint: amount of time and energy investing subsistence activities. I argue that intensive farming could lead to higher time-stress and energy costs with higher risks for less returns than simple foraging (e.g., Lee 1979). Foragers did not have to work hard all day to feed themselves. On the contrary, farmers did, and they even possibly faced the possibility of crop failure. This alternative time-stress should therefore be considered one of the characteristics of early farming.

If people have to work hard at farming, investing more time and energy in crop-growing, then their investment of time and energy in other aspects of economic or social production would be reduced (Kleindienst pers. comm.). Generalized lithic production may have resulted from such time- or energy-saving strategies. The specialized production (e.g., bifacial or blade production) would have required people with high skill and energy to produce standardized items, which to some extent wastes more raw materials (Hayden et al. 1996; or see chapter 6 for discussion). If raw materials were not handy on site, this would have required people to make more trips to sourcing areas for acquisition. Under the constraints of time-stress, it would be expected that the number of people traveling to chert sources would be minimal. As a result, people at Grand Banks likely adopted the "transformed" core reduction as an economizing strategy for raw material conservation of on-site use.

Although Torrence has attempted to test her hypothesis using Oswalt's (1976) data, she is aware that her hypothesis has never truly been directly tested (Torrence 1983:14). Moreover, it is not possible to directly test the positive correlation between high time-stress and generalized production in the earlier farming societies in the study region. However, the notion that constraints on time and energy would have caused simplified procedures in lithic manufacture and a generalized use of flake tools in diversified economic activities, deserves serious consideration in future studies.

Productive Organization

If there are reasons to assume that generalized stone tool production relates to increased population and time-stress, then some aspects of social relations also need to be taken into account. I suggest that changes in the means of production and economic organization during the transition to food production may also influence the pattern of lithic production in the Eastern Woodland as a *long-term* change. Land-use and individualized production -- two aspects of productive organization determined by changes in subsistence, are likely the internal factors responsible for any changes in lithic production.

First, the shift to food production, especially field agriculture, causes a fundamental change in the use of land (Wolf 1982; Southall 1988; Woodburn 1982). Land in hunter-gatherer societies may be an object of labor, which means that people use the resources of the land, not the land itself. In a food-producing society, land as such becomes a means of production. Territorialized farm-land is the center of food production, as higher energy input is involved in cultivated land. Second, "a change from community to household levels of economic organization may have accompanied the transition to agriculture, including a shift from communal sharing to familial or individual accumulation" (Price and Gebauer 1995:8). The appearance of individualized farming at this time may result from the higher labor costs invested in the land-use, which would increase the value of land and perpetuate ideas about land-ownership. Of course, land-ownership and individualized production are relativist phenomena in a society whose economy was mixed horticulture and hunting-gathering. However, such changes in social relations related to food production have been observed ethnographically.

In his description of Huron horticulture, Trigger (1969:28) noted: "any man could clear as much land as he wished, and this land remained in the possession of his family members as long as they wished to cultivate it. Once abandoned, however, a field could be planted by anyone who wished to do so." Exactly the same situation is found in the Dulong society of southwestern China, which represents a small-scale farming and hunting-gathering society (Song and Shen 1999).

I have noted earlier that there is similarity between foraging collectors and early farmers in settlement organization and resource procurement. However, the major difference between the two forms of society is probably the productive organization. With the emergence of field horticulture, households in the farming societies became basic economic units and started developing the ownership of land and setting productive boundaries between and within their societies. This phenomenon unlikely existed in the foraging (collectors) societies. The individualized production (household economic activity) could be responsible for different requirements of stone tool designs and use.

This is not to say that generalized lithic production can only be achieved through individualized production. Nor do I suggest that in early hunting-gathering society, individual foragers did not produce specialized tools; stone tool production is in fact a matter of individual performance. Lithic production by individuals and individualized production have completely different meanings; the former is a form of stone tool manufacture whereas the latter refers to economic organization in food procurement. While hunting or gathering took place in a co-operative manner (in a general sense for example, communal hunting-drives or gathering), horticultural activity was organized by household units. Co-operative production in foraging societies may likely have required formalized toolkits that resulted in standardized production. However, with the establishment of individualized production in food-producing societies, the members of the household productive units were likely responsible for their own production needs.

Although few researchers have demonstrated the development of lithic production in households in an archaeological context, Parry's (1987) study of Formative household lithic production may provide some evidence. He states:

> It appears that occupants of every household manufactured flake tools for their own use. Every excavated Early or Middle Formative household unit, even those with small sample sizes, had evidence of flake production, including decortication flakes, unused flakes, exhausted flake cores, angular fragments, hammerstones, or other debris. In every case, the quantities of debitage in each household units exceed the number of tools, and seem to be sufficient to account for all of the flake tools present in that unit. Thus, it would seem that flake production was not a specialized craft; on the contrary, individuals in every household manufactured flake tools (Parry 1987:57).

Other archaeological evidence for household lithic production in northeastern North American comes from the Labras Lake site in the American Bottom (Yerkes 1987, 1989; Phillips et al. 1980). Yerkes has demonstrated, through studies of lithic technology and tool use, that Mississippian household production constitutes the basic economic foundation in these communities. Further, he implies that such strong household production would prompt social differentiation through accumulation or redistribution of "trade" goods and ceremonial items. The archaeological facts are that, among the six house structures examined, "all of the houses seem to have about the same number of deep 'storage/refuse' pits" (Yerkes 1989:202). Microwear analysis, performed by Yerkes, indicates "all six of the Mississippian houses contained domestic tools that were used to cut meat, scrape hide, and work wood, bone, and antler" (Yerkes 1989:205). These facts suggest that households became essential units for the manufacture and use of stone tools.

Tools produced in household economic units were probably needed for the performance of a wide range of use-tasks responding to different working systems such as horticulture, hunting, fishing, gathering, food preparation, etc.. Thus, it should be expected that there are many more tool use-tasks in the farming context than in the foraging context. A comparison of stone tools use-tasks from the three lithic assemblages suggest that the more recent the site is, the more tool use-tasks may likely be present (Fig. 9.4)*. In the HH Middle Woodland site, only 21 use-tasks are indicated by use-wear. A dramatic increase of stone tool use-tasks occurred in the Grand Banks and Lone Pine assemblages; 32 and 36 kinds, respectively. The new use-tasks at both later period sites are plant-working, bone-working, or soil-digging use-related activities. However, the difference in the number of use-tasks between the early and later assemblages could have been affected by sample sizes. The samples selected from the HH assemblage for microscopic analysis are much smaller than those from both the Grand Banks and Lone Pine assemblages. Nevertheless, the patterns presented here should not be ignored and need to be confirmed by other data.

A similar pattern of tool use has also been evidenced through the study of the Labras Lake lithic assemblages. Philips (1980) and Yerkes (1987, 1989) observed a different tool use pattern from Late Archaic, through Late Woodland, to the Mississippian components:

> Microwear analysis revealed the same sort of parallel pattern in the functional classifica-

* Given unrepresentative samples of Young 1, I excluded the Young 1 assemblage from the comparison.

Tool Motion \ Use-Task \ Contact Material	soft animal (SA) meat, fresh hide, etc	soft vegetal (SV) plant, grass, etc.	medium-soft vegetal (1M) fresh wood, hard plan, etc	medium animal (MA) dried hide, fat, fish, etc.	medium-hard vegetal (2M) dried, seasonal wood	hard animal (1H) fresh bone, antler, etc	very-hard animal (2H) dried bone, antler	hard inorganic (3H) stone, ceramic, etc.
cut/saw	HH GB LP	HH GB LP	HH GB LP	GB LP	HH LP	HH GB LP	LP	
slice/curve	HH GB LP		LP	HH GB LP		GB		
scrape/shove	HH GB LP	GB LP	HH GB LP	HH GB LP	HH GB LP	HH GB LP	HH GB LP	
plane/whittle	LP		HH GB LP	GB	GB LP	HH GB LP	GB	
projection	GB			HH GB LP	LP	GB		
bore/drill	LP		GB LP		HH GB LP	GB LP		
grave	LP		GB		GB	HH GB LP	LP	
chop					GB LP			
hoe/dig								GB LP
pound/grind					HH	LP		LP
wedge					HH GB LP			
pick			LP			HH		

Total use-task: HH 21 UT; Grand Bank (GB) 32 UT; Lone Pine (LP) 36 UT

Fig. 9.1: Comparison of Use-Tasks at the Three Sites.

tion of the 308 utilized artifacts in the Labras Lake microwear sample. However, there were some differences. Plant knives were found in the Late Woodland and Mississippian samples, but they were absent in the Late Archaic microwear assemblage, and shell drills were only found in Mississippian contexts (Yerkes 1987:185).

They also noted that at Labras Lake, bifacial tools produced by standardized technology were used for a variety of tasks that were not often related to their form. "Bifacial tools in the microwear assemblages were used as bone, antler, or wood saws as often as they were used to cut meat, so the designation 'bifacial knife' is something of a misnomer" (Yerkes 1987:186). This observation is, not surprisingly, consistent with the findings of bifacial tool use at Grand Banks (Chapter 5).

Summary and Future Research

In the above discussion, I have considered both external and internal factors that possibly influenced the shift to generalization of stone tool production. Increased sedentism, the use of local resources, increase of localized population, time-stress, and the establishment of land-ownership and individual production, are all associated with the origins of food production. Since there is no single factor that results in the emergence of food production, it can be said that the shift to the generalization of stone tool production can only be caused by an integrated relationship of external conditions (sedentism, raw material variability, increase of population) and internal elements (e.g., time-stress, individual production, land-use).

In summary, the model proposed here contends that there is a plausible relationship between the establishment of generalized lithic production and the emergence of food production during the Princess Point period in the study region. I do not suggest, however, that specialized lithic production occurs only in association with mobile hunter-gatherers. Shott (1986) has demonstrated that small hunting-gathering groups would likely use generalized toolkits in response to the various tasks needed in more highly mobile settlements. The lithic technology of small and highly mobile hunter-gatherer groups, however, is commonly characterized by highly standardized reduction sequences and a greater degree of diversity and complexity of formalized tool types. It is indeed mobility and other relevant variables (e.g., raw material availability) that could cause or, more accurately, influence the pattern of generalized lithic production among hunter-gathering groups. Mechanisms for this may differ a great deal from those that exist in early farming societies. A detailed study of technological variability among small, mobile hunting-gathering groups from a lithic production system perspective would yield many more insightful results, but such a task is beyond the scope of this work.

It should be noted that complex hunter-gatherers (collectors) are also characterized by large populations, intensification of subsistence and technology, sedentism, and some level of social circumscription (Price and Brown 1985). It would not be surprising to see the trends towards the shift from specialized to generalized lithic production in logistic foraging societies. In recent studies of agricultural origins, it is suggested that agriculture emerges initially among more complex groups of hunter-gatherers in areas with substantial resources (Price and Gebauer 1995; Gebauer and Price 1992). Therefore, if the generalization of stone tool production appeared in the complex hunter-gatherer groups, it would also be ultimately associated with the emergence of food production.

It is unfortunate that at present this study is not able to independently support the model using the Grand Banks case since our understanding of the Princess Point complex is preliminary. However, the model is proposed on the basis of our understanding of the Princess Point lithic production system. The study of the Grand Banks lithic assemblage demonstrates for the first time that the Princess Point lithic industry indeed represents a generalized production. Use-wear analysis indicates that a great variety of use-tasks were performed by simplified flake tools in Princess Point, rather than through bifacial tools. Nevertheless, to test this model significantly more evidence is needed. Therefore, this alternative explanation of the change in lithic technology during the Middle to Late Woodland in Southern Ontario should serve as an exploratory hypothesis for further study, rather than as a conclusive generalization.

As expected, there are two main problems with testing this model in future studies. First, although social relations are taken into account in this model, an understanding of social relations in lithic production systems needs to be further enhanced through an extended ethnographic analogy. Future lithic study should not only investigate the characteristics of lithic technology, but also should explore the social dynamics that lithic production may reflect. Archaeologically, the study of lithic artifacts from household and mortuary contexts should be appropriate to test the model. An in-depth study of Princess Point settlement patterns is needed. It is unfortunate that the lack of household settlement data from the Princess Point site does not enable us to support the assumption that the shift from co-operative production to individualized production took place in the study region. However, the relationships between social relations and patterns of lithic production are based on comparative data and logical interpretations, and could be tested when more lithic data becomes available.

Second, to test the proposed model requires a more comprehensive data set for intersite comparisons on a regional scale. Although three of the four lithic assemblages selected for this study have been interpreted as inhabitant occupations (except Young 1), the interpretation of tool use pattern can still be biased by possible difference in site functions and by sample sizes. A comparative study of a few lithic assemblages of the Middle Woodland sites in the study region would be quite helpful to clarify the relationship between lithic production and food production. Considerably more data from the Princess Point sites in the Lower Grand River Valley or other areas will also be necessary in future studies. I suggest that the study of lithic production systems should be carried out on a regional scale to achieve a comprehensive understanding of the transformation of lithic production.

In summary, this study has carried out the first detailed analysis of lithic materials from a Middle to Late Woodland transitional culture in southwestern Ontario. The two general goals that address the lithic production system of the Princess Point Complex were achieved in this study. At the descriptive level, I have characterized the pattern of Princess Point lithic production. At the interpretive level, I have offered a model for explaining the transformation of lithic production during the transition to food production in the study region. Both of these goals are directed towards a comprehensive understanding of the Princess Point lithic production system.

According to the research design outlined for an archaeological study of lithic production systems (Fig 1.2), the highest level of interpretations could be attained through a reconstruction of the organization of lithic production. Given the data available and our preliminary understanding of this cultural complex, I am unable to provide comprehensive interpretations on this issue. However, as I have argued earlier, lithic production systems will be better understood when a series of social variables are taken into account. Based upon lithic data drawn from this study, it is plausible to assume that the organization of lithic production within the Princess Point Complex represents a communal mode of production (Lee 1981, 1988, 1990; Wolf 1982; Patterson 1993). According to this mode of production, no individual at the site was able to control the lithic resources and labor; such evidence does not exist at the Grand Banks site. The resources necessary for production were widely accessible to anyone. In a communal society like Princess Point, the stone tool production would have been directed towards making tools for a variety of daily-life use-tasks, rather than for the purposes of exchange or of surplus accumulation. The reconstruction of lithic production organization should provide significantly more information for the retrieval and reconstruction of the political-economic structure of Princess Point, but at present the archaeological data is far from achieving this goal. Any further investigations of the Princess Point Complex, especially addressing its settlement pattern, catchment analysis, subsistence pattern, and site function, will provide insightful data for testing the hypotheses proposed in this study.

Regardless of our limited data, the analysis of lithic artifacts in this study has yielded fruitful data for understanding the Princess Point lithic industry. It is hoped that the methodology developed for this study will also be useful to others conducting lithic analyses of comparative assemblages in this region. At the very least, this study has demonstrated that the results from an integrated study of typo-technological and use-wear analysis are informative. The study of flaked stone artifacts within the framework of the lithic production system is significant in order for us to understand both the basic characteristics of lithic industries and the changes in lithic technology over time.

Photo 1: Flaked stone artifacts recovered from Lone Pine (AfGx-113)

Photo 2: Flaked stone artifacts recovered from Lone Pine (AfGx-113)

Photo 3: Flaked stone artifacts recovered from Grand Bank (AfGx-3)

Photo 4: Flaked stone artifacts recovered from Grand Bank (AfGx-3)

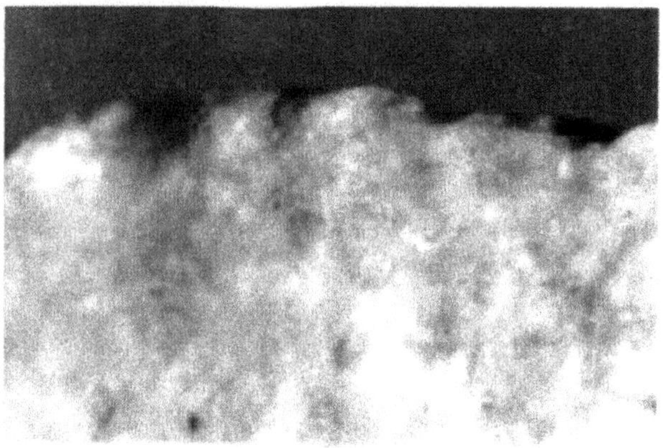

Photo 5: ES#18, used for cutting dried antler, very hard material. Uneven, denticulated distribution, snap and crushed scarring. Note use-polish near the edge of the center. 20x.

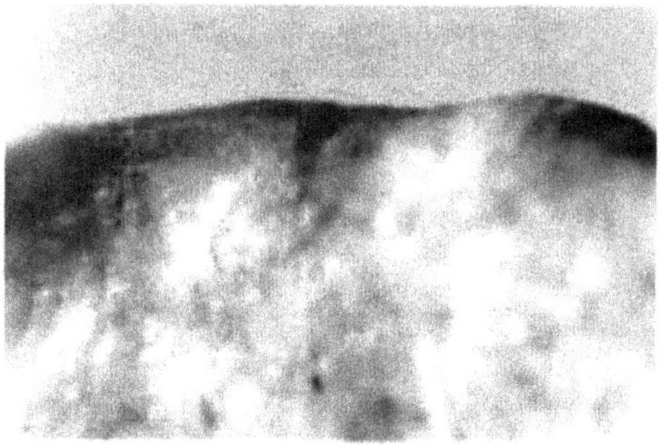

Photo 6: ES#18, same use location as Photo 5, magnified to show the use-polish. Note the lineal appearance of polish on the left part of the edge. 50x.

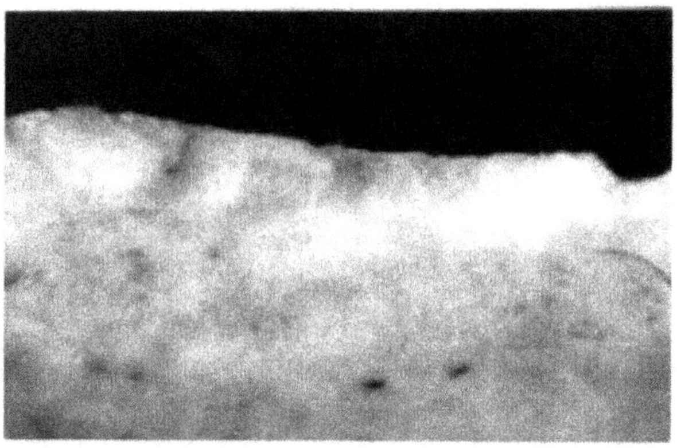

Photo 7: ES#02, used for sawing dried bone. Medium to large feathered scarring, heavy rounding, and bright polish. 20x.

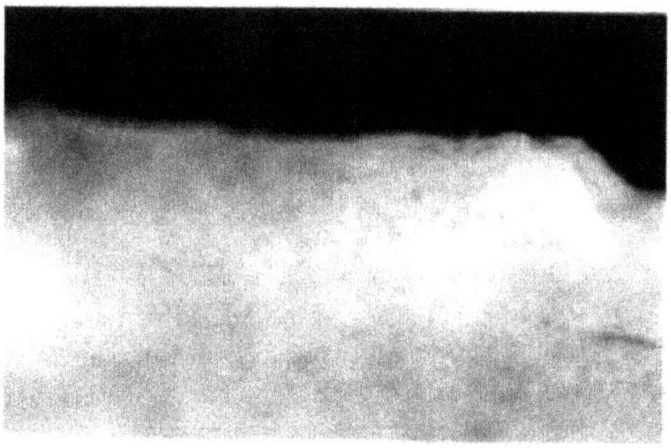

Photo 8: ES#02, same used location as the one of photo 7. Magnified to show the use-polish on the right side corner. 45x.

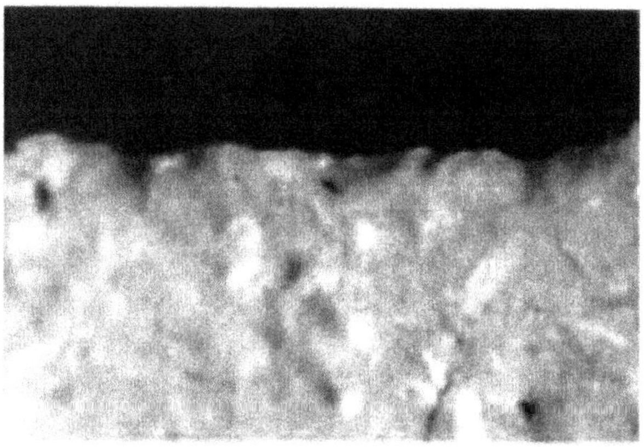

Photo 9: ES#16 used for scraping soft vegetal (potato). Small sized, feathered scars appear on the dorsal side only. 35x.

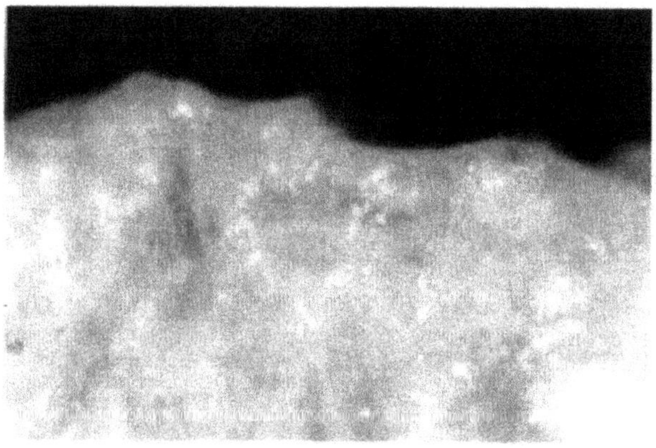

Photo 10: ES#16, same use location as Photo 9, ventral side. Note polish that is formed perpendicular on the working edge. 100x.

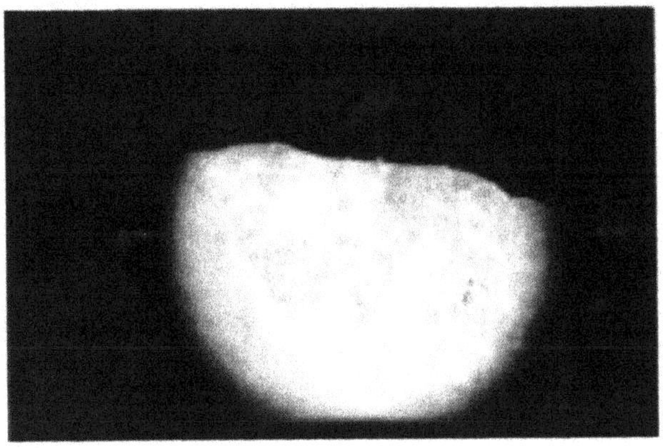

Photo 11: ES#38, used for cutting grass. Very small-sized feather scars unevenly distributed along the working edge. Rounding and polish are observed. 63x.

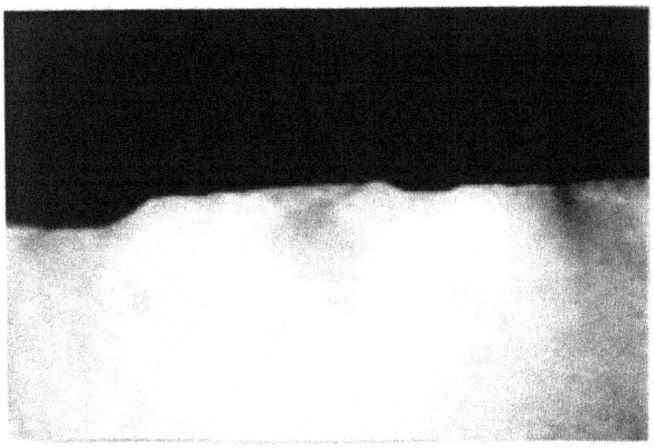

Photo 12: TU#A11, used for sawing dried bone. Heavy rounding, matt polish and large hinged and feather scarring on the edge. 20x.

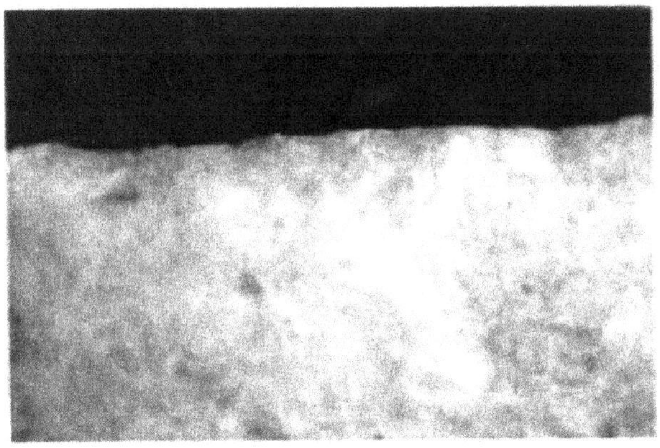

Photo 13: TU#B137 shows meat-cutting wear: rough rounding edge, bifacially incipient polish, and small sized and feather-terminated scarring. 30x.

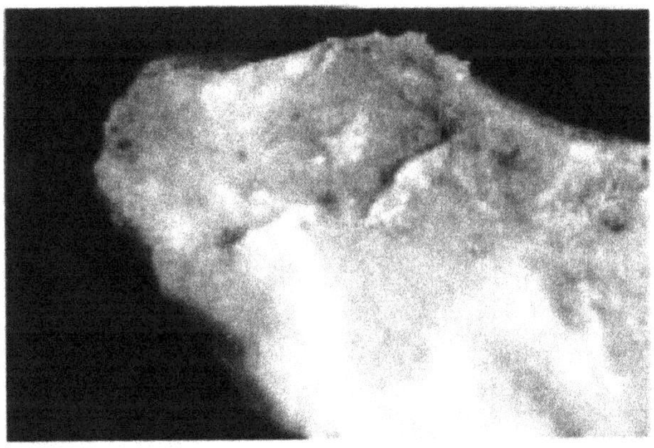

Photo 14: ES#12, a drilling tip shows heavy rounding and crushed scarring on all edges, a little polish can be seen on the worked ridge. 20x.

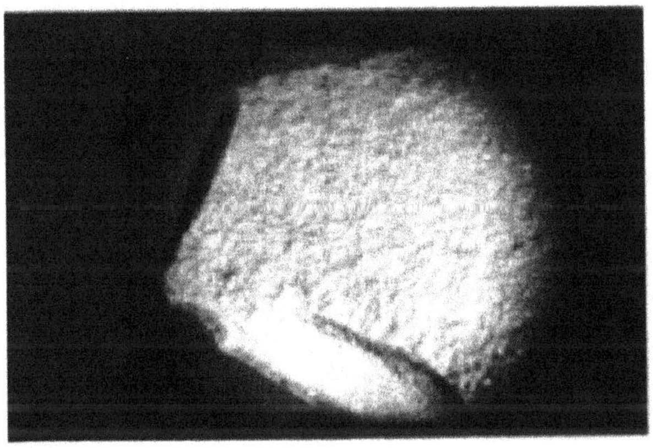

Photo 15: ES#33, large snap scars are the result of penetrating a medium animal substance (meat with bone contact). 20x.

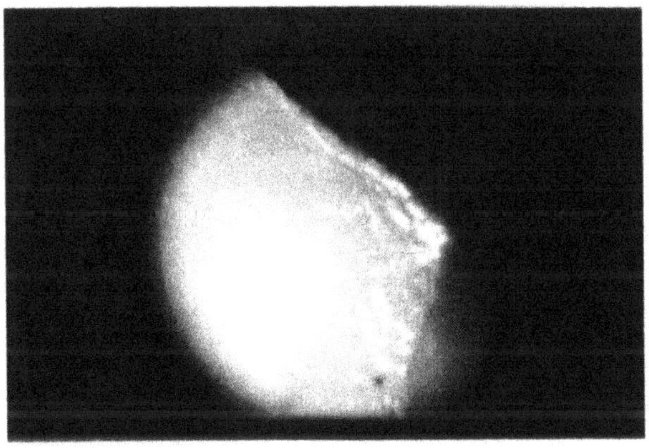

Photo 16: ES#40, heavy rounding and bright polish on the contact surface of the graving tip. 120x.

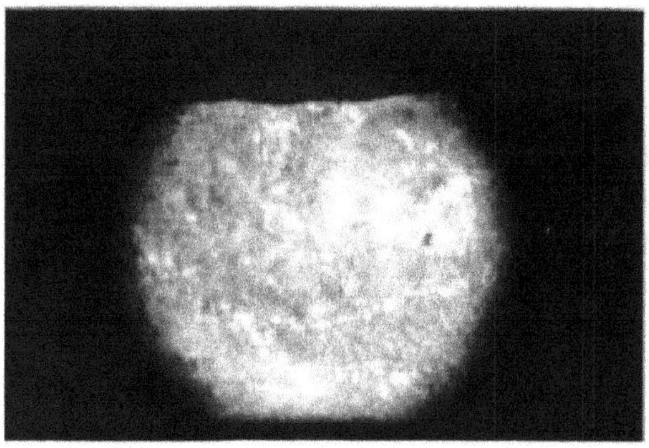

Photo 17: ES#41, used for chopping tree in the experimentation. Polish apears on the working edge. 48x.

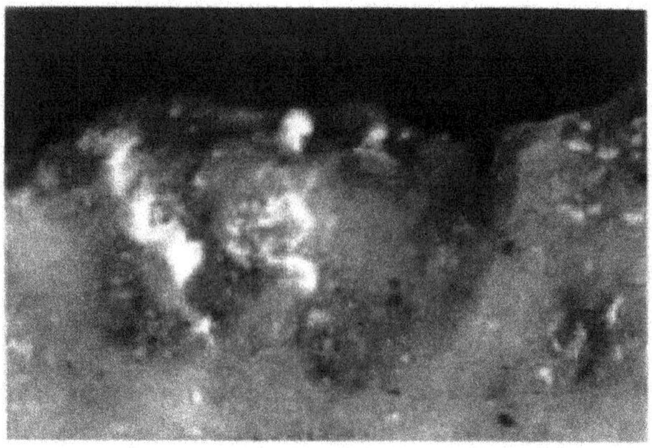

Photo 18: ES#19, used for scraping cooked pig bone. Large feather and step-terminated scar perpendicular to the edge, rounded and crushed edge. 20x.

Photo 19: ES#36, used for cutting hard wood, noting large step and feather scars in directional distribution, and rounded edge. 20x.

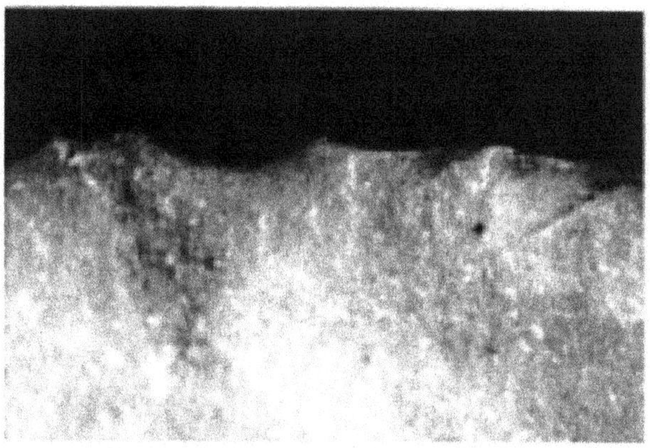

Photo 20: TU#B145, used for sawing soft animal substance (leather), noting small-to-medium sized snap and feather scars in directional distribution. 20x.

Photo 21: Lone Pine projectile point (AfGx-113:909), crushed edge and medium sized, step and feather scarring, interpreted as sawing hard wood. 15x.

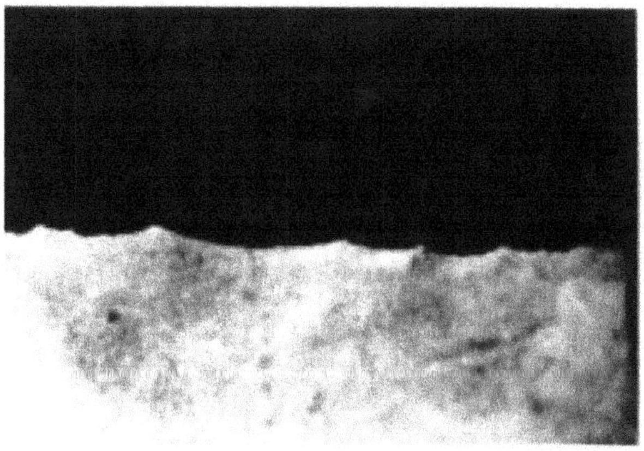

Photo 22: Lone Pine projectile point (AfGx-113:3239), rolled-over scarring pattern with medium feather scars, interpreted as cutting/sawing soft wood materials. 10x.

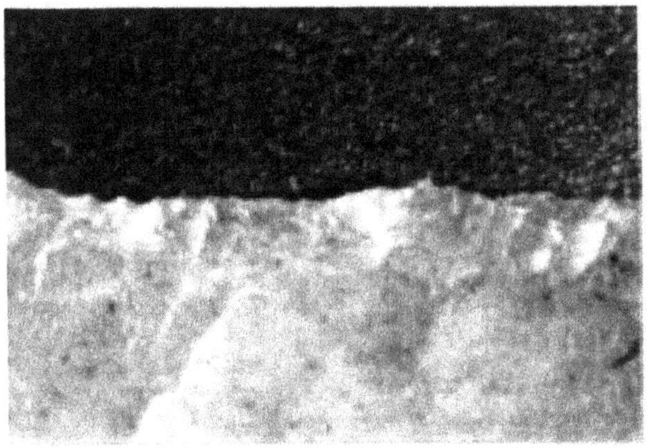

Photo 23: Lone Pine projectile point (AfGx-113:960): small scarring on the worn edge, interpreted as haft wear. 13x.

Photo 24: Lone Pine projectile point (AfGx-113:372), worn edge showing hand-hold wear pattern: blunt edge with a row of small regular scars. 13x.

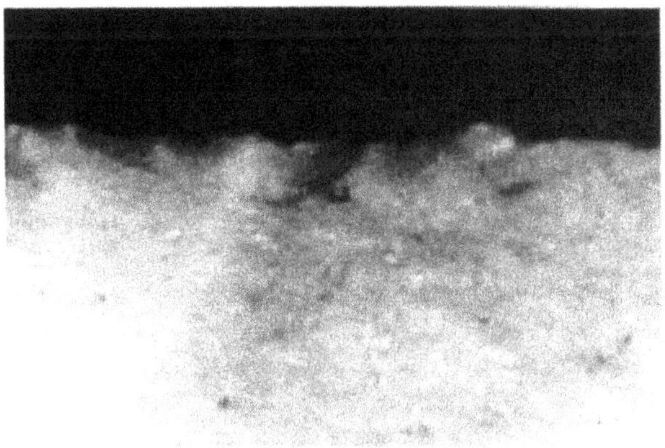

Photo 25: Utilized core (AfGx-3:5963): small to medium sized feather-terminated scars and rounded working edge, interpreted as meat cutting wear on location A. 20x.

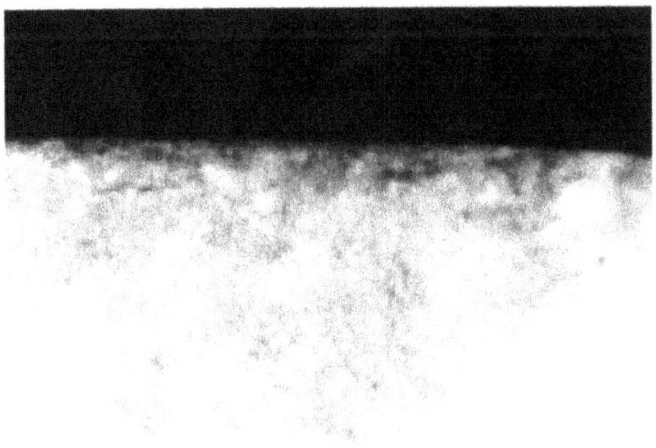

Photo 26: Utilized core, same as the one of photo 25. Blunting and small crushed scars, interpreted as hand-holding wear. 16x.

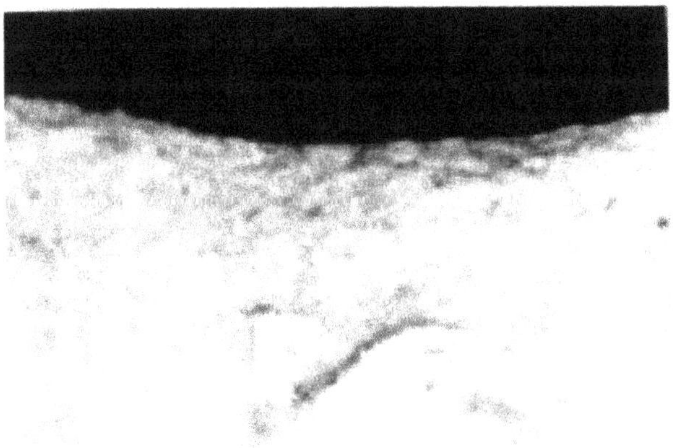

Photo 27: Utilized core (AfGx-3:6321): unifacial distributed feather and step scars and light polish, interpreted as a wood planing tool. 16x.

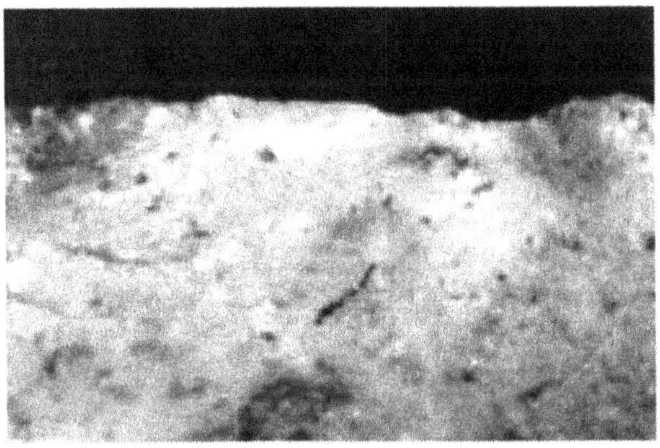

Photo 28: Utilized unretouched-flake (AfGx-3:5353.1A) rolled-over scarring, feather-terminated, and incipient polish on ventral side, interpreted as a wood scraping working edge. 28x.

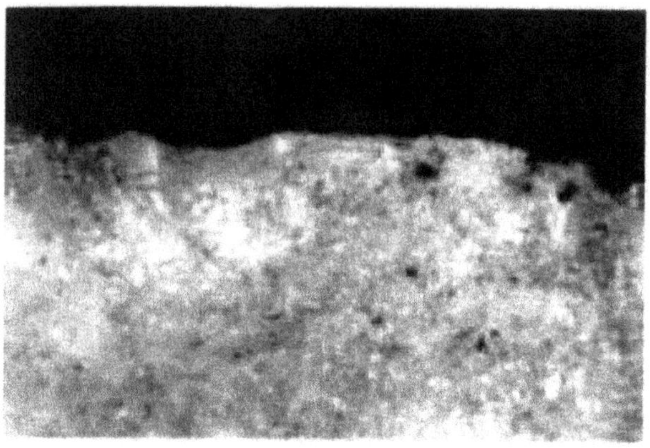

Photo 29: Utilized retouched-flake (AfGx-3:5353.1B): denticulated scarring, medium sized feather-terminated scars, rounding edge, interpreted as an edge employed for cutting medium-soft wood. 15x.

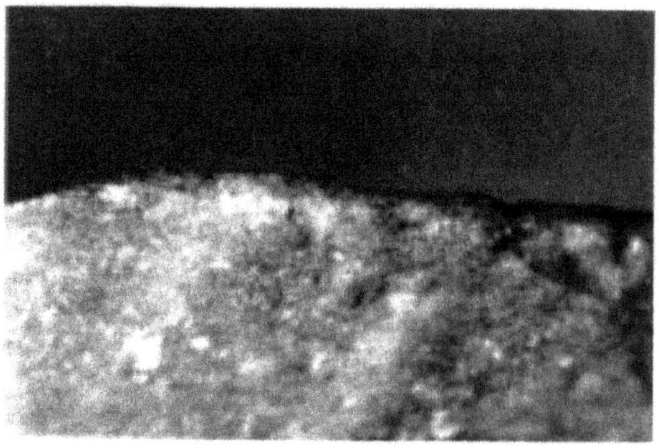

Photo 30: Utilized retouched-flake, same as the above on C: heavy rounding on the working edge and bright polish along the edge that was probably caused by scraping of animal skins or hides. 15x.

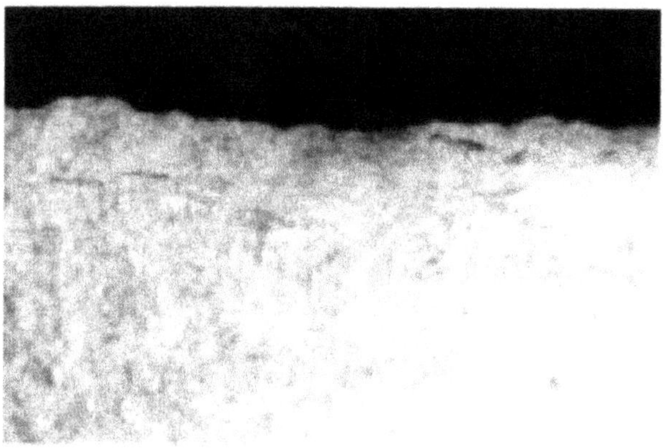

Photo 31: Utilized unretouched-flake (AfGx-3:6886.1A): shallowed stepped and feathered scars on the ventral side and bright polish. Used for planning wood. 18x.

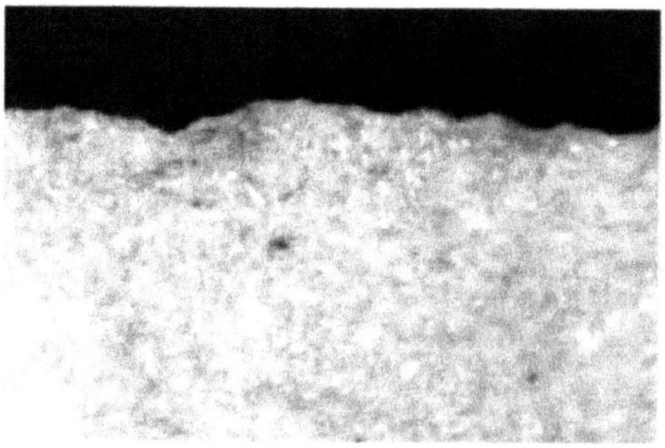

Photo 32: Utilized unretouched-flake (AfGx-3:6886.1B) bright polish on the ventral side only and small feathered and stepped scars, interpreted as wood planning working edge. 18x.

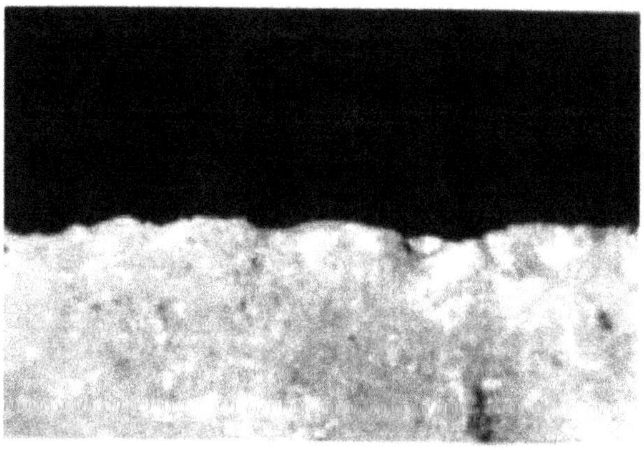

Photo 33: Utilized unretouched-flake (AfGx-113:1252.1) location A, dorsal side showing denticulated and small feather scars, probably causing by slicing meat with bone contact. 30x.

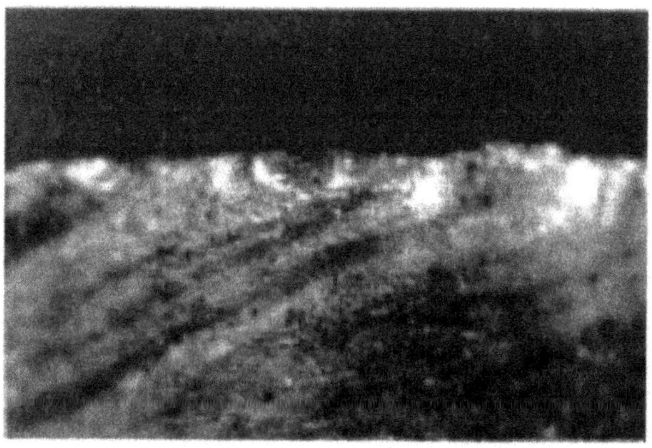

Photo 34: Utilized unretouched-flake (AfGx-113:1252.1) same working edge as the above on ventral side where the surface has mostly contacted with the material, bright polish and small scars appear. 30x.

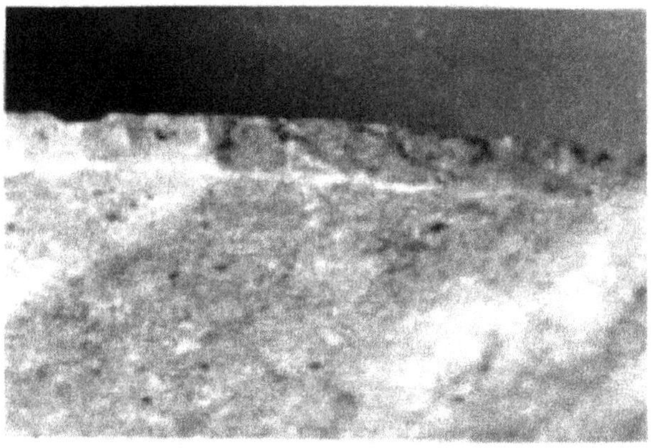

Photo 35: Utilized unretouched-flake (AfGx-113:1252.1) location B, where there are intentional micro-retouch scars to dull the edge for hand-holding. Prehensile wear is also indicated by small utilized scars along the edge. 11x.

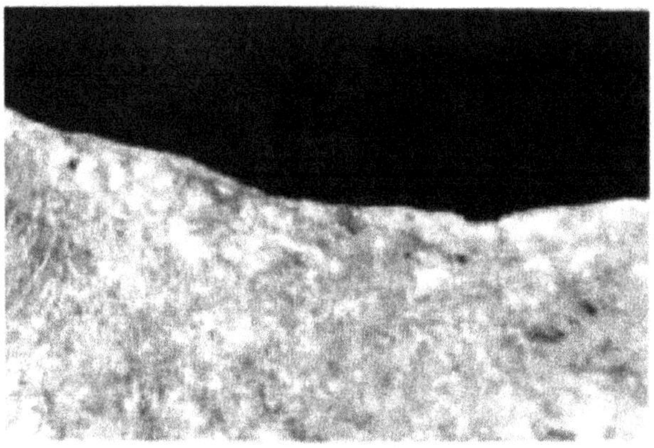

Photo 36: Utilized unretouched-flake (AfGx-113:1252.1) same working edge as the left on ventral side where the surface has mostly contacted with material, bright plish and small scars appear. 30x.

References

Adams, W. Y. and E. W. Adams 1991 *Archaeological Typology and Practical Reality: A Dialectical Approach to Artifact Classification and Sorting*. Cambridge University Press, Cambridge.

Ahler, S. A. 1971 *Projectile Point Form and Function at Rodgers Shelter, Missouri*. Missouri Archaeological Society Research Series 8.

Ahler, S. A. 1989a Experimental Knapping With KRF and Midcontinent Cherts: Overview and Application. In *Experiments in Lithic Technology*, edited by Daniel S. Amick and Raymond P. Mauldin, pp. 199-234. BAR International Series 528. Oxford.

Ahler, S. A. 1989b Mass Analysis of Flaking Debris: Studying the Forest Rather Than the Trees. In *Alternative Approaches to Lithic Analysis*, edited by Donald O. Henry and George H. Odell, pp. 85-118. Archaeological Paper of the American Anthropological Association No. 1.

Aikens, C. M. 1970 *Hogup Cave*. University of Utah Anthropological Papers, No. 93.

Amick, D. S. and R. P. Mauldin (editors) 1989 *Experiments in Lithic Technology*. BAR International Series 528, Oxford.

Amick, D. S., R. P. Mauldin and L. R. Binford 1989 The Potential of Experiments in Lithic Technology. In *Experiments in Lithic Technology*, edited by Daniel S. Amick and Raymond P. Mauldin, pp. 1-14. BAR International Series 528, Oxford.

Andrefsky, W. Jr. 1994 Raw-material Availability and the Organization of Technology. *American Antiquity* 59(1):21-34.

Arnold, J. 1987 *Craft Specialization in the Prehistoric Channel Islands, California*. University of California Publications in Anthropology No. 18. University of California at Berkeley, Berkeley, CA.

Arnold, J. 1992 Complex Hunter-Fishers of Prehistoric California: Chiefs, Specialists, and Maritime Adaptations of the Channel Islands. *American Antiquity* 57(1):60-84.

Baerreis, D. and R. A. Bryson 1965 Climatic episodes and the dating of Mississippian cultures. in *Wisconsin Archaeology* 46(4):203-220

Bamforth, D. B. 1986 Technological Efficiency and Tool Curation. *American Antiquity* 51:38-50.

Bamforth, D. B 1991 Technological Organization and Hunter-Gatherer Land Use: A California Example. *American Antiquity* 56(1):216-234.

Bar-Yosef, O. 1970 *The Epipaleolithic Cultures of Palestine*. Unpublished PhD dissertation, Hebrew University.

Bar-Yosef, O. (editor) 1998 *The Transition to Agriculture in the Old World*: Special issue of *The Review of Archaeology*

Bar-Yosef, O. and A. Belfer-Cohen 1989 The Origins of Sedentism and Farming Communities in the Levant. *Journal of World Prehistory* 3:447-498.

Bar-Yosef, O. and A. Belfer-Cohen 1991 Form Sedentary Hunter-Gatherers to Territorial Farmers in the Levant. In *Between Bands and States*, edited by S. A. Gregg, pp. 181-202. Center for Archaeological Investigations, Occasional Paper No. 9, Carbondale, IL.

Bar-Yosef, O. and A. Belfer-Cohen 1992 From Foraging to Farming in the Mediterranean Levant. In *Transitions to Agriculture in Prehistory*, edited by Anne B. Gebauer and T. Douglas Price, pp. 21-48. Mono-graphs in World Archaeology No.4, Pre-history Press, Madison, Wisconsin.

Bar-Yosef, O. and R. H. Meadow 1995 The Origins of Agriculture in the Near East. In *Last Hunters-First Farmers: New Perspectives on the Prehistoric Transition to Agriculture*, edited by T. Douglas Price and Anne Birgitte Gebauer, pp. 39-94. School of American Research Press, Santa Fe, New Mexico.

Barham, L. S. 1987 The Bipolar Technique in Southern Africa: A Replicative Experiment. *South African Archaeological Bulletin* 42:45-50.

Baumler, M. F. and C. E. Bownum 1989 Between Micro and Macro: A Study in the Interpretation of Small-Sized Lithic Debitage. In *Experiments in Lithic Technology*, edited by Daniel S. Amick and Raymond P. Mauldin, pp. 101-116. BAR International Series 528, Oxford.

Bekerman, A. 1995 *Relative Chronology of Princess Point Sites*. Unpublished M.Sc. Thesis, Department of Anthropology, University of Toronto.

Bender, B.1975 *Farming in Prehistory: From Hunter -Gatherer to Food Producer*. John Baker, London.

Bender, B.1978 Gatherer-Hunter to Farmer: A Social Perspective. *World Archaeology* 10:204-222.

Bender, B. 1981 Gatherer-Hunter Intensification. In *Economic Archaeology*, edited by A. Sheridan and G. Bailey. pp. 149-158 BAR International Series 96, Oxford.

Bernabo, J. C. 1981 Quantitative Estimates of Temperature Changes over the Last 2,700 Years in Michigan Based on Pollen Data. *Quaternary Research* 15:143-159.

Biggar, H. P. (editor) 1930 *A Collection of Documents Relating to Jacques Cartier and the Sieur de Roberval.* Publications of the PAC, No. 14, Ottawa.

Binford, L. R. 1977 Forty-Seven Trips: A Case Study in the Character of Archaeological Formation Process. In *Stone Tools as Cultural Markers: Change, Evolution and Complexity,* edited by R. V. S. Wright, pp. 24-36. Australian Institute of Aboriginal Studies, Canberra.

Binford, L. R. 1979 Organization and Formation Process: Looking at Curated Technologies. *Journal of Anthropological Research* 35:255-275.

Binford, L. R. 1980 Willow Smoke and Dog's Tails: Hunter-Gatherer Settlement Systems and Archaeological Site Formation. *American Antiquity* 45(1):4-20.

Binford, L. R. and G. I. Quimby 1963 Indian Sites and Chipped Stone Materials in the Northern Lake Michigan Area. *Fieldiana Anthropology* 36(12):227-307.

Bleed, P. 1986 The Optimal Design of Hunting Weapons: Maintainability or Reliability. *American Antiquity* 51:737-747.

Blitz, J. H. 1988 Adoption of the Bow in Prehistoric North America. *North American Archaeologist* 9:123-145.

Bordes, F. 1961 *Typologie du Paleolithique Ancien st Moyen 2.* Bordeaux:Delmas.

Bordes, 1968 *The Old Stone Age*. McGraw-Hill, New York.

Bowyer, V. E 1995 *Paleoethnobotanical Analysis of Two Princess Point Sites: Grand Banks (AfGx-3) and Lone Pine (AfGx-113) in the Grand River Area, Ontario.* Unpublished M.Sc. Thesis, Department of Anthropology, University of Toronto.

Brink, J. W. 1978 *An Experimental Study of Microwear Formation on Endscrapers*. National Museum of Man, Mercury Series No. 83, Ottawa.

Brose, D. S. 1994 *The South Park Village Site and the Late Prehistoric Whittlesey Tradition of Northeast Ohio.* Monographs in World Archaeology No. 20. Prehistory Press, Madison, Wisconsin.

Brown, D. M., G. A. Mckay and L. J. Chapman 1974 *The Climate of Southern Ontario.* Climatological Studies No. 5, Department of Transport, Toronto, Canada, Toronto.

Bryson, R. and W. M. Wendland 1967 Tentative climatic patterns for some late glacial and post-glacial episodes in central North America. In *Life, Land and Water* edited by W. J. Mayer-Oakes. pp. 271-298. University of Manitoba Press. Winnipeg.

Burden, E. T., J. McAndrews and G. Norris 1986 Palynology of Indian and European Forest Clearance and Farming in Lake Sediment Cores from Awaenda Provincial Park, Ontario. *Canadian Journal of Earth Sciences* 23:55-65.

Bursey, J. A. 1994 Chaingate (AhGw-11): A Late Archaic Perkiomen Site in Burlington, Ontario. *Ontario Archaeology* 57:45-63.

Bursey, J. A. 1995 The Transition from the Middle to Late Woodland Periods: A Re-Evaluation. In *Origins of the People of the Longhouse*, edited by Andre Bekerman and Gary Warrick, pp. 43-54. The Ontario Archaeological Society, Toronto.

Cabeceiras, D. 1994 Faunal Report: the Lone Pine site (AfGx-113). report on file Anthropology Department, University of Toronto.

Callahan, E. 1979 The Basics of Biface Knapping in the Eastern Fluted Point Tradition: A manual for flintknappers and lithic analysis. *Archaeology of Eastern North America* 7(1):1-180.

Campbell, C. and I. D. Campbell 1988 The Little Ice Age and Neutral Faunal Assemblages. *Ontario Archaeology* 49:13-33.

Carr, P. J. (editor) 1994 *The Organization of North American Prehistoric Chipped Stone Tool Technologies.* International Monographs in Prehistory, Ann Arbor, MI.

Casey, J. L. 1993 *The Kintampo Complex in Northern Ghana: Late Holocene Human Ecology on the Gambaga Escarpment.* Unpublished PhD dissertation, Depart-ment of Anthropology, University of Toronto.

Chang, K. C. 1967 *Rethinking Archaeology*. Random House, New York.

Chang, K. C. 1986 *The Archaeology of Ancient China.* Yale University Press, New Haven.

Chapman, L. J. and D. F. Putnam 1969 *The Physiography of Southern Ontario,* (Second

Edition). University of Toronto Press, Toronto.

Chapman, R. C. 1977 Analysis of the Lithic Assemblages. In *Settlement and Subsis-tence Along the Lower Chaco River: The CGP Survey*, edited by Charles A. Reher, pp. 371-452. University of New Mexico Press, Albuquerque.

Cheng, T. and R. E. M. Hedges 1994 AMS Radiocarbon Dating of Pottery from Pengtoushan and Hujiawuchang sites and the Earliest Rice Remains in China [in Chinese with English abstract]. *Wenuu* 1994(3):88-94.

Christenson, A. L. 1986 Projectile Point Size and Projectile Aerodynamics: An Exploratory Study. *Plains Anthropologist* 31:109-128.

Clark, J. D. and M. R. Kleindienst 1974 The Stone Age Cultural Sequence: Terminology, Typology and Raw Material. In *Kalambo Falls Prehistoric Site*, edited by J. Desmond Clark, pp. 71-106. vol. II. Cambridge University Press, Cambridge.

Clark, J. D., J. L. Philips and P. S. Staley 1974 Interpretations of Prehistoric Technology from Ancient Egyptian and Other Sources, Part I: Ancient Egyptian Bows and Arrows and Their Relevance for African Prehistory. *Paleorient* 2:323-388.

Clark, J. E. 1987 Politics, Prismatic Blades, and Mesoamerican Civilization. In *The Organization of Core Technology*, edited by J. K. Johnson and C. A. Morrow. Westview Press, Boulder.

Clarke, D. L. 1968 *Analytic Archaeology*. Methuen & Co. Ltd., London.

Close, A. E. 1978 The Identification of Style in Lithic Artefacts. *World Archaeology* 10:223-237.

Close, A. E. 1978 1989 Identifying Style in Stone Artefacts: A Case Study from the Nile Valley. In *Alternative Approaches to Lithic Analysis*, edited by Donald O. Henry and George H. Odell, pp. 3-26. Archaeological Papers of the American Anthropological Association No. 1.

Cowan, C. W. and P. J. Watson 1993 Some Concluding Remarks. In *Origins of Agriculture: An International Perspective*, edited by P. J. Watson and C. W. Cowan, pp. 207-212. Smithsonian Institution, Washington, D. C.

Crabtree, D. E. 1972 *An Introduction to Flintworking*. Occasional Papers of the Idaho State University Museum No. 28, Pocatello.

Crawford, G. W. 1993 Prehistoric Plant Domestication in East Asia. In *The Origins of Plant Domestication in World Perspective,* edited by Patty Jo Watson and C. Wesley Cowan, pp. 7-38. Smithsonian Institution Press, Washington, D.C.

Crawford, G. W. 1999 Northeast Palaeoethnobotany: How are We Doing? In *Palaeoethnobotany in the Northeast*, pp.227- 236, edited by John Hart, New York State Museum, Albany.

Crawford, G. W. and C. Shen 1998 The Origins of Rice Agriculture: Recent Progress in East Asia. *Antiquity* 72(278):858-866

Crawford, G. W. and D. G Smith 1996 Migration in Prehistory: Princess Point and The Northeastern Iroquoain Case. *American Antiquity* 61(4):782-790

Crawford, G. W., D. G. Smith and V. Bowyer 1997 Dating the Entry of Corn (Zea Mays) into the Lower Great Lakes Region. *American Antiquity* 62(1):112-119

Crawford, G. W., D. G. Smith, J. Desloges and A. Davis 1998 Floodplains and Agricultural Origins: A Case Study from South-Central Ontario, Canada. *Journal of Field Archaeology.* 25(2):123-137

Custer, J. F. 1987 Core Technology at the Hawthorn Site, New Castle County, Delaware: A Late Archaic Hunting Camp. In *The Organization of Core Technology*, edited by Jar K Johnson and Carol A. Morrow, pp. 45-62. Westview Press, Inc., Boulder, Colorado.

Davis, D. D. (editor) 1978 Lithics and Subsistence: the Analysis of Stone Tool Use in *Prehistoric Economies*. Vanderbilt University Publications in Anthropology No. 20, Nashville, Tennessee.

Deller, D. B. 1976 Paleo-Indian Locations on Late Pleistocene Shorelines, Middlesex County, Ontario. *Ontario Archaeology* 26:3-19.

Deller, D. B. and C. J. Ellis 1992 *Thedford II: A Paleo-Indian Site in the Ausable River Watershed of Southwestern Ontario.* University of Michigan Museum of Anthropology, Memoir 24, Ann Arbor.

Dice, L. R. 1943 *The Biotic Provinces of North America.* University of Michigan Press, Ann Arbor, Michigan.

Dodd, C. F., D. R. Poulton, P. A. Lennox, D. G. Smith and G. A. Warrick 1990 The Middle Ontario Iroquoian Stage. In *The Archaeology of Southern Ontario to A.D. 1650,* edited by Chris J. Ellis and Neal Ferris, pp.

321-360. vol. 5. Occasional Publication of the London Chapter, Ontario Archaeological Society.

Dunnell, R. C. 1978 Style and Function: A Fundamental Dichotomy. *American Antiquity* 43:192-202.

Dunnell, R. C. 1986 Methodological Issues in Americanist Artifact Classification. In *Advances in Archaeological Method and Theory Vol. 9*, edited by Michael B. Schiffer, pp. 149-207. Academic Press, New York.

Ebert, J. I. 1979 An Ethnoarchaeological Approach to Reassessing the Meaning of Variability in Stone Tool Assemblages. In *Ethnoarchaeology: Implications of Ethnography for Archaeology*, edited by C. Kramer, pp. 59-74, New York, Columbia University Press.

Eley, B. E. and P. H. von Bitter 1989 *Cherts of Southern Ontario*. Royal Ontario Museum, Toronto.

Ellis, C. J. 1984 *Paleo-Indian Lithic Technological Structure and Organization in the Lower Great Lakes Region: A First Approximation*. Unpublished PhD Dissertation, Department of Archaeology, Simon Fraser University.

Ellis, C. J. and D. B. Deller 1988 Some Distinctive Paleo-Indian Tool Types From the Lower Great Lakes Region. *MIdcontinemtal Journal of Archaeology* 13(2):111-158.

Ellis, C. J. and D. B. Deller 1990 Paleo-Indians. In *The Archaeology of Southern Ontario to A.D. 1650*, edited by Chris J. Ellis and Neal Ferris, pp. 37-64. Occasional Publication of the London Chapter, Ontario Archaeological Society.

Ellis, C. J., I. T. Kenyon, and M. W. Spence 1990 The Archaic In *The Archaeology of Southern Ontario to A.D. 1650*, edited by Chris J. Ellis and Neal Ferris, pp. 65-124. Occasional Publication of the London Chapter, Ontario Archaeological Society.

Ellis, C. J. and N. Ferris (editors) 1990 *The Archaeology of Southern Ontario to A.D. 1650*. Occasional Publication of the London Chapter, Ontario Archaeological Society.

Ericson, J. E. 1982 *Production for Obsidian Exchange in California*. In *Contexts of Prehistoric Exchange*, edited by J. E. Ericson and T. K. Earle, pp. 129-148. Academic Press, New York.

Ericson, J. E. 1984 Toward the Analysis of Lithic Production System. In *Prehistoric Quarries and Lithic production*, edited by Jonathon E Ericson and Barbara A. Purdy, pp. 1-9. Cambridge University Press, Cambridge.

Evens, Sir John 1872 *Ancient Stone Implements, Weapons and Ornaments of Great Britain*. Longmans, Green, Reader and Dyer, London.

Fagan, B. M. 1991 *Ancient North America: The Archaeology of a Continent*. Thames and Hudson, London.

Finlayson, W. D. 1977 *The Saugeen Culture: A Middle Woodland Manifestation in Southwestern Ontario*. National Museum of Man, Archaeological Survey of Canada, Mercury Series 61.

Fitting, J. E. 1970 *The Archaeology of Michigan: A Guide to the Prehistory of the Great Lakes Region*. Natural History Press, New York.

Flenniken, J. H. and A. W. Raymond 1986 Morphological Projectile Point Typology: Replication Experimentation and Technological Analysis. *American Antiquity* 51(3):603-614.

Ford, J. A. 1954a Comment on A. C. Spaulding, "Statistical Techniques for the Discovery of Artifact Types." *American Antiquity* 19(4):390-391

Ford, J. A. 1954b On the Concept of Types. *American Anthropologist* 56:42-53.

Ford, R. I. 1974 Northeastern Archaeology: Past and Future Directions. *Annual Review of Anthropology* 3:385-413.

Forsman, M. R. A. 1975 Bipolar Stone Working Technology. In *Primitive Art and Technology, Proceedings of the Seventh Annual Chacmool Conference*, edited by J. S. Raymond, B. Loveseth, C. Arnold and G. Reardon, pp. 16-26. University of Calgary, Calgary.

Fox, W. A. 1979 An Analysis of an Historic Huron Attignawantan Lithic Assemblage. *Ontario Archaeology* 32:61-88.

Fox, W. A. 1981 The Foliate Biface. *Kewa* 81(3):2-4.

Fox, W. A. 1990 The Middle Woodland to Late Woodland Transition. In *The Archaeology of Southern Ontario to A.D. 1650*, edited by Chris J. Ellis and Neal Ferris, pp. 171-188. Occasional Publication of the London Chapter, Ontario Archaeological Society.

Fritz, G. J. 1990 Multiple Pathways to Farming in Precontact Eastern North America. *Journal of World Prehistory* 4(4):387-435.

Grace, R. 1989 *Interpreting the Function of Stone*

Tools: the Quantification and Computerisation of Microwear Analysis. BAR International Series 474, Oxford.

Grace, R. 1990 The Limitations and Applications of Use-Wear Analysis. In *The Interpretative Possibilities of Microwear Studies*, edited by B. Graslund, H. Knutsson, K. Knutsson and J. Taffinder, pp. 9-14. Societas Archaeologica Upsaliensis, AUN 14, Uppsala.

Grace, R. 1993 The Use of Expert Systems in Lithic Analysis. In *Traces et Function; les gestes retrouves*, edited by P. C. Anderson, S. Beyries, M. Otte, and H. Plisson. pp. 389-400. Etudes et Recherches Archaeologiques de l'University de Liege (ERAUL), Vol 50.

Grace, R., K. Ataman, R. Fabregas and C. M. B. Haggren 1988 A Multi-variate Approach to the functional Analysis of Stone Tools. In *Industries Lithiques: Traceologie et Technologie,* edited by Beyries S., pp. 217-230. BAR International Series 411, Oxford.

Greber, N., R. S. Davis and A. S. Dufresne 1981 The Micro Component of the Ohio Hopewell Lithic Technology: Bladelets. In *The Research Potential of Anthropological Museum Collections,* edited by Anne-Marie Cantwell, James B. Griffin and Nan A. Rothschild, pp. 486-528. Annals 376, New York Academy of Sciences, New York.

Green, S. W. and K. E. Sassaman 1983 The Political Economy of Resource Management: A General Model and Application to Foraging Societies in the Carolina Piedmont. In *Ecological Models in Economic Prehistory,* edited by Gordon Bronitsky, pp. 261-290. Arizona State University, Anthropological Research Papers No. 29.

Griffin, J. B. 1983 The Midlands. In *Ancient North Americans,* edited by Jesse D. Jennings, pp. 243-301. W. H. Freeman and Company, New York.

Hall, R. L. 1980 An Interpretation of the Two-Climax Model of Illinois Prehistory. In *Early Native Americans: Prehistory, Demography*, Economy and Technology, edited by D. L. Browman, pp. 401-462. The Hague, Mouton.

Harris, D. R. 1996 *The Origins and Spread of Agriculture and Pastoralism in Eurasia*, Smithsonin Institute Press, Washington D.C.

Hayden, B. (editor) 1979 *Lithic Use-Wear Analysis* Academic Press, New York.

Hayden, B. 1980 Confusion in the Bipolar World: Bashed Pebbles and Splintered Pieces. *Lithic Technology* 9:2-7.

Hayden, B. 1990 Nimrods, Piscators, Pluckers and Planters: The Emergence of Food Production. *Journal of Anthropological Archaeology* 9:31-69.

Hayden, B. 1992 Contrasting Expectations in Theories of Domestication. In *The Transitions to Agriculture in Prehistory*, edited by Anne Birgitte Gebauer and T. Douglas Price, pp. 11-20. Prehistory Press, Madison, Wisconsin.

Hayden, B. and W. K. Hutchings 1989 Whither the Billet Flakes? In *Experiments in Lithic Technology*, edited by Daniel S. Amick and Raymond P. Mauldin, pp. 235-258. BAR International Series 528., Oxford.

Hayden, B., N. Franco and J. Spafford 1996 Evaluating Lithic Strategies and Design Criteria. In *Stone Tools: Theoretical Insights into Human Prehistory,* edited by G. H. Odell, pp. 9-43. Plenum Press Inc., New York.

Heidenreich, C. E. 1971 *Huronia: A History and Geography of the Huron Indians 1600-1650,* McClelland and Stewart Limited

Heidenreich, C. E. 1978 Huron. In *Handbook of North American Indians, Vol. 15, Northeast*, edited by Bruce G. Trigger, pp. 368-388. Smithsonian Institution, Washington.

Heidenreich, C. E. 1990 History of the St. Lawrence-Great Lakes Area to A.D. 1650. In *The Archaeology of Southern Ontario to A.D. 1650*, edited by Chris J. Ellis and Neal Ferris, pp. 475-492. vol. 5. Occasional Publication of the London Chapter, Ontario Archaeological Society.

Henry, D. O. 1973 *The Natufian of Palestine: Its Material Culture and Ecology.* Unpublished PhD dissertation, Department of Anthropology, Southern Methodist University.

Henry, D. O. 1989a *From Foraging to Agriculture: the Levant at the End of the Ice Age.* University of Pennsylvania Press, Philadelphia.

Henry, D. O. 1989b Correlations Between Reduction Strategies and Settlement Patterns. In *Alternative Approaches to Lithic Analysis*, edited by Donald O. Henry and George H. Odell, pp. 139-156. Archaeological Papers of the American Anthropological Association NO. 1.

Henry, D. O. 1995 *Prehistoric Cultural Ecology and Evolution: Insights from Southern Jordan.* Plenum Press, New York and London.

Henry, D. O., C. V. Haynes and B. Bradley 1976 Quantitative Variations in Flaked Stone Debitage. *Plains Anthropologist* 21(71): 57-61.

Hester, T. R. 1973 *Chronological Ordering of Great Basin Prehistory.* Contributions of the University of California, Archaeological Research Facility, No.17.

Hofman, J. L. 1987 Hopewell Blades from Twenhafel: Distinguishing Local and Foreign Core Technology. In *The Organization of Core Technology*, edited by Jar K Johnson and Carol A. Morrow, pp. 87-118. Westview Press, Inc., Boulder, Colorado.

Hole, F. 1986 Chipped-Stone Tools. In *Guila Naquitz: Archaic Foraging and Early Agriculture in Oaxaca*, edited by Kent V. Flannery, pp. 97-139. Academic Press, Inc., Orlando.

Inizan, M., H. Roche and J. Tixier 1992 *Technology of Knapped Stone.* Publie avec le Concours du Centre National de la Recherche Scientifique, Meudon: CREP.

Jelinek, A. J. 1976 Form, Function, and Style in Lithic Analysis. In *Cultural Change and Continuity*, edited by C. E. Cleland. Academic Press, Inc., New York.

Jelks, E. B. 1993 Observations on the Distributions of Certain Arrow-Point Types in Texas and Adjoining Regions. *Lithic Technology* 18 (1-2):9-15.

Jensen, H. J. 1994 Flint Tools and Plant Working: Hidden Traces of Stone Age Technology. Aarhus University Press, Aarhus.

Jeske, R. 1989 Economies in Raw Material Use by Prehistoric Hunter-Gatherers. In *Time, Energy and Stone Tools*, edited by Robin Torrence, pp. 34-45. Cambridge University Press, Cambridge.

Jeske, R. 1992 Energetic Efficiency and Lithic Technology: an Upper Missisippian example. *American Antiquity* 57(3):467-481.

Jochim, M. A. 1981 *Strategies for Survival: Cultural Behavior in an Ecological Context.* Academic Press, New York.

Johnson, J. K. 1986 Amorphous Core Technology in the Mid-South. *Midcontinental Journal of Archaeology* 11:135-151.

Johnson, J. K. 1987 Introduction. In *The Organization of Core Technology*, edited by Jay K. Johnson and Carol A. Morrow, pp. 1-12. Westview Press, Boulder and London.

Johnson, J. K. 1989 The Utility of Production Trajectory Modeling as a Framework for Regional Analysis. In *Alternative Approaches to Lithic Analysis*, edited by Donald O. Henry and George H. Odell, pp. 119-138. Archaeological Papers of the American Anthropological Association No. 1.

Johnson, J. K. 1996 Lithic Analysis and Questions of Cultural Complexity: the Maya. In *Stone Tools: Theoretical Insights into Human Prehistory*, edited by George H. Odell, pp. 159-180. Plenum Press, New York and London.

Johnson, J. K. and C. A. Morrow (editors) 1987 *The Organization of Core Technology.* Westview Press, Boulder and London.

Johnson, A. W. and T. Earle 1987 *The Evolution of Human Societies: from Foraging Group to Agrarian State.* Standford University Press, Stanford, Callifornia.

Justice, N. D. 1987 *Stone Age Spear and Arrow Points of the Midcontinental and Eastern United States.* Indiana University Press, Bloomington & Indianapolis.

Kamminga, J. 1982 *Over the Edge: Functional Analysis of Australian Stone Tools.* Anthropology Museum, University of Queensland, Occasional Papers in Anthropology No. 12.

Karrow, P. F. and B. G. Warner 1990 The Geological and Biological Environment for Human Occupation in Southern Ontario. In *The Archaeology of Southern Ontario to A.D. 1650*, edited by Chris J. Ellis and Neal Ferris, pp. 5-36. Occasional Publication of the London Chapter, Ontario Archaeological Society.

Keeley, L. H. 1974 Technique and Methodology in Microwear Studies. *World Archaeology* 5:323-326.

Keeley, L. H. 1980 *Experimental Determination of Stone Tool Uses.* The University of Chicago Press, Chicago.

Keeley, L. H. and M. H. Newcomer 1977 Microwear Analysis of Experimental Flint Tools: A Test Case. *Journal of Archaeological Science* 4(1):29-62.

Kelly, R. L. 1983 Hunter-Gatherer Mobility Strategies, *Journal of Anthropological Research* 39:277-306.

Kelly, R. L. 1988 The Three Sides of A Bifaces. *American Antiquity* 53(4):717-734.

Kelly, R. L. 1992 Mobility/Sedentism: Concepts, Archaeological Measures, and Effects.

Annual Review of Anthropology 21:43-66.

Kelly, J. E. , F. A. Finney, D. L. McElrath and S. J. Ozuk 1984 Late Woodland Period. In *American Bottom Archaeology: A Summary of the FAL-270 Project*, Contribution to the Culture History of the Mississippi River Valley, edited by C. Bareis and J. Porter, pp. 104-127. University of Illinois Press, Urbana.

Klejn, L. S. 1982 *Archaeological Typology*. BAR International Series 153, Oxford.

Knight, G. C. and J. D. Keyser 1983 A Mathematical Technique for Dating Projectile Points Common to the Northwestern Plains. *Plains Anthropologist* 28:199-207.

Knight, M. 1993 The Lone Pine site Faunal Report: Emphasis on Mammalian Remains. report on file Anthropology Department, University of Toronto.

Krieger, A. D. 1944 The Typological Concept. *American Antiquity* 9(3):271-288.

Kuijt, I., W. C. Prentiss and D. L. Pokotylo 1995 Bipolar Reduction: An Experimental Study of Debitage Variability. *Lithic Technology* 20(2):116-127.

Knudson, R. 1973 *Organizational Variability in Late Paleo-Indian Assemblages*. Unpublished PhD dissertation, Department of Anthropology, Washington State University.

Latta, M. A. 1976 *The Iroquoian Cultures of Huronia: A Study of Acculturation Through Archaeology*. Unpublished Ph.D. Dissertation, Department of Anthropology, University of Toronto.

Lee, R. B. 1979 *The !Kung San: Men, Women and Work in a Foraging Society*. Cambridge University Press, Cambridge.

Lee, R. B. 1981 Is there a Foraging mode of production?. *Canadian Journal of Anthropology* 2(1):13-19.

Lee, R. B. 1988 Reflections on Primitive Communism. In *Hunters and Gatherers 1: History, Evolution, and Social Change*, edited by T. Ingold, D. Riches and J. Woodburn, pp. 252-268. St. Martin's Press, New York.

Lee, R. B. 1990 Primitive Communism and the Origin of Social Inequality. In *The Evolution of Political Systems: Sociopolitical in Small-Scale Sedentary Societies*, edited by Steadman Upham, pp. 225-246. Cambridge University Press, Cambridge.

Lennox, P. A. 1986 The Innes Site: A Plow-Disturbed Archaic Component, Brant County, Ontario. *Midcontinental Journal of Archaeology* 11(2):221-268.

Lennox, P. A. 1993 The Kassel and Blue Dart Sites: Two Components of the Early Archaic, Bifurcate Base Projectile Point Tradition, Waterloo County, Ontario. *Ontario Archaeology* 56:1-31.

Lennox, P.A. and B. Morrison n.d. The Ramsey Site: A Princess Point Camp, Brant County, Ontario. report on file, Ontario Ministry of Transportation.

Leudtke, B. E. 1992 *The Archaeologists Guide to Chert and Flint*. Archaeological Research Tools 7, UCLA, Institute of Archaeology.

Levi-Sala, I. 1993 Use-Wear Traces: Processes of Development and Post-Depositional Alterations. In *Traces et Function; les gestes retrouves*, edited by P. C. Anderson, S. Beyries, M. Otte, and H. Plisson. pp. 401-416. Etudes et Recherches Archaeologiques de l'University de Liege (ERAUL), Vol 50.

Levi-Sala, I. 1996 *A Study of Microscopic Polish on Flint Implements*. BAR International Series 629, BAR Publishing.

Lewenstein, S. M. 1987 *Stone Tool Use at Cerros: the Ethnoarchaeological and Use-Wear Evidence*. University of Texas Press, Austin.

Linton, R. 1936 *The Study of Man*. Appleton-Century-Crofts, Inc., New York.

MacDonald, J. D.A. 1986a The Varden Site: A Multi-Component Fishing Station on Long Point, Lake Erie. Report on file, Ontario Ministry of Culture and Communications.

MacDonald, J. D.A. 1986b New Dates for Old Chronologies: Radiocarbon Dates from the Varden Site. *Kewa* 86(9):8-22.

Magne, M. P. R. 1985 *Lithic and Livelihood: Stone Tools Technologies of Central and Southern Interior British Columbia*. Mercury Series No 133. National Museum of Man, Ottawa.

Marks, A. (editor) 1976 *Prehistoric and Paleoenvironments in the Central Nagev, Isreal. Vol I.* Southern Methodist University Press, Dallas.

Marks, A. (editor) 1977 *Prehistoric and Paleoenvironments in the Central Nagev, Isreal. Vol II.* Southern Methodist University Press, Dallas.

Mason, R. J. 1981 *Great Lakes Archaeology*. Academic Press, New York.

Mauldin, R. P. and D. S. Amick 1989 Investigating Patterning in Debitage from Experimental Bifacial Core Reduction. In *Experiments in Lithic Technology*, edited by Daniel S. Amick and Raymond P. Mauldin, pp. 67-88. BAR International Series 528, Oxford.

McAndrews, J. H. 1988 Human Disturbance of North American Forests and Grasslands: the Fossil Pollen Record. In *Vegetation History*, edited by B. Huntley and T. III Webb, pp. 673-695.

McAndrews, J. H. and M. Boyko-Diakonow 1989 Pollen Analysis of the Varved Lake Sediment at Crawford Lake, Ontario: Evidence of Indian and European Farming. In *Quaternary Geology of Canada and Greenland*, edited by R. J. Fulton, J. A. Heginbottom and R. J. Funder, pp. 528-530. Geological Survey of Canada, Ottawa, Ontario.

McDonald, G. F. 1968 *Debert: A Paleo-Indian Site in Central Nova Scotia*. National Museum of Canada, Anthropological Papers, 16, Ottawa.

Mendenhall, W. 1968 *Introduction to Probability and Statistics*. Wadsworth Publishing, Co., Belmont, California.

Michlovic, M. G. 1976 Social Interaction and Points in the Eastern U.S. *Pennsylvania Archaeologist* 46(1-2):13-16.

Minnis, P. and C. L. Redman (editors) 1990 *Perspectives on Southwestern Prehistory*. Westview Press, Boulder.

Morgan, L. H. 1901 *League of the Iroquois. Buty Franklin* (reprint from 1851), New York.

Morrow, C. A. 1987 Blades and Cobden Chert: A Technological Argument for their Role as Markers of Regional Identification During the Hopewell Period in Illinois. In *The Organization Core Technology*, edited by Jar K Johnson and Carol A. Morrow, pp. 119-150. Westview Press, Inc., Boulder, Colorado.

Morse, P. and D. F. Morse 1990 The Zebree Site: An Emerged Early Mississippian Expression of Northeast Arkansas. In *The Mississippian Emergence*, edited by B. Smith, pp. 51-66. Smithsonian Institution Press, Washington, D.C.

Muller, J. 1986 *Archaeology of the Lower Ohio Valley*. Academic Press, Orlando.

Muller, J. 1987 Lower Ohio Valley Emergent Horticulture and Mississippian. In *Emergent Horticultural Economies of the Eastern Woodlands*, edited by William F. Keegan, pp. 243-273. vol. 7. Southern Illinois University at Carbondale, Center for Archaeological Investigations, Occasional Paper.

Murphy, C. n.d. *An Archaeological Survey of Highway 54 York to Caledonia*. Report on file, Ontario Ministry of Transportation.

Murphy, C. and N. Ferris 1990 The Late Woodland Western Basin Tradition in Southwestern Ontario. In *The Archaeology of Southern Ontario to A.D. 1650*, edited by Chris J. Ellis and Neal Ferris, pp. 189-279. Occasional Publication of the London Chapter, Ontario Archaeological Society.

Nash, S. E. 1996 Is Curation a Useful Heuris-tic? In *Stone Tools: Theoretical Insights into Human Prehistory*, edited by G. H. Odell, pp. 81-100. Plenum Press, Inc., New York.

Nassaney, M. S. 1992 Communal societies and the emergence of elites in the prehistoric American Southeast. In *Lords of the Southeast: Social Inequality and the Native Elites of Southeastern North America*, edited by Alex W. Barker and Timothy R. Pauketat. Archaeological Papers of he American Anthropological Association No. 3.

Nassaney, M. S. 1996 The Role of Chipped Stone in the Political Economy of Social Ranking. In *Stone Tools: Theoretical Insights into Human Prehistory*, edited by George H. Odell, pp. 181-228. Plenum Press, New York.

Nelson, M. C. 1991 The Study of Technological Organization. In *Archaeological Method and Theory*, Vol. 3, edited by M. Schiffer. University of Arizona Press, Tucson.

Neusius, P. 1996 Lithic Assemblage Variability in the Parker Collections. In *Reanalyzing the Ripley Site: Earthwork and Late Prehistory on the Lake Erie Plain*, edited by L. P. Sullivan. New York State Museum Bulletin No.489. New York State Museum, Albany

Noble, W. C. and I. T. Kenyon 1972 Porteous (AgHb -1): A Probable Early Glen Meyer Village in Brant County, Ontario. *Ontario Archaeology* 19:11-38.

Noon, J. A. 1949 *Law and Government of the Grand River Iroquois*. Viking Fund Publications in Anthropology Number 12, New York.

Odell, G. H. 1975 Micro-wear in Perspective: A sympathetic Response to Lawrence H. Keeley. *World Archaeology* 7:226-240.

Odell, G. H. 1977 *The Application of Micro-wear Analysis to the Lithic Component of an Entire Prehistoric Settlement: Methods, Problems and Functional Reconstructions.*

Unpublished PhD dissertation, Department of Anthropology, Harvard University.

Odell, G. H. 1979 A New Improved System for the Retrieval of Functional Information from Microscopic Observations of Chipped Stone Tools. In *Lithic Use-Wear Analysis*, edited by Brian Hayden, pp. 329-344. Academic Press, Inc.

Odell, G. H. 1981a The Morphological Express at Function Junction: searching for meaning in lithic tool type. *Journal of Anthropological Research* 37(4):319-342.

Odell, G. H. 1981b The Mechanics of Use-Breakage of Stone Tools: Some Testable Hypotheses. *Journal of Field Archaeology* 8:197-209.

Odell, G. H. 1985 On Evaluation "Blind Tests" in Lithic Use-Wear Research. *Western Canadian Archaeologist* 2:26-30.

Odell, G. H. 1988 Addressing Prehistoric Hunting Practices Through Stone Tools Analysis. *American Anthropologist* 90(2):335-356.

Odell, G. H. 1989a Experiments in Lithic Reduction. In *Experiments in Lithic Technology*, edited by Daniel S. Amick and Raymond P. Mauldin, pp. 163-198. BAR International Series 528, Oxford.

Odell, G. H. 1989b Fitting Analytical Techniques to Prehistoric Problems with Lithic Data. In *Alternative Approaches to Lithic Analysis*, edited by O. Henry and G. H. Odell, pp. 159-182. Archaeological Papers of the American Anthropological Association No. 1.

Odell, G. H. 1990 Brer Rabit Seeks True Knowledge. In *The Interpretative Possibilities of Microwear Studies*, edited by B. Graslund, H. Knutsson, K. Knutsson and J. Taffinder, pp. 125-134. Societas Archaeologica Upsaliensis, AUN 14, Uppsala.

Odell, G. H. 1993 A North American Perspective on Recent Archaeological Stone Tool Research. *Palimpsesto, Revista de Arqueologia* 3:109-122.

Odell, G. H. 1994a The Role of Stone Bladelets in Middle Woodland Society. *American Antiquity* 59(1):102-120.

Odell, G. H. 1994b Assessing Hunting-Gathering mobility in the Illinois Valley: exploring ambiguous results. In *The Organization of North American Prehistoric Chipped Stone Tool Technologies*, edited by Philip J. Carr, pp. 70-86. International Monographs in Prehistory, Ann Arbor, MI.

Odell, G. H. 1994c Prehistoric Hafting and Mobility in the North America Midcontinent: examples from Illinois. *Journal of Anthropological Archaeology* 13:51-73.

Odell, G. H. 1996a *Stone Tools and Mobility in the Illinois Valley: from hunter-gatherer camps to agricultural villages.* International Monographs in Prehistory, Ann Arbor, Michigan.

Odell, G. H. 1996b Economizing Behavior and the Concept of "Curation". In *Stone Tools: Theoretical Insights into Human Prehistory*, edited by G. H. Odell, pp. 51-80. Plenum Press, Inc., New York.

Odell, G. H. (editor) 1996c *Stone Tools: Theoretical Insights into Human Prehistory*. Plenum Press, Inc., New York.

Odell, G. H. 1998 Investigating Correlates of Sedentism and Domestication in Prehistoric North American. *American Antiquity* Vol63(4):553-576

Odell, G. H. and F. Odell-Vereecken 1980 Verifying the Reliability of Lithic Use-Wear Assessments by `Blind Tests': the Low-Power Approach. *Journal of Field Archaeology* 7(1):87-120.

Odell, G. H., B. D. Hayden, J. K. Johnson, M. Kay, T. Morrow, S. E. Nash, M. S. Nassaney, J. W. Rick, M. F. Rondeau, S. A. Rosen, M. J. Shott and P. T. Thacker 1996 Some Comments on a Continuing Debate. In *Stone Tools: Theoretical Insights into Human Prehistory*, edited by G. H. Odell, pp. 377-392. Plenum Press, Inc., New York.

Ormerod, T. 1994 *The Lone Pine Flaked Lithic Aggregate: Behavioral Implications For A Late Transitional Woodland Site.* Unpublished M.Sc. Thesis, Department of Anthropology, University of Toronto.

Oswalt, W. H. 1976 *An Anthropological Analysis of Food-getting Technology*. John Wiley & Sons, New York.

Parker, R. L. 1986 Haldimand Chert: A Preferred Raw Material in Southwestern Ontario During the Early Holocene Period. *Kewa* 86(4):4-21.

Parker, R. L. 1994 Archaeological Assessment Stage 2 & 3 field Assessment Unit Gas Limited Project Caledonia Pipeline (NPS 12) Grand River to Mckenzie Road Town of Haldimand R.M. of Haldimand-Norfolk. Report on file, Ontario Ministry of Culture, Tourism, and Recreation.

Parry, W. J. 1987 *Chipped Stone Tools in Formative*

Oaxaca, Mexico: Their Procurement, Production and Use. Memoirs of the Museum of Anthropology University of Michigan, Ann Arbor.

Parry, W. J. and A. L. Christenson 1986 *Prehistoric Stone Technology on Northern Black Mesa, Arizona. Center for Archaeological Investigations*, Southern Illinois University, Carbondale.

Parry, W. J. and R. L. Kelly 1987 Expedient Core Technology and Sedentism. In *The Organization of Core Technology*, edited by Jar K Johnson and Carol A. Morrow, pp. 285-304. Westview Press, Inc., Boulder, Colorado.

Patterson, L. W. 1981 The Importance of Flake Size Distribution. *Contract Abstracts and CRM Archaeology* 3:70-72.

Patterson, L. W. 1982 Replication and Classification of Large Size Lithic Debitage. *Lithic Technology* 11(3):50-58.

Patterson, L. W. 1985 Distinguishing Between Arrow and Spear Points n the Upper Texas Coast. *Lithic Technology* 14:81-89.

Patterson, L. W. 1987 Amorphous Cores and Utilized Flakes: A Commentary. *Lithic Technology* 16(2-3):51-52.

Patterson, L. W. 1990 Characteristics of Bifacial Reduction Flake-Size Distribution. *American Antiquity* 55:550-558.

Patterson, L. W. 1995 Thermal Damage to Chert. *Lithic Technology* 20(1):72-80.

Patterson, L. W. and J. B. Sollberger 1978 Replication and Classification of Small Size Lithic Debitage. *Plains Anthropologist* 23(80):103-112.

Patterson, T. C. 1992 *Archaeology: The Historical Development of Civilizations.* Prentice Hall, Englewood Cliffs, NJ.

Philips, P. 1988 Traceology (Microwear) Studies in the USSR. *World Archaeology* 19:349-356.

Phillips, J. L., R. L. Hall and R. W. Yerkes 1980 *Investigations at the Labras Lake Site, Vol.1:* Archaeology. Department of Anthropology, University of Illinois at Chicago.

Presant, E. W. and C. J. Acton 1984 The Soils of the Regional Municipality of Haldimand-Norfolk. Report No.57 of the Ontario Institute of Pedology, Land Research Institute.

Price, T. D. and J. A. Brown (editors) 1985 *Prehistoric Hunter-Gatherers: The Emergence of Complex Societies.* Academic Press, Inc., New York.

Price, T. D. and G. M. Feinman (editor) 1995 *Foundations of Social Inequality.* Plenum Press, New York.

Price, D. T. and A. B. Gebauer (editors) 1995 New Perspectives on the Transition to Agriculture. In *Last Hunters-First Farmers: New Perspectives on the Prehistoric Transition to Agriculture*, edited by T. Douglas Price and Anne Birgitte Gebauer, pp. 3-20. School of American Research Press, Santa Fe, New Mexico.

Quin, J. 1996 *Chert Sourcing in the Lower Grand River Valley: A Study in Methods and Application.* Unpublished M.Sc. Thesis, Department of Anthropology, University of Toronto.

Railey, J. A. 1992 Chipped Stone Artifacts. In *Fort Ancient Cultural Dynamics in the Middle Ohio Valley*, edited by A. G. Henderson, pp. 137-170. Monographs in World Archaeology No. 8. Prehistory Press, Madison, Wisconsin.

Ramsden, P. G. 1990 The Hurons: Archaeology and Culture History. In *The Archaeology of Southern Ontario to A.D. 1650*, edited by Chris J. Ellis and Neal Ferris, pp. 361-384. Occasional Publication of the London Chapter, Ontario Archaeological Society.

Read, D. W. and G. Russell 1996 A Method for Taxonomic Typology Construction and an Example: Utilized Flakes. *American Antiquity* 61(4):663-685.

Reid, K. C. 1976 Prehistoric Trade in the Lower Missouri River Valley: An Analysis of Middle Woodland Bladelets. In *Hopewellian Archaeology in the Lower Missouri River Valley*, edited by Alfred E. Johnson, pp. 55-66. Publications in Anthropology No. 8 University of Kansas.

Richards, T. 1984 Searching High and Low: A Review and Comparison of Microwear Analysis Methodologies. *Western Canadian Anthropologist* 1:18-25.

Richards, T. 1988 *Microwear Patterns on Experimental Basalt Tools.* BAR International Series 460, Oxford.

Ritchie, W. A. 1961 *Typology and Nomenclature for New York Projectile Points.* New York State Museum and Science Service, Bulletin 384, Albany, New York.

Ritchie, W. A. 1965 *The Archaeology of New York State.* Natural History Press, Garden City, New York.

Rosen, S. A. 1996 The Decline and Fall of Flint. In *Stone Tools: Theoretical Insights into*

Human Prehistory, edited by George H. Odell, pp. 129-154. Plenum Press, New York.

Rousseau, M. K. 1992 *Integrated Lithic Analysis: the Significance and Function of Key-shaped Formed Unifaces on the Interior Plateau of Northwestern North America.*Department of Anthropology Publication No.20, Simon Fraser University.

Sackett, J. R. 1973 Style, Function and Artifact Variability in Palaeolithic Assemblages. In *The Explanation of Culture Change: Models in Prehistory*, edited by Colin Renfrew, pp. 317-328. Gerald Duckworth and Co. Ltd.

Sanger, D. 1970 Mid-Latitude Core and Blade Traditions. *Arctic Anthropology* 7(2):106-114.

Sassaman, K. E. 1994 Changing Strategies of Biface Production and Evolution: Notes from the Continent. In *The Organization of North American Prehistoric Chipped Stone Tool Technologies*, edited by Philip J. Carr, pp. 99-117. International Monographs in Prehistory, Ann Arbor, MI.

Scarry, M. 1993 *Foraging and Farming in the Eastern Woodlands.* University of Florida, Gainesville.

Schiffer, M. B 1975 Classification of Chopped-Stone Tool Use. In *The Cache River Archaeological Project: an experiment in contract archaeology*, edited by Michael B. Schiffer and John H. House, pp. 249-251. Arkansas Archaeological Survey, Research Series 8.

Schiffer, M. B 1976 *Behavioral Archeology*. Academic Press, New York.

Schiffer, M. B 1979 The Place of Lithic Use-Wear Studies in Behavioral Archaeology. In *Lithic Use-Wear Analysis*, edited by Brian Hayden, pp. 15-26. Academic Press, Inc., New York.

Schiffer, M. B 1992 *Technological Perspectives on Behavioral Change*. University of Arizona Press, Tucson.

Seeman, M. F. 1992 The Bow and Arrow, the Intrusive Mound Complex, and a Late Woodland Jack's Reef Horizon in the Mid -Ohio Valley. In *Cultural Variability in Context: Woodland Settlements of the Mid-Ohio Valley*, edited by M. Seeman, pp. 41-51. MCJA Special Paper No.7, Kent.

Semenov, S. A. 1964 *Prehistoric Technology*. Cory, Adams and Mackay, London.

Shanghai Culture Relics Management Committee 1987 An Important Discovery of Third Season Excavation at Fuqunshan Site, Shanghai [in Chinese with English abstract]. *Dongnan Wenhua* 3:86-103.

Shea, J. J. 1987 On Accuracy and Relevance in Lithic Use-Wear Analysis. *Lithic Technology* 16(2-3):44-50.

Shea, J. J. 1991 *The Behavioral Significance of Lecantine Mousterian Industrial Variability.* Unpublished PhD Dissertation, Department of Anthropology, Harvard University.

Shea, J. J. 1995 Lithic Microwear Analysis of Tor Faraj Rockshelter. In *Prehistoric Cultural Ecology and Evolution: Insights from Southern Jordan*, edited by Donald O. Henry, pp. 85-106. Plenum Press, New York.

Sheets, P. D. 1978 From Craftsman to Cog: Quantitative Views of Mesoamerican Lithic Technology. In *Papers on the Economy and Architecture of the Ancient Maya*, edited by R. Sidrys, pp. 40-71. University of California, Los Angeles, Institute of Archaeology, monograph 8, Los Angeles.

Shen, C. 1995 Lithic Analysis of the Princess Point Complex from Southwestern Ontario. *Annual Archaeological Report Ontario* 6:148-153.

Shen, C. 1999 Were 'Utilized Flakes' Utilized? An Issue of Lithic Classification in Ontario Archaeology. *Ontario Archaeology* 68:63-73.

Shen, C. 2000 Tool Use-Patterning at the Grand Banks site of the Princess Point Complex, Southwestern Ontario *Northeast Anthropology* 60:63-87

Shennan, S. 1988 *Quantifying Archaeology*. Academic Press, Edinburgh.

Shott, M. J. 1986 Technological Organization and Settlement Mobility: An Ethnographic Examination. *Journal of Anthropological Research* 42(1):15-53.

Shott, M. J. 1989 On Tool-Class Use Lives and the Formation of Archaeological Assemblages. *American Antiquity* 54(1):9-30.

Shott, M. J. 1993 Spears, Darts, and Arrows: Late Woodland Hunting Techniques in The Upper Ohio Valley. *American Antiquity* 58(3):425-443.

Shott, M. J. 1994 Size and Form in the Analysis of Flake Debris: Review and Recent Approaches. *Journal of Archaeological Method and Theory* 1(1):69-110.

Shott, M. J. 1996a Stage Versus Continuum in the Debris Assemblage From Production of A

Fluted Biface. *Lithic Technology* 21(1):6-22.

Shott, M. J. 1996b Innovation and Selection in Prehistory: A Case Study from the American Bottom. In *Stone Tools: theoretical insights into Human Prehistory*, edited by G. H. Odell, pp. 279-309. Plenum Press, Inc., New York.

Shott, M. J. 1997 Stones and Shafts Redux: The Metric Discrimination of Chipped-Stone Dart and Arrow Points. *American Antiquity* 62(1):86-102

Sievert, A. K. 1992 *Maya Ceremonial Specialization: Lithic Tools from the Sacred Cenote at Chichen Itza, Yucatan.* Monographs in World Archaeology No.12. Prehistory Press, Madison, Wisconsin.

Smith, B. D. 1987 The Independent Domestication of Indigenous Seed-Bearing Plants in Eastern North America. In *Emergent Horticultural Economies of the Eastern Woodlands*, edited by William F. Keegan, pp. 1-47. vol. 7. Southern Illinois University at Carbondale, Center for Archaeological Investigations, Occasional Paper.

Smith, B. D. 1989 Origins of Agriculture in Eastern North America. *Science* 246:1566-1571.

Smith, B. D. 1993 Prehistoric Plant Husbandry in Eastern North America. In *The Origins of Agriculture: An International Perspective,* edited by C. Wesley Cowan and Patty Jo Watson, pp. 101-119. Smithsonian Institution Press, Washington.

Smith, B. D. 1995 Seed Plant Domestication in Eastern North America. In *Last Hunters-First Farmers: New Perspectives on the Prehistoric Transition to Agriculture*, edited by T. Douglas Price and Anne Birgitte Gebauer, pp. 193-214. School of American Research Press, Santa Fe, New Mexico.

Smith, B. D. (editor) 1992 *Rivers of Change: Essays on Early Agriculture.* Smithsonian Institution Press, Washington, D.C.

Smith, D. G. 1996 The Early Development of Iroquoian Smoking Pipes in Ontario. Paper presented at the 61st annual meeting of the Society for American Archaeology, New Orleans.

Smith, D. G. 1997 Radiocarbon Dating the Middle to Late Woodland Transition and Earliest Maize in Southern Ontario. *Northeast Anthropology* 54:37-73

Smith, D. G. and G. W. Crawford 1993 The Origins of Agriculture in Ontario. Paper presented at the 26th Annual Meeting of the Canadian Archaeological Association, Montreal.

Smith, D. G. and G. W. Crawford 1994 The Princess Point Project: 1993 Field Season. *Annual Archaeological Report, Ontario* 5:147-152.

Smith, D. G. and G. W. Crawford 1995 The Princess Point Complex and the Origins of Iroquoian Societies. In Origins of the People of the Longhouse: proceedings of the 21st annual symposium of the Ontario Archaeological Society, edited by Andre Bekerman and Gary Warrick, pp. 55-70. Ontario Archaeological Society Inc., Toronto.

Smith, D. G. and G. W. Crawford 1997 Recent Developments in the Archaeology of the Princess Point Complex in Southern Ontario. *Canadian Journal of Archaeology* 21:9-32

Snow, D. R. 1980 *The Archaeology of New England.* Academic Press, New York.

Song, E. and C. Shen 1999 The Dulong of Southwestern China. in *Foraging Peoples: an Encyclopaedia of Contemporary Hunter-Gatherers*, edited by Richard B. Lee and Richard Daly, Cambridge University Press. Pp303-306

Southall, A. 1988 On Mode of Production Theory: the Foraging Mode of Production and the Kinship Mode of Production. *Dialectical Anthropology* 12:165-192.

Spaulding, A. C. 1953 Statistical Techniques for the Discovery of Artifact Types. *American Antiquity* 18:305-314.

Spaulding, A. C. 1960 The Dimensions of Archaeology. In *Essays in the Science of Culture*, edited by G.E. Dole and R.L. Carneiro, pp. 437-456. Thomas Y. Crowell, New York.

Spaulding, A. C. 1980 Structure in Archaeological Data: Nominal Variables. In *Essays on Archaeological Typology*, edited by R. Whallon and J.A. Brown, pp. 1-20. Center for American Archeology Press, Evanston.

Staffford, M. 1999 *From Forager to Farmer in Flint: A Lithic Analysis of the Prehistoric Transition to Agriculture in Southern Scandinavia* Aarhus University Press, Aarhus.

Stahle, D. W. and J. E. Dunn 1982 An Analysis and Application of the Size Distribution of Waste Flakes from the Manufacture of Bifacial Stone Tools. *World Archaeology* 14(1):84-97.

Stahle, D. W. and J. E. Dunn 1984 An Experimental Analysis of the Size Distribution of Waste Flakes from Biface Reduction. Arkansas Archaeological Survey, Technical Paper No. 2.

Storck, P. 1974 Two Probable Shield Archaic Sites in Killarney Provincial Park, Ontario. *Ontario Archaeology* 21:3-36.

Storck, P. 1978 The Coates Creek Site: A Possible Late Paleo Indian-Early Archaic Site in Simcoe County, Ontario. *Ontario Archaeology* 30:25-46.

Storck, P. 1983 The Fisher Site, Fluting Techniques, and Early Paleo-Indian Cultural Relationships. *Archaeology of Eastern North America* 11:80-97.

Storck, P. 1984 Research into the Paleo-Indian Occupations of Ontario: A Review. *Ontario Archaeology* 41:3-28.

Storch, P. 1997 *The Fisher Site* Memoirs Museum of Anthropology University of Michigan No. 30, Ann Arbor, Michigan.

Stothers, D. M. 1970 The Princess Point Complex and Its Relationship to the Owasco and Ontario Iroquois Traditions. *Newsletter of the Ontario Archaeological Society* 70(3):4-6.

Stothers, D. M. 1973 *Early Evidence of Agriculture in the Great Lakes.* Canadian Archaeological Association Bulletin 5:62-76.

Stothers, D. M. 1974a An Archaeological Survey of the Lower Grand River Valley 1974: Field Notes. Report on file,Ontario Ministry of Culture, Tourism, an Recreation.

Stothers, D. M. 1974b The Glass Site (AgHb-5) Oxbow Tract, Brantford Township, Brant County, Ontario. *Ontario Archaeology* 21:37-45.

Stothers, D. M. 1976 The Princess Point Complex: A Regional Representative of an Early Late Woodland Horizon in the Great Lakes Area. In *The Late Prehistory of the Lake Erie Drainage Basin*, edited by David S. Brose, pp. 137-161. Cleveland Museum of Natural History, Cleveland.

Stothers, D. M. 1977 *The Princess Point Complex.* National Museum of Man, Archaeological Survey of Canada, Mercury Series 58. Ottawa.

Stothers, D. M. and R. A. Yarnell 1977 An Agricultural Revolution in the Lower Great Lakes. In *Geobotany*, edited by R.D. Romans, pp. 209-232. Plenum Press, New York.

Sullivan, A. P., III and K. C. Rozen 1985 Debitage Analysis and Archaeological Interpretation. *American Antiquity* 50(4):755-779.

Thomas, D. H. 1978 Arrowheads and Atlatl Darts: How he Stones Got the Shaft. *American Antiquity* 43(461-472).

Timmins, P. A. 1985 The Analysis and Interpretation of Radiocarbon Dates in Iroquoian Prehistory. Museum of Indian Archaeology, Research Report.

Timmins, P. A. 1992a The Alder Creek Site (AiHd-75): Interior Paleo-Indian and Princess Point Settlement Patterns in the Region of Waterloo, Ontario. Ministry of Transportation Southwestern Region Environmental Unit.

Timmins, P. A. 1992b *An Interpretive Framework for the Early Iroquoian Village.* Unpublished Ph.D. Dissertation, McGill University.

Tixier, J. 1974 Glossary for the Description of Stone Tools, with Special Reference to the Epipalaeolithic of the Maghreb. *News-letter of Lithic Technology* Special Publication 1. Newsletter of Lithic Technology Special Publication 1.

Tomenchuk, J. 1985 *The Development of A wholly Parametric Use-Wear Methodology and It's Applications to Two Selected Samples of Epipaleolithic Chipped Stone Tools from Hayonim Cave, Israel.* Unpublished PhD. Dissertation, Department of Anthropology, University of Toronto.

Tomka, S. A. 1989 Differentiating Lithic Reduction Techniques: an experimental approach. In *Experiments in Lithic Technology*, edited by Daniel S. Amick and Raymond P. Mauldin, pp. 137-162. BAR International Series 528., Oxford.

Tooker, E. 1964 *An Ethnography of the Huron Indians 1615 - 1649.* American Bureau of Ethnology, Bulletin.

Tooker, E. (editor) 1985 *An Iroquois Source Book.* Garland Publication, New York.

Torrence, R. 1989 Retooling: Towards a Behavioral Theory of Stone Tools. In *Time, Energy and Stone Tool*, edited by Robin Torrence, pp. 57-66. Cambridge University Press, Cambridge.

Torrence, R. 1983 Time Budgeting and Hunter-Gatherer Technology. In *Hunter-Gatherer Economy in Prehistory*, edited by G. Bailey, pp. 11-22. Cambridge University Press, Cambridge.

Torrence, R. 1994 Strategies for Moving on in Lithic Studies. In *The Organization of North American Prehistoric Chipped Stone Tool Technologies*, edited by Philip J. Carr, pp.

123-131. International Monographs in Prehistory, Ann Arbor, MI.

Trigger, B. G. 1969 *The Huron: farmers of the north.* Holt, Rinehart and Winston, New York.

Trigger, B. G. 1976 *The Children of Aataentsic: A History of the Huron People to 1660.* McGill-Queen's University Press, Montreal.

Trigger, B. G. 1981 Prehistoric Social and Political Organization: An Iroquoian Case Study. In *Foundations of Northeast Archaeology,* edited by D.R. Snow, pp. 1-50. Academic Press, New York.

Trigger, B. G. 1985 *Natives and Newcomers: Canada's 'Heroic Age' Reconsidered.* McGill-Queen's University Press, Montreal.

Trigger, B. G. 1987 *The Children of Aataentsic: A History of the Huron People to 1660.* McGill-Queen's University Press (reprint from 1976), Montreal.

Trigger, B. G. (editor) 1978 *Handbook of North American Indians, Vol. 15, Northeast.* Smithsonian, Washington.

Tringham, R. E., G. Cooper, G; Odell, B. Voytek and A Whitman 1974 Experimentation in the Formation of Edge Damage: a new approach to lithic analysis. *Journal of Field Archaeology* 1:171-196.

Unger-Hamilton, R. 1988 *Method in Microwear Analysis: Prehistoric Sickles and Other Stone Tools from Arjoune, Syria.* BAR International Series 435, Oxford.

Unger-Hamilton, R. 1992 Experiments in Harvesting Wild Cereals and Other Plants. In *Prehistoire de L'agriculture: Nouvelles Approaches Experimentales et Ethnographiques*, edited by P. Anderson, pp. 211-224. Monographie du Centre de Recherches Archeologiques No.6 Centre National de la Recherche Scientifique, Paris.

Upham, S. 1983 Intensification and Exchange: An Evolutionary Model of Non-Egalitarian Socio-Political Organization for the Prehistoric Plateau Southwest. In *Ecological Models in Economic Prehistory*, edited by Gordon Bronitsky, pp. 219-245. Arizona State University, Anthropological Research Papers No. 29.

Upham, S. (editor) 1989 *The Sociopolitical Structure of Prehistoric Southwestern Societies.* Westview Press, Boulder.

Upham, S. (editor) 1990 *The Evolution of Political Systems: Sociopolitical in Small-Scale Sedentary Societies.* Cambridge University Press, Cambridge.

Vaughan, P. C. 1985 *Use-Wear Analysis of Flaked Stone Tools.* The University of Arizona Press, Tucson, Arizona.

Vierra, B. J. 1995 *Subsistence and Stone Tool Technology: An Old World Perspective.* Arizona State University, Anthropological Research Papers No. 47.

Walker, I. 1995 Lateral Bar Formation and Stability on the Grand River near Cayuga, Ontario. Report on file, Anthropology Department, University of Toronto.

Walker, I, J.R. Desloges, G.W. Crawford and D.G. Smith 1997 Floodplain Formation Processes and Archaeological Implications. *Geoarchaeology* 12(8):865-887.

Watson, P. J. 1985 The Impact of Early Horticulture in the Upland Drainages of the Midwest and Midsouth. In *Prehistoric Food Production in North America*, edited by Richard I. Ford, pp. 99-148. vol. 75. University of Michigan, Museum of Anthropology, Anthropological Papers.

Watson, P. J. 1989 Early Plant Domestication in the Eastern Woodlands of North America. In *Foraging and Farming: the Evolution of Plant Exploitation*, edited by D. R. Harris and G. C. Harris, pp. 555-571. Unwin Hyman.

Waugh, F.W. 1916 Iroquois Foods and Food Preparation. In *Canada Department of Mines, Geological Section, Memoir Vol. 86.*

Webster, G. S. 1980 Recent Data Bearing on the Question of the Origins of the Bow and Arrow in the Great Basin. *American Antiquity* 45:63-66.

Whallon, R. and J. A. Brown 1982 *Essays on Archaeological Typology.* Center for American Archeology Press, Evanston.

White, A. M. 1963 Analytic Description of the Chipped Stone Industry from Snyders Site, Calhoun County, Illinois. In *Miscellaneous Studies in Typology and Classification*, edited by Anta M. White, Lewis R. Binford and Mark L. Papworth. Anthropological Papers No. 19. Museum of Anthropology, University of Michigan, Ann Arbor.

Williamson, R. F. 1990 The Early Iroquoian Period of Southern Ontario. In *The Archaeology of Southern Ontario to A.D. 1650*, edited by Chris J. Ellis and Neal Ferris, pp. 291-320. vol. 5. Occasional Publication of the

London Chapter, Ontario Archaeological Society.

Wills, W. H. 1991 Organizational Strategies and the Emergence of Prehistoric Villages in the American Southwest. In *Between Bands and States*, edited by Susan A. Gregg. Center for Archaeological Investigations, Occasional Paper No.9, Carbondale, IL.

Wills, W. H. 1995 Archaic Foraging and the Beginning of Food Production. In *Last Hunters-First Farmers: New Perspectives on the Prehistoric Transition to Agriculture*, edited by T. Douglas Price and Anne Birgitte Gebauer, pp. 215-242. School of American Research Press, Santa Fe, New Mexico.

Winterhalder, B. 1981 Optimal Foraging Strategies and Hunter-Gatherer Research in Anthropology: Theory and Models. In *Hunter-Gatherer Foraging Strategies: Ethnographic and Archaeological Analyses*, edited by B. Winterhalder and E. Smith, pp. 13-35. University of Chicago Press, Chicago.

Wolf, E. 1982 *Europe and the People Without History*. University of California, Berkley and Los Angeles.

Woodburn, J. 1982 *Egalitarian Societies*. Man 17:431-451.

Woodley, P. J. 1996 The HH Site (AgGw-81). QEW Highway and Redhill Creek Expressway, Regional Municipality of Hamilton-Wentworth. Report on file, Ministry of Transportation, Ontario.

Wright, J. V. 1972 *Ontario Prehistory: An Eleven-Thousand-Year Archaeological Outline*. National Museums of Canada, Ottawa.

Wright, J. V. 1973 *The Ontario Iroquois Tradition*. National Museum of Man, Ottawa.

Wright, K. 1991 The Origins and Development of Ground Stone Assemblages in Late Pleistocene Southwest Asia. *Paleorient* 17(1):19-45.

Wright, K. 1992 *Ground Stone Assemblage Variations and Subsistence in the Levant, 22,000-5,500 B.P.* Unpublished PhD dissertation, Department of Anthropology, Yale University.

Wright, K. 1994 Ground Stone Tools and Hunter-Gatherer Subsistence in Southwest Asia: Implications for the Transition to Farming. *American Antiquity* 59(2):238-263.

Yerkes, R. W. 1983 Microwear, Microdrills, and Mississippian Craft Specialization. *American Antiquity* 48(3):499-518.

Yerkes, R. W. 1987 *Prehistoric Life on the Mississippi Floodplain: Stone Tool Use, Settlement Organization, and Subsistence Practices at the Labras Lake Site, Illinois. Prehistoric Archaeology and Ecology.* The University of Chicago Press, Chicago.

Yerkes, R. W. 1989 Lithic Analysis and Activity Patterns at Labras Lake. In *Alternative Approaches to Lithic Analysis*, edited by Donald O. Henry and George H. Odell, pp. 183-212. Archaeological Papers of the American Anthropological Association No. 1.

Yerkes, R. W. 1994 A Consideration of the Function of Ohio Hopewell Bladelets. *Lithic Technology* 19(2):109-127.

Yerkes, R. W. and P. N. Kardulias 1993 Recent Developments in the Analysis of Lithic Artifacts. *Journal of Archaeological Research* 1(2):89-166.

You, X. 1995 The Research Progress and New Perspectives of Rice Remains from Hemudu [in Chinese]. *Nongye Kaogu* (1):66-70.

Young, D. and D. B. Bamforth 1990 On the Macroscopic Identification of Used Flakes. *American Antiquity* 55(2):403-409.

www.ingramcontent.com/pod-product-compliance
Lightning Source LLC
LaVergne TN
LVHW071207210926
848482LV00015B/747